Handbook of Electrical and Computer Engineering

Volume I

Handbook of Electrical and Computer Engineering
Volume I

Edited by **Dewayne Hopper**

NY RESEARCH PRESS

New York

Published by NY Research Press,
23 West, 55th Street, Suite 816,
New York, NY 10019, USA
www.nyresearchpress.com

Handbook of Electrical and Computer Engineering: Volume I
Edited by Dewayne Hopper

International Standard Book Number: 978-1-63238-234-4 (Hardback)

Printed in China.

Contents

Preface

The field of engineering that deals with the study and application of electronics and electricity is termed as electrical engineering. This field gained recognition as one of the biggest fields of engineering during the 19th century. Electric power has gained a foothold in the most basic activities of our lives. There is not a part of our daily lives that is not involved with electrical power in some way or the other. There are many subdivisions of electrical engineering like telecommunications, control systems, instrumentation and electronics, though electronics can be termed as a field in its own right. The skills required of an electrical engineer are variable and they work in a huge variety of industries. Computer engineering on the other hand is a discipline that combines various fields of electrical engineering and computer science and specializes in the development of computer hardware and software. Computer engineers can be involved in the many hardware and software facets of computing, like the design of microprocessors and personal computers to circuit design. It can be said that this field of study focuses on both how computer systems work as well as their integration in the larger picture. Both electrical and computer engineering are disciplines that can be counted as the biggest and fastest growing in the range of engineering fields that are there in the industry.

This book is an attempt to collate all current data and research on computer and electrical engineering. I am thankful to all the contributing authors for the hard work and effort put in these researches. I also wish to acknowledge the efforts of the publishing team who provided excellent technical assistance, whenever needed. Lastly, I wish to thank my friends and family who have supported me at every step in my life.

Editor

1

Evaluating Subpixel Target Detection Algorithms in Hyperspectral Imagery

Yuval Cohen,[1] Yitzhak August,[1] Dan G. Blumberg,[2] and Stanley R. Rotman[3]

[1] *The Unit of Electro-Optics Engineering and the Earth and Planetary Image Facility, Ben-Gurion University of the Negev, P.O. Box 653, 84105 Beer-Sheva, Israel*
[2] *The Department of Geography and Environmental Development and the Earth and Planetary Image Facility, Ben-Gurion University of the Negev, P.O. Box 653, 84105 Beer-Sheva, Israel*
[3] *Department of Electrical and Computer Engineering and the Earth and Planetary Image Facility, Ben-Gurion University of the Negev, P.O. Box 653, 84105 Beer-Sheva, Israel*

Correspondence should be addressed to Stanley R. Rotman, srotman@ee.bgu.ac.il

Academic Editor: James Theiler

Our goal in this work is to demonstrate that detectors behave differently for different images and targets and to propose a novel approach to proper detector selection. To choose the algorithm, we analyze image statistics, the target signature, and the target's physical size, but we do not need any type of ground truth. We demonstrate our ability to evaluate detectors and find the best settings for their free parameters by comparing our results using the following stochastic algorithms for target detection: the constrained energy minimization (CEM), generalized likelihood ratio test (GLRT), and adaptive coherence estimator (ACE) algorithms. We test our concepts by using the dataset and scoring methodology of the Rochester Institute of Technology (RIT) Target Detection Blind Test project. The results show that our concept correctly ranks algorithms for the particular images and targets including in the RIT dataset.

1. Introduction

Ideally, one would like to choose a hyperspectral detection algorithm for use in a particular scenario with the assurance that it would be "optimal," that is, that the type of algorithm to be used and its free parameters would be optimized for the particular task for which it is being considered. Of course, in such cases, the complexity of real-world scenarios and the difficulties of predicting the exact target signature *in situ*, make it hard to believe that we can predict the optimal target detection algorithm ahead of time. Because the responses of these algorithms can vary depending on target placement, we adapted the Rotman-Bar Tal Algorithm (RBTA) [1] for comparing point target detection algorithms, used for infrared broadband images, to the analysis of hyperspectral imagery [2–4]. The RBTA implants targets and evaluates the response of the detecting algorithm to their presence in every pixel in the dataset. Indeed, our development of new algorithms based on this tool has been validated by results obtained by other researchers in actual field tests [5, 6].

An inherent weakness of the RBTA method is its assumption that subpixel targets will each be contained within a single pixel. In light of our recent work [7], which showed that even very small targets can affect several pixels, here we fine-tuned the RBTA method to account for this possibility.

Sections 2–6 describes the RBTA in detail. We show how the simulation of target detection performance is dependent on the spatial correlation of the pixels present in the target.

Sections 7–12 analytically considers the expected performances of several detection algorithms under conditions of "pixel phasing," that is, a small target located simultaneously in several adjacent pixels. Our improved RBTA (IRBTA) takes into account target blurring and pixel phasing. The results presented in Sections 13–16 show that the superiority of the ACE algorithm and the importance of accounting for target blurring are validated in a real data analysis based on the

RIT target detection blind test experiment. Conclusions are presented in Section 17.

2. Determining the "Best Algorithm" for Target Detection

Manolakis et al. [8] claimed that to identify the best algorithm for target detection, we need datasets with reliable ground truth spanning a diverse range of targets and backgrounds under various scenarios, target fill factors, and atmospheric conditions. Statistically significant numbers of target and background pixels are necessary to construct reliable ROC curves. Because in many cases this degree of data confirmation is unavailable, we suggest an alternative approach for estimating the best algorithm from among several detectors for specific backgrounds and targets. We start by presenting the RBTA [1]. The algorithm was originally developed for broadband infrared images with subpixel targets, but we altered it to account for pixel blur (atmospheric and system effects which would cause the emitted power of the target to be spread over several pixels) and multipixel targets in hyperspectral imagery.

To estimate detector performance, Rotman and Bar-Tal proposed a multistep process that begins with an analysis of the unmodified reflectance image that is available in the website without any embedded targets. (We assume that ideally no targets are present in the datacube being analyzed; if one were present, it would slightly distort the histogram of the image. We trust that such a distortion will not disturb the overall analysis of the image statistics). The algorithm being tested is evaluated for each pixel, and the results are summarized in what we call a false-alarm histogram. Next we embed targets into every pixel and evaluate each of the algorithms. This is done independently for each pixel (rather than simultaneously) so that surrounding pixels are not changed prior to algorithm evaluation. The results are arranged in a target detection histogram. Each histogram (false-alarm and target detection) is then normalized; a variable threshold is set and the area of the false-alarm and target histograms to the right of the threshold are measured. For any particular threshold, a pair of P_{FA} and P_D (probability of false alarm and probability of detection) values are generated. The threshold is swept through all possible detector outputs, generating a set of these pairs. When graphed, these points produce the ROC (receiver operating characteristic) curves.

We note that the target implantation mechanism as given here has ignored several possibly significant effects which would affect the values found of PD. In particular, the target spectrum is a nearly noiseless lab spectrum that does not have the same artifacts, noise, and degradation as the real imagery. Additionally, this approach assumes the data has been perfectly atmospherically compensated by RIT's algorithm, which is not necessarily true. In our opinion, this seems to limit the use of our method rather than to invalidate it. Since the atmospheric conditions at the time of the measurement were not known, we cannot implant atmospherically corrected signatures or validate the reflectance dataset that is available in RIT website. Instead, we are testing the response of the algorithms to an implanted nonatmospherically corrected target which has been substituted in the reflectance dataset as described above; in each examined pixel, the fraction of the laboratory signature replaced the fraction of the background signal. While inaccurate atmospheric correction may result in an unknown decrease in the target detection, we note that the final comparisons are for variations in algorithm selection for a given target signature. The method should not be used to calculate absolute values for the probability of detection of a particular target which indeed has been altered by atmospheric and other effects. Rather, we are attempting to determine which algorithms will have a superior probability of detecting a target of this type in the scenario. Future work should include a quantitative determination to what degree atmospheric effects change the ranking of different algorithms.

This methodology can be used for the following reasons: as a rule, the ROC curve, which are generated tend to have probabilities of detection which range from 0 to 1; the value of probabilities of false alarm, on the other hand, vary from 0 to some chosen threshold $P_{FA\text{-}MAX}$. This threshold is normally set quote low; a standard value would be 0.01. This is appropriate since the acceptable use of most detection algorithms could only be in the range where a small percentage of the pixels in the image would be false alarms.

Now, the exact distribution of the background pixels is crucial for the analysis of our detection algorithms; it will indeed be the exceptional pixels in the tail of the distribution which will determine the ROC curve. However, since the probability of detection is being determined by the entire P_D scale from 0 to 1, all pixels contribute. In other words, the target detection scheme in this paper is extremely sensitive to a few false alarms; it is much less sensitive to a few pixels with missed "synthetic" target signatures. As such, subtle effects affecting the exact form of the target signature *in situ* are not being measured; rather the average response of the algorithm to the target signatures placement in all the pixels is the key factor. For our above goal, that is, the comparison of different target algorithms, we believe our method to be reliable.

To summarize, ROC curve evaluation entails the following steps as demonstrated in Figure 1.

3. Subpixel Target Detection: Global Methods

3.1. CEM. In many cases, it is convenient to scale the matched filter such that it has a value of 1 when the target signature fills the pixel being examined. This scaling can be achieved by normalizing the matched filter to its value when operating on the designated target spectrum:

$$\mathrm{CEM}(\overline{x}) = \frac{\overline{s}^T \overline{\overline{R}}^{-1} \overline{x}}{\overline{s}^T \overline{\overline{R}}^{-1} \overline{s}}, \qquad (1)$$

where s is the reference signature of the target, R is the background correlation matrix, that is, an [L × L] matrix, L is the number of bands, and x is the observed pixel. Geometrically speaking, the CEM algorithm measures the projection of x onto s normalized by the length of s in the

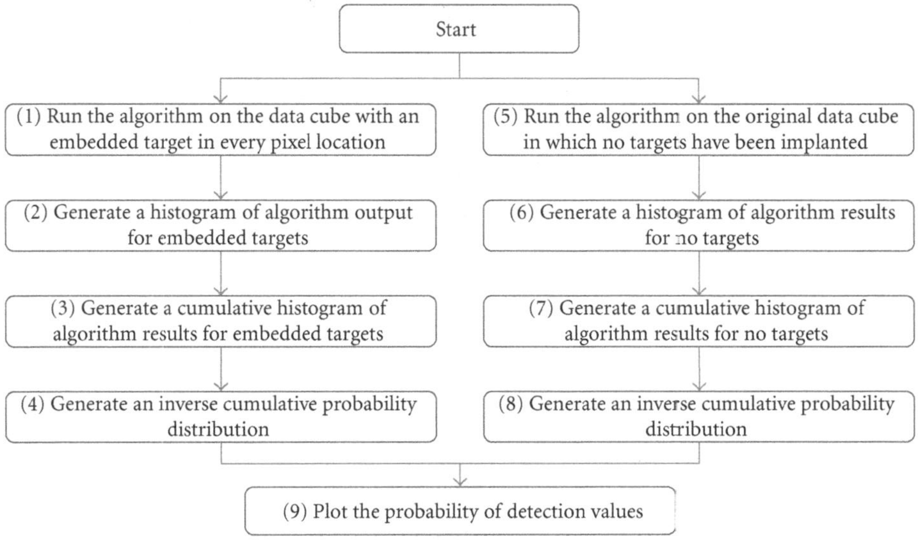

Figure 1: RBTA flow chart.

whitened space and thus leads to planar decision surfaces in that space. An important characteristic of the CEM algorithm is that its output is correlated to the target's fractional abundance in signature x, assuming the target signature is well isolated from the other endmembers, mixing is linear, and the relative abundances of the endmembers follow a Dirichlet distribution [9].

3.2. GLRT and ACE. Manolakis and his group [10–14] have described a number of stochastic target detection algorithms, including that attributed to Kelly [15] for solving to the Neyman-Pearson decision/detection theory for maximizing the probability of detection of a target with a fixed probability of false alarms. The solution uses a GLRT expressed as

$$\begin{aligned}\mathrm{GLRT}(\overline{x}) = & \left(\left[\left(\overline{s}-\overline{m_g}\right)^T \overline{\overline{G}}^{-1}\left(\overline{x}-\overline{m_g}\right)\right]^2\right) \\ & /\left(\left[\left(\overline{s}-\overline{m_g}\right)^T \overline{\overline{G}}^{-1}\left(\overline{s}-\overline{m_g}\right)\right]\right. \\ & \left.\cdot\left[1+(1/M)\cdot\left(\overline{x}-\overline{m_g}\right)^T \overline{\overline{G}}^{-1}\left(\overline{x}-\overline{m_g}\right)\right]\right),\end{aligned} \tag{2}$$

where s and x are the same as for (1), m_g is the global mean, G is the background covariance matrix, and M is the total number of samples.

The ACE algorithm, a variation of the GLRT algorithm, is expressed as

$$\begin{aligned}\mathrm{ACE}(\overline{x}) = & \left(\left[\left(\overline{s}-\overline{m_g}\right)^T \overline{\overline{G}}^{-1}\left(\overline{x}-\overline{m_g}\right)\right]^2\right) \\ & /\left(\left[\left(\overline{s}-\overline{m_g}\right)^T \overline{\overline{G}}^{-1}\left(\overline{s}-\overline{m_g}\right)\right]\right. \\ & \left.\cdot\left[\left(\overline{x}-\overline{m_g}\right)^T \overline{\overline{G}}^{-1}\left(\overline{x}-\overline{m_g}\right)\right]\right),\end{aligned} \tag{3}$$

with a maximum value of 1 for the case of $x = s$ and a minimum value of 0 when $x = m_g$.

In the context of target detection, the sign of $(s - m_g)^T G^{-1}(x - m_g)$ is important, as only positive abundances are of interest. (In contrast, this would not be the case for thermal gas detection, for example, where the target could be either absorptive or emissive in nature). Thus, in practice, a signed version of the GLRT algorithm is used as follows:

$$\mathrm{GLRT}_{\mathrm{sign}}(\overline{x}) = \mathrm{sign}\left[\left(\overline{s}-\overline{m_g}\right)^T \overline{\overline{G}}^{-1}\left(\overline{x}-\overline{m_g}\right)\right]\cdot \mathrm{GLRT}(\overline{x}). \tag{4}$$

The corresponding ACE algorithm for target detection, also a variation of the GLRT algorithm, is expressed as

$$\mathrm{ACE}_{\mathrm{sign}}(\overline{x}) = \mathrm{sign}\left[\left(\overline{s}-\overline{m_g}\right)^T \overline{\overline{G}}^{-1}\left(\overline{x}-\overline{m_g}\right)\right]\cdot \mathrm{ACE}(\overline{x}). \tag{5}$$

Because real data does not necessarily match the assumptions from which the above algorithms are derived, that is, a background probability distribution function assumed to be multivariate Gaussian with zero mean bias and an additive target model, we generally cannot expect that any of the algorithms will be optimal or even that one will consistently outperform another [8]. Nevertheless, it was shown by Manolakis [13] that for a limited dataset, although each of the algorithms exhibited some degree of success in target detection, the ACE algorithm performed best on the limited dataset tested.

In Figure 1, step 1, note that the target is not in all the positions simultaneously; rather, the result is obtained sequentially. Steps 4 and 8 are generated by one minus the cumulative histogram using the results from step 3 and 7, respectively, (these are the probability of detection-PD). In Step 9, we plot PD values (step 4) versus the PFA (step 8).

4. Subpixel Target Detection Using Local Spatial Information

Improving target detection involved replacing the global mean with the local mean. Using the local mean is definitely double edged: on one hand, we would expect that the closer the points used to evaluate the background are to the suspected target, the more likely it is that the estimate will be accurate. On the other hand, the noise in the estimate will decrease given more points entering into the estimation, assuming that the background is stationary and the noise is linearly added to the background and independent thereof. Our empirical experience confirmed by several studies (4) and (5) is that the closer we choose the pixels the better, with the condition that we do not have target contamination of the background pixels. It is this proviso that we wish to test here.

We note that we are not dealing here with a "local" covariance matrix which would change when evaluating each pixel in the image. Rather, we use the same covariance matrix throughout the image; it will simply be based on the difference of the sample pixels and their "local" background.

Since we are dealing with a subpixel target, which in the physical domain can affect only pixels in a limited spatial area surrounding the center of the target, we used the eight nearest neighbors approach to estimate the value of the test pixels. The CEM algorithm does not use the mean and will therefore be unaffected by the above changes. The GLRT can be improved as follows:

$$\mathrm{GLRT}_{\mathrm{local}}(\overline{x}) = \left(\left[\left(\overline{s}-\overline{m_g}\right)^T\overline{\overline{G}}^{-1}(\overline{x}-\overline{m_8})\right]^2\right) \Big/ \left(\left[(\overline{s}-\overline{m_8})^T\overline{\overline{G}}^{-1}(\overline{s}-\overline{m_8})\right] \cdot \left[1+(1/M)\cdot(\overline{x}-\overline{m_8})^T\overline{\overline{G}}^{-1}(\overline{x}-\overline{m_8})\right]\right), \quad (6)$$

and for target detection

$$\mathrm{GLRT}_{\mathrm{sign-local}}(\overline{x}) = \mathrm{sign}\left[(\overline{s}-\overline{m_8})^T\overline{\overline{G}}^{-1}(\overline{x}-\overline{m_8})\right] \cdot \mathrm{GLRT}_{\mathrm{local}}(\overline{x}), \quad (7)$$

with m_8, the mean of the eight nearest neighbors, replacing the global mean m_g. For the ACE detector, the same procedure (replacing m_g by m_8) may be followed.

Segmentation [16–18] or even more local covariance matrices [2, 4, 6, 19] can be used to improve the covariance matrix. Common to all these methods is an increased need for high performance computational resources, while the corresponding influence each method has on detection ability is uncertain and highly dependent on the pictures being analyzed. Used in parallel, the algorithms create new difficulties through the combination of results from different segments. We used a global covariance matrix, but adapted it to local variations by using the local rather than the global mean, that is,

$$\overline{\overline{G}}_{\mathrm{global}} = \frac{\left[\overline{X}-\overline{\overline{M_g}}\right]^T\left[\overline{X}-\overline{\overline{M_g}}\right]}{M}, \qquad \overline{\overline{G}}_{\mathrm{local}} = \frac{\left[\overline{X}-\overline{\overline{M_8}}\right]^T\left[\overline{X}-\overline{\overline{M_8}}\right]}{M}, \quad (8)$$

where X is a two-dimensional matrix ($M \times L$), in which M is the number of pixels and L is the number of bands, m_g $[1 \times L]$ is the mean vector of X, and M_g *is* m_g replicate M times. When we use M_8 for the covariance matrix, we do not need to replicate the mean, because M_8 is also of size $[M \times L]$, and this is the appropriate covariance matrix for whitening $X - M_8$.

5. Data

We tested our algorithms on the online reflectance data sets and the hyperspectral data collected over Cooke City. The Cooke City imagery was acquired on 4 July 2006 using the HyMap VNIR/SWIR sensor with 126 spectral bands. Two hyperspectral scenes are provided with the dataset, one intended to be used for development and testing (the "Self Test" scene, where the positions of some targets are known) and the other intended to be used for detection performance evaluation (the "Blind Test" scene, where the position of targets is unknown). The data was corrected for atmospheric effects and available in the website but the exact atmospheric condition and the atmospheric correction algorithm are not available in the website and we assume that the reflectance dataset is good but not perfect. In Figure 2, we present the image in false color.

The target signatures, used both in the algorithm for detection and in the implantation of the synthetic targets in the RBTA method were laboratory measured and in reflectance units. The GSD is approximately 3 m. In Figure 3, we present the spectral signature of the targets in the blind test image.

The list of all targets is presented in Table 1 below.

6. Spatial Effect

6.1. Analytical and Simulated Performances of GLRT and ACE

6.1.1. Simple Case. The general form for local target detection as described in Section 3 is

$$D_{\mathrm{Local}}(\overline{x}) = \left(\left[(\overline{s}-\overline{m_8})^T\overline{\overline{G}}^{-1}(\overline{x}-\overline{m_8})\right]^2\right) \Big/ \left(\left[(\overline{s}-\overline{m_8})^T\overline{\overline{G}}^{-1}(\overline{s}-\overline{m_8})\right] \cdot \left[\Psi_1+\Psi_2\cdot(\overline{x}-\overline{m_8})^T\overline{\overline{G}}^{-1}(\overline{x}-\overline{m_8})\right]\right), \quad (9)$$

Table 1: Targets description.

Target ID	Target description	size (m²) No. 1	size (m²) No. 2	Self test ground truth	Blind test ground truth
F1	Red cotton fabric panel	3 × 3	N/A	Yes	No
F2	Yellow nylon fabric panel	3 × 3	N/A	Yes	No
F3	Blue cotton fabric panel	2 × 2	1 × 1	Yes	No
F4	Red nylon fabric panel	2 × 2	1 × 1	Yes	No
F5	Maroon nylon fabric panel	2 × 2	1 × 1	No	Web score
F6	Gray nylon fabric panel	2 × 2	1 × 1	No	Web score
F7	Green cotton fabric panel	2 × 2	1 × 1	No	Web score
V1	Chevy Blazer, green	4 × 2	N/A	Yes	Web score
V2	Toyota T100, white with black plastic liner	3 × 1.7	N/A	Yes	Web score
V3	Subaru GL Wagon, Red	4.5 × 1.6	N/A	Yes	Web score

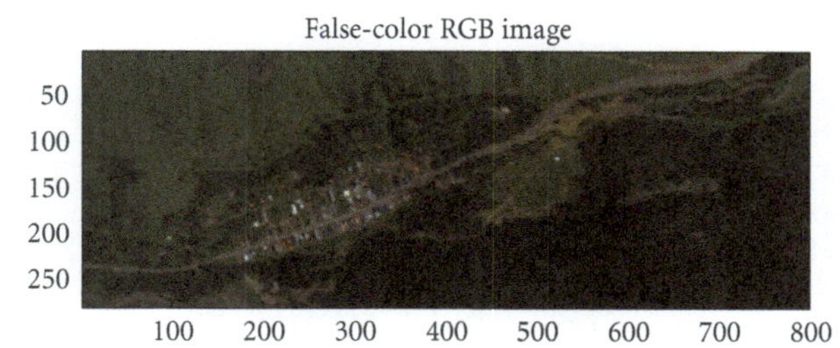

Figure 2: False-color RGB of the Cooke City imagery.

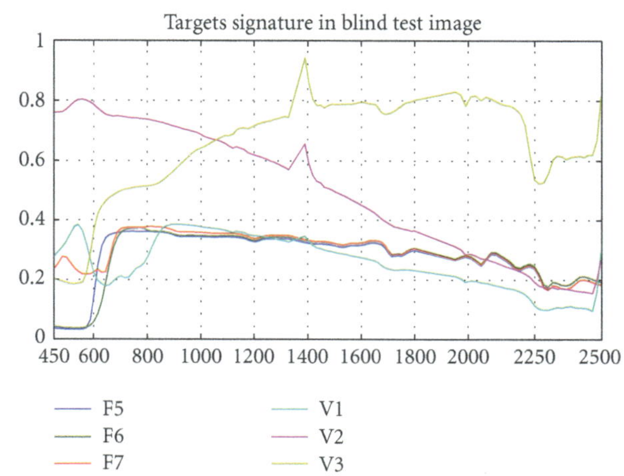

Figure 3: Spectral signatures of the targets that are present in the Blind test image x-axis is the wavelength [nm] and y-axis is the reflectance unit less.

with m_8 as the mean of eight neighbors. G, the global-local covariance matrix, is computed as

$$G_{\text{global\_local}} = \frac{\left(\overline{\overline{X}} - \overline{\overline{M_8}}\right)^T \cdot \left(\overline{\overline{X}} - \overline{\overline{M_8}}\right)}{L}, \tag{10}$$

where we can get GLRT and ACE as functions of $\Psi_1 + \Psi_2$:

$$\begin{aligned} \text{GLRT}: \Psi_1 &= M, \qquad \Psi_2 = 1, \\ \text{ACE}: \Psi_1 &= 0, \qquad \Psi_2 = 1. \end{aligned} \tag{11}$$

For the case in which the PUT (pixel under testing) x is exactly s, we obtain the following results:

$$\begin{aligned} D_{\text{Local}}(\overline{x}) = {} & \left(\left[(\overline{s} - \overline{m_8})^T \overline{\overline{G}}^{-1} (\overline{s} - \overline{m_8})\right]^2\right) \\ & / \left(\left[(\overline{s} - \overline{m_8})^T \overline{\overline{G}}^{-1} (\overline{s} - \overline{m_8})\right]\right. \\ & \left. \cdot \left[(\Psi_1 + \Psi_2) \cdot (\overline{s} - \overline{m_8})^T \overline{\overline{G}}^{-1} (\overline{s} - \overline{m_8})\right]\right). \end{aligned} \tag{12}$$

Let us define the scalar C as−

$$C = (\overline{s} - \overline{m_8})^T \overline{\overline{G}}^{-1} (\overline{s} - \overline{m_8}). \tag{13}$$

Therefore, when $\overline{x}$ is exactly $\overline{s}$, GLRT and ACE can be written as

$$\begin{aligned} \text{GLRT}(\overline{s}) &= \frac{C}{[M + C]} = \frac{1}{M/C + 1}, \\ \text{ACE}(s) &= 1. \end{aligned} \tag{14}$$

Assuming that the data is normally distributed, C is chi-square distributed with $E(C) = \text{L}$, where L is the number of bands. For the case in which $M \gg E(C) = \text{L}$, we can assume that

$$\text{GLRT}(s) \cong \frac{C}{M}. \tag{15}$$

7. Pixel Phasing Case

When imaging, the target can often fall across several pixels even if its total size is only a single pixel; we will call this effect pixel phasing even though it is a natural consequence of imaging system quantization. The pixel phasing effect can be demonstrated by a target one pixel in size, the imaging of which leads to pixel phasing registration defined by p,

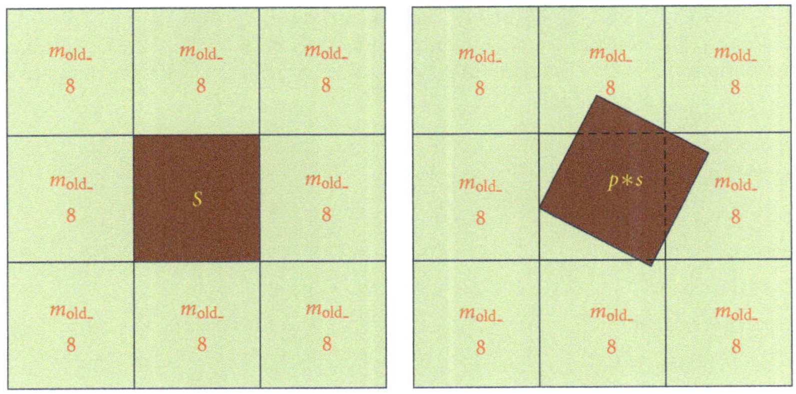

(a) Simple case (b) Pixel phasing case

FIGURE 4: Pixel phasing schema.

such that $0 < p < 1$ (0 corresponds to perfect sampling, with the target completely replacing the background) (Figure 4). From the point of view of the central pixel, it is not important the spatial location of the fraction within the pixel nor the location of the remainder of the target signature. Assuming uniform backgrounds of $m_{\text{old\_8}}$ for both center pixels, they can now be given as in Figure 4.

We obtain the following:

$$\overline{x_{\text{New}}} = p \cdot \overline{s} + (1 - p) \cdot \overline{m_{\text{old\_8}}}, \tag{16}$$

where x_{new} is the new PUT for the pixel phasing case and

$$m_{\text{new\_8}} = \frac{7+p}{8} \cdot \overline{m_{\text{old\_8}}} + \frac{1-p}{8} * \overline{s}, \tag{17}$$

where $m_{\text{new\_8}}$ is the new mean for the background.

We now evaluate the terms $(\overline{s} - \overline{m_{\text{new8}}})$ and $(\overline{x} - \overline{m_{\text{new8}}})$ as follows:

$$\begin{aligned}(\overline{s} - \overline{m_{\text{New8}}}) &= \overline{s} - \left[\frac{7+p}{8} \cdot \overline{m_{\text{old\_8}}} + \frac{1-p}{8} * \overline{s}\right] \\ &= \frac{(7+p)}{8} \cdot (\overline{s} - \overline{m_{\text{old\_8}}}),\end{aligned} \tag{18}$$

$$\begin{aligned}(\overline{x_{\text{New}}} - \overline{m_{\text{New8}}}) &= (p \cdot \overline{s} + (1-p) \cdot \overline{m_{\text{old\_8}}}) \\ &\quad - \left(\frac{7+p}{8} \cdot \overline{m_{\text{old\_8}}} + \frac{1-p}{8} * \overline{s}\right) \\ &= \frac{9 \cdot p - 1}{8} \cdot (\overline{s} - \overline{m_{\text{old\_8}}}).\end{aligned} \tag{19}$$

The GLRT result now becomes

$$\begin{aligned}D_{\text{Local}}(\overline{x}) = &\left([((9 \cdot p - 1)/8) \cdot (((7+p)/8) \cdot C)]^2\right) \\ &/([((9 \cdot p - 1)/8) \cdot ((9 \cdot p - 1)/8) \cdot C] \\ &\cdot [\Psi_1 + \Psi_2((7+p)/8) \cdot ((7+p)/8) \cdot C]),\end{aligned} \tag{20}$$

where $D_{\text{Local}}(\overline{x})$ is the general local detector for the pixel phasing case.

For the case in which $N \gg C$, we calculate that

$$\begin{aligned}\text{GLRT}_{\text{miss\_sampling}}(\overline{x}) &\cong \frac{(p^2 + 14p + 49)}{64} \cdot \frac{C}{M} \\ &= \frac{(p^2 + 14p + 49)}{64} \cdot \text{GLRT}_{\text{local}},\end{aligned} \tag{21}$$

where $\text{GLRT}_{\text{miss\_sampling}}(x)$ is the expected GLRT value for the pixel phasing case and $M \gg L$. The GLRT expected value degrades as a function of p. But for ACE ($\Psi_1 = 0, \Psi_2 = 1$) we still get expected values of 1:

$$\text{ACE}_{\text{miss\_sampling}}(x) = 1 = \text{ACE}_{\text{local}}(x). \tag{22}$$

In this model, the complete lack of ACE degradation as a function of pixel phasing may explain why ACE is a more robust detector than GLRT in many test cases, as noted in the literature [8, 20].

8. Ranking the Algorithms by RBTA

The difficult task of synthesizing a synthetic image to help predict which algorithm to select is simplified and detector selection is facilitated if we synthesize only the target signature of our real image. Suppose we want to determine the proper detector for a specific target. We have already selected our method (e.g., CEM, GLRT, or ACE), and now we want to select the size of the local window. One approach is to assume that the best size for the local window is that under which the PUT value can be predicted with minimum error vis-à-vis the real PUT (in which we normalize each band by the mean values of the pixels in that band).

The approach outlined above depends only on the background image, not on the target signature, and it entails two assumptions: first, estimating signature values will improve our detector results independent of the different target signatures and second, the target has no effect on its neighbors. Address these assumptions in the following sections.

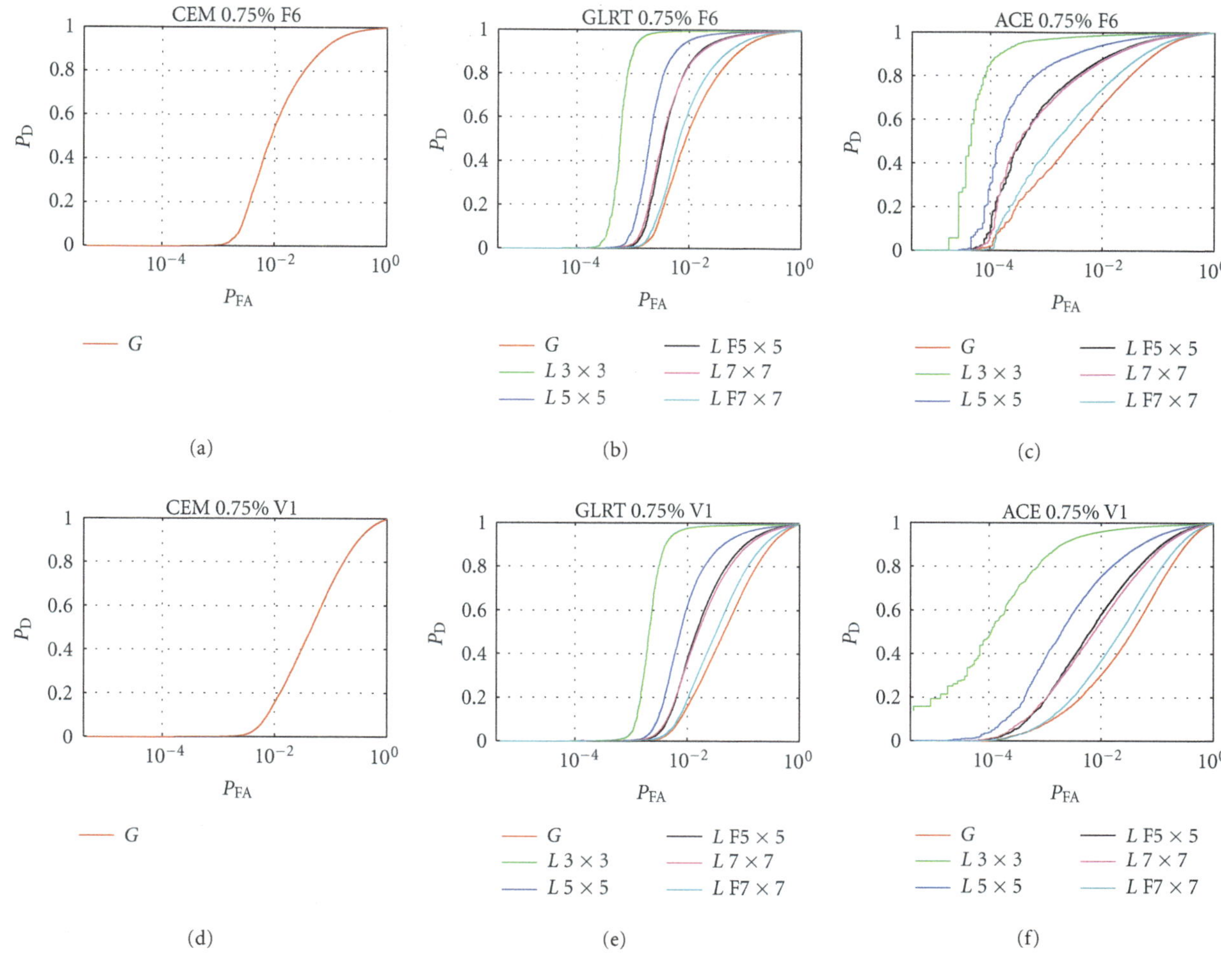

Figure 5: RBTA results for different size of local windows.

9. How to Use RBTA

The implementation of RBTA, which depends on our ability to implant realistic signals into backgrounds and measure detector response, should be done carefully. We cannot expect the real signature to be identical to a library signature, but we can hope for a high level of similarity. The low percentage of the target signature that actually enters any particular pixel is demonstrated in Figure 6; the response of the CEM filter, which responds proportionally to the percentage of the target fill in the tested filter, was maximum at 0.06.

As a rule, to test and challenge our algorithms by examining the area under the ROC curve, we need to test targets which neither "saturate" the ROC curve (with a probability of detection close to one with no false alarms detected) nor result in a "diagonal" ROC curve (in which the probability of detection equals the probability of false alarms. As the allowable false alarm rate decreases, the strength of our synthetic implanted target would need to increase; if we know what the acceptable false alarm rate is, we can select the target percent that will demonstrate the dynamic range around this rate and get results for our detectors (Figure 5).

For the values found experimentally for p, the target was easily detectable and saturated our ROC curve. Thus, we only embedded 0.0075 of the target signature in the background pixels to generate the target detection histogram. In our results, we found that for GLRT and ACE, the best local window size is 3×3 pixels (CEM has no local form). We also see that using bigger windows to estimate the pixel signature value gets us closer to the performance of global detectors that use a global mean. In this case, it is clea that local detectors are superior to global detectors.

In terms of real data, we must expect each target to affect more than one pixel even if its total physical size is at the subpixel level. A discussion of this point follows below and leads to improvement of the RBT algorithm.

10. Improvements to RBTA

10.1. Target Size. As will be discussed in Sections 11-12 the apparent target size in the final digital image is related both to its physical size and to various atmospheric and sensor effects, for example, its point spread function (PSF), Gibbs effect, crosstalk between pixels, spatial sampling, band-to-

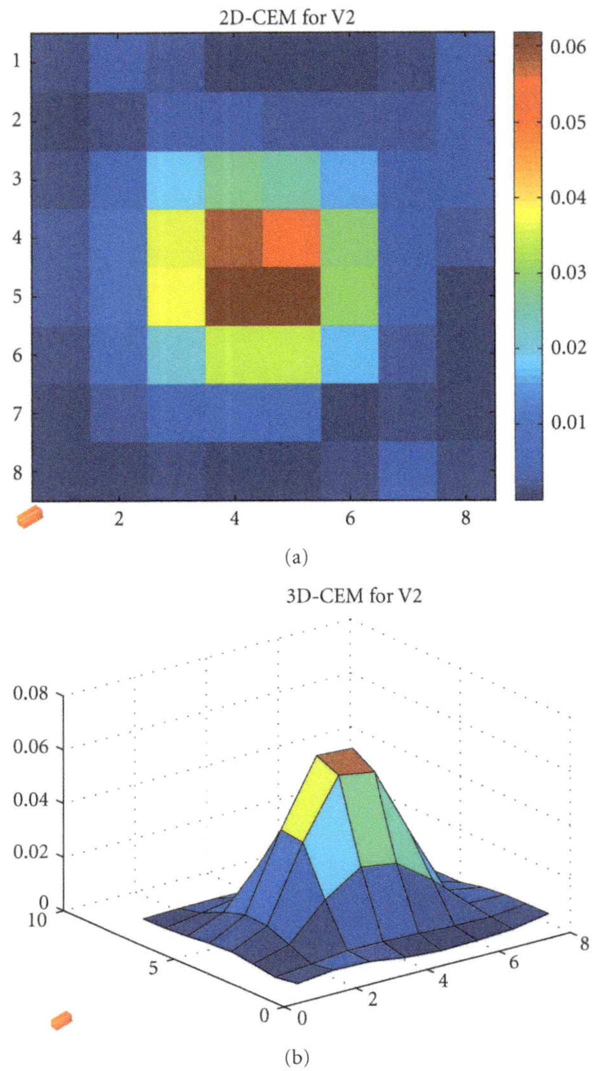

FIGURE 6: CEM results (2D and 3D) for target with pixel size.

band misregistration [5], and motion compensation. Thus, a target of a single pixel could actually occupy several pixels. In the RIT blind test, there are two 3×3-m targets, that is, exactly the size of the ground sample resolution (GSD) for the self and blind test images. Figure 6 shows a sample target of this size.

11. Spatial Sampling Effect

If we take into account only the spatial sampling effect, we can estimate the percent of pixel area partially occupied by the target. Notice that even targets of subpixel size often spread over neighboring pixels (Figure 7).

Put formally, the percent of pixel area covered as a function of target size, target location, and target orientation is

$$S_{\text{target\_in\_pixel}} = \iint_{-0.5}^{0.5} \frac{4}{\pi} \cdot (\alpha \cdot \beta) \cdot \text{UnitBox}[\tilde{x}, \tilde{y}] \cdot dx\, dy,$$

$$\tilde{x} = [\alpha \cdot [(x - \Delta x) \cdot \cos(\theta) - (y - \Delta y) \cdot \sin(\theta)]],$$
$$\tilde{y} = [\beta \cdot [(x - \Delta x) \cdot \sin(\theta) + (y - \Delta y) \cdot \cos(\theta)]], \tag{23}$$

where θ represents the clockwise rotation of the target relative to the pixel grid, and α, β represent the proportions of target length and width, respectively, relative to pixel physical dimension, Δx, Δy are the transition of the target origin relative to the pixel origin, as demonstrated in Figure 8

$$\text{UnitBox}[x, y] = \begin{cases} 1 & -\frac{1}{2} < x < \frac{1}{2} \text{ and}, -\frac{1}{2} < y < \frac{1}{2}, \\ 0 & \text{else}. \end{cases} \tag{24}$$

The expected value $E[S\alpha, \beta]$ is

$$E\left[S_{\alpha,\beta}\right] = \iiint_D S_{\text{target\_in\_pixel}}\, d\Delta x \cdot d\Delta y \cdot d\theta,$$
$$D = \begin{Bmatrix} 0 \le \Delta x \le 0.5 \\ 0 \le \Delta y \le 0.5 \\ 0 \le \theta \le \pi \end{Bmatrix}. \tag{25}$$

If we set θ to a constant value of zero and the target length and width are half the size of a pixel, we can simulate all the locations of the target where it's covering the same percent of pixel area (Figure 9).

Calculating (25) for a different size target using a numerical example produced the results shown in Figure 10.

In the graphs depicting pixel coverage as a function of physical target size, the x-axis is the ratio either between the target area and the pixel area (Figure 10(a)) or between the target length and the pixel length (Figure 10(b)). The blue line in both figures represents the percent of target within the pixel, while the green line is the percent of the pixel expected to be covered. It is intuitive that a very small target will be located in only one pixel, covering a small percent of that pixel. It is less intuitive, however, that the expected pixel to be covered will be entirely covered only by a target with an area four times that of the pixel.

12. Point Spread Function Effect

The PSF effect, present in any optical system, is not always known. Let us assume that the PSF is a typical, rotationally symmetric Gaussian filter of size 3×3 with standard deviation sigma 1/2.

Figure 11 demonstrates the synthetic spread effect that emerges from the convolution of the optical PSF (Figure 11(a)), and the physical pixels phasing due to target size (Figure 11(b)). For the spatial sampling we took the mean case representing the average pixel phasing that we could expect. Figure 11(c) represents the total effect, for example, convolution between (a) and (b). We devised an improved RBTA (IRBTA) and embedded the pixel signature and its neighbors with ratios as shown in Figure 11(c).

A comparison of Figures 5 and 12 shows that global detector and local detectors that both use only 7×7 frames

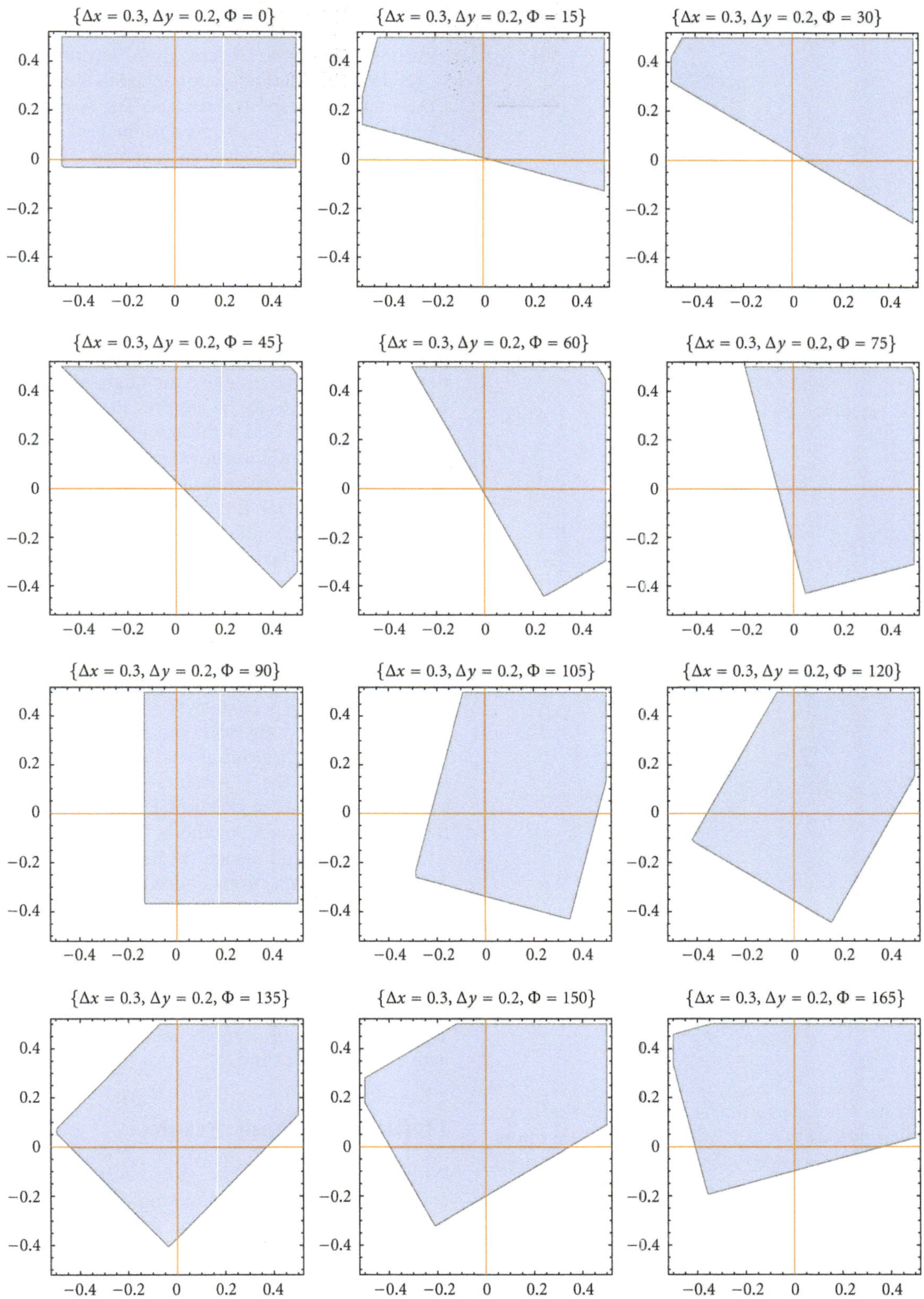

FIGURE 7: Demonstration for target size of 4/3 over 2/3 with origin at (0.3,0.2) and different rotation angle.

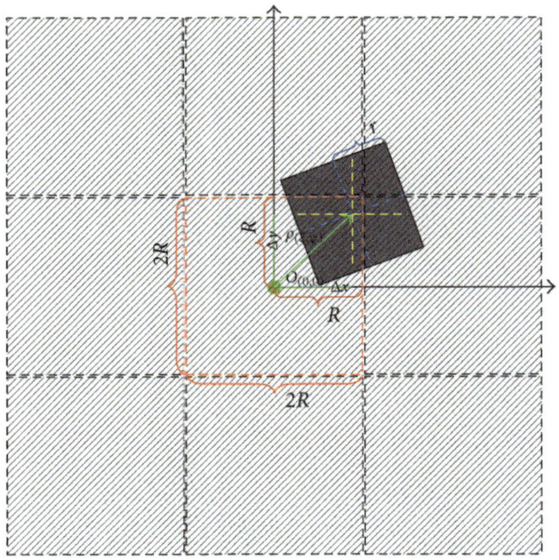

FIGURE 8: Variable demonstration.

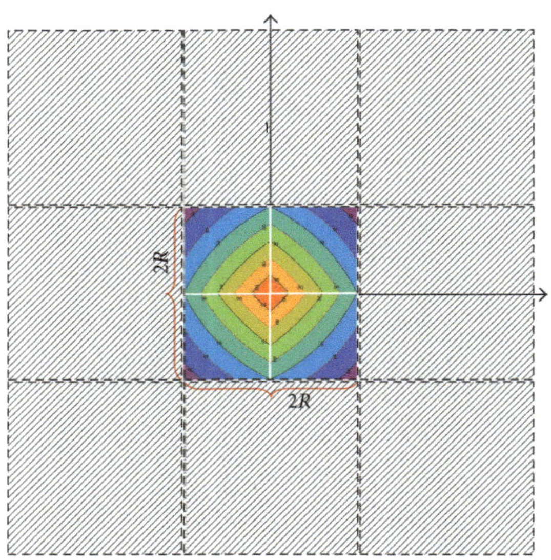

FIGURE 9: Pixel covered as function of target size for $\theta = 0$, $\alpha = 2$, and $\beta = 2$.

have exactly the same performance. Evidently, the target spread effect is only localized in a 5×5 window. Furthermore, the most significant effect is for a local 3×3 window. Indeed, the local 3×3 seems robust, and it outperformed all other detectors. We expected these detectors to be the best for this target in this image.

For our next stage of proof of concept, we need to compare real detection performances to these simulated results. We obtain this by using the self-test dataset and by submitting our algorithms to the RIT target detection blind test and comparing the ranking of our algorithms for each of the target signatures available in the set; we can then see if the IRBTA predictions of the preferred algorithms are true.

13. Detection Performance: Experimental Results

13.1. Scoring Methodology and Result Presentation. Detection algorithm performances are given according to the RIT target detection methodology applied to the aforementioned Cooke City hyperspectral dataset. The score is based on a comparison of the values given to the background pixels in the image to the value given to the target. The target value is defined as the maximum value given the pixels in the target area; the metric then counts how many pixels there are in the overall image greater than or equal to the target value. Since the threshold needed to detect the target would have to be less than or equal to the target value, all points above this value are false alarms. The score given for any algorithm/target combination would thus be the number of pixels above the target value. Perfect detection would equal the value 1, since the only pixel equal or above the target value would be the target itself; no false alarms are present.

In Tables 2, 3, and 4 the scores for the self-test targets were calculated, and those for the blind test were obtained by uploading our results to the RIT website. We have colored coded the results for the ease of the reader. A value of 1 (italic background) indicates perfect performance (no false positives). We marked all scores greater than 448 (e.g., PFA = $2 * 10^{-3}$) with bold italic backgrounds. We will associate a false alarm score greater than this as a fundamentally undetected target; ranking above these values is irrelevant All results (i.e., scores) were divided among the three tables: Table 2 presents the global method (CEM, GLRT and ACE) while Tables 3 and 4 contain the results for the local GLRT and ACE, respectively, with different sized windows. The best score for each group is marked with a bold. Ignoring (as stated above) the PFA values $> 2 * 10 - 3$ (bold italic background), then the global ACE is clearly the best method from among the global methods. The local GLRT and ACE with the 3×3 windows are the best of the local GLRT and ACE methods, respectively. Indeed, the overall performance of the local ACE algorithm with 3×3 windows was superior to all other algorithms. All these results were obtained using the IRBTA without any need of ground truth. Some targets, for example, V3 in the blind test and V2 and V3 in the self test, degraded with the 3×3 window of the local ACE algorithm, but those targets were effectively undetectable with all three algorithms.

14. Global Methods: Results

Within the results for the global methods (Table 2), we marked detector scores indicating a significant advantage relative to other detectors in bold. Our results clearly show that the ACE detectors outperformed the other global detectors.

15. Local GLRT and Local ACE: Results

Similar to the global methods analysis, here (Tables 3 and 4) we also applied **yellow** highlights to cells with exceptional

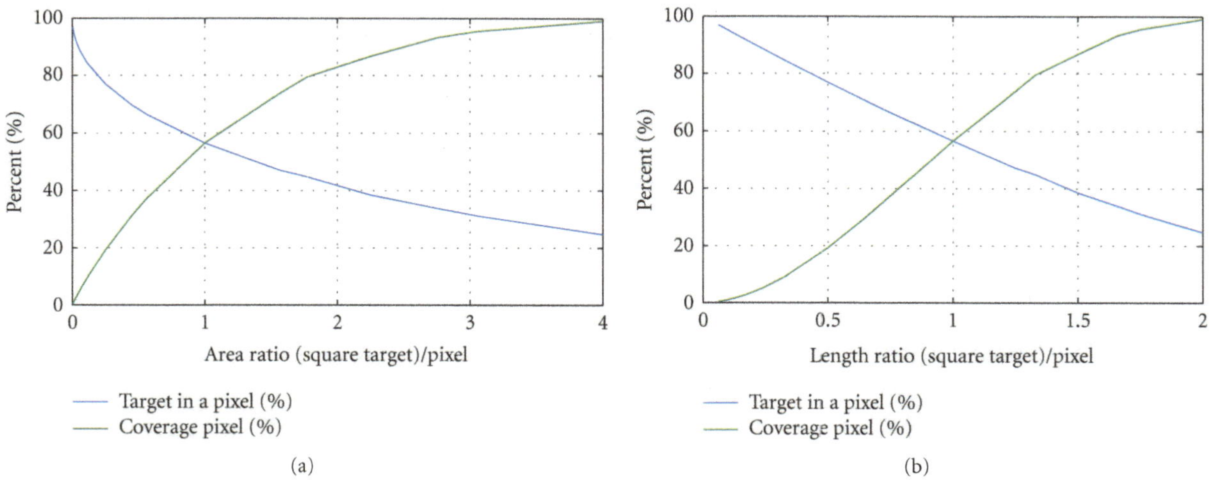

Figure 10: Pixel coverage as a function of target size.

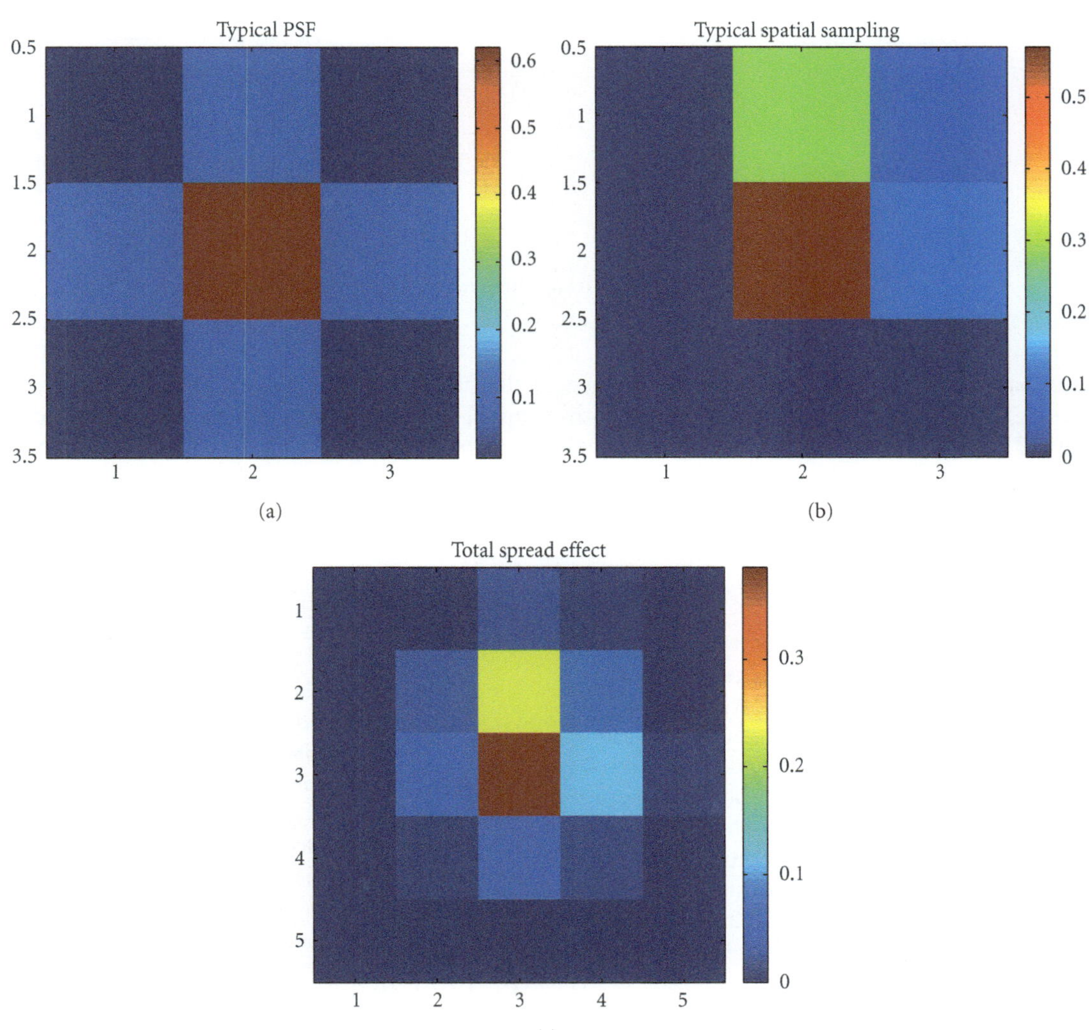

Figure 11: Synthetic spread emulation.

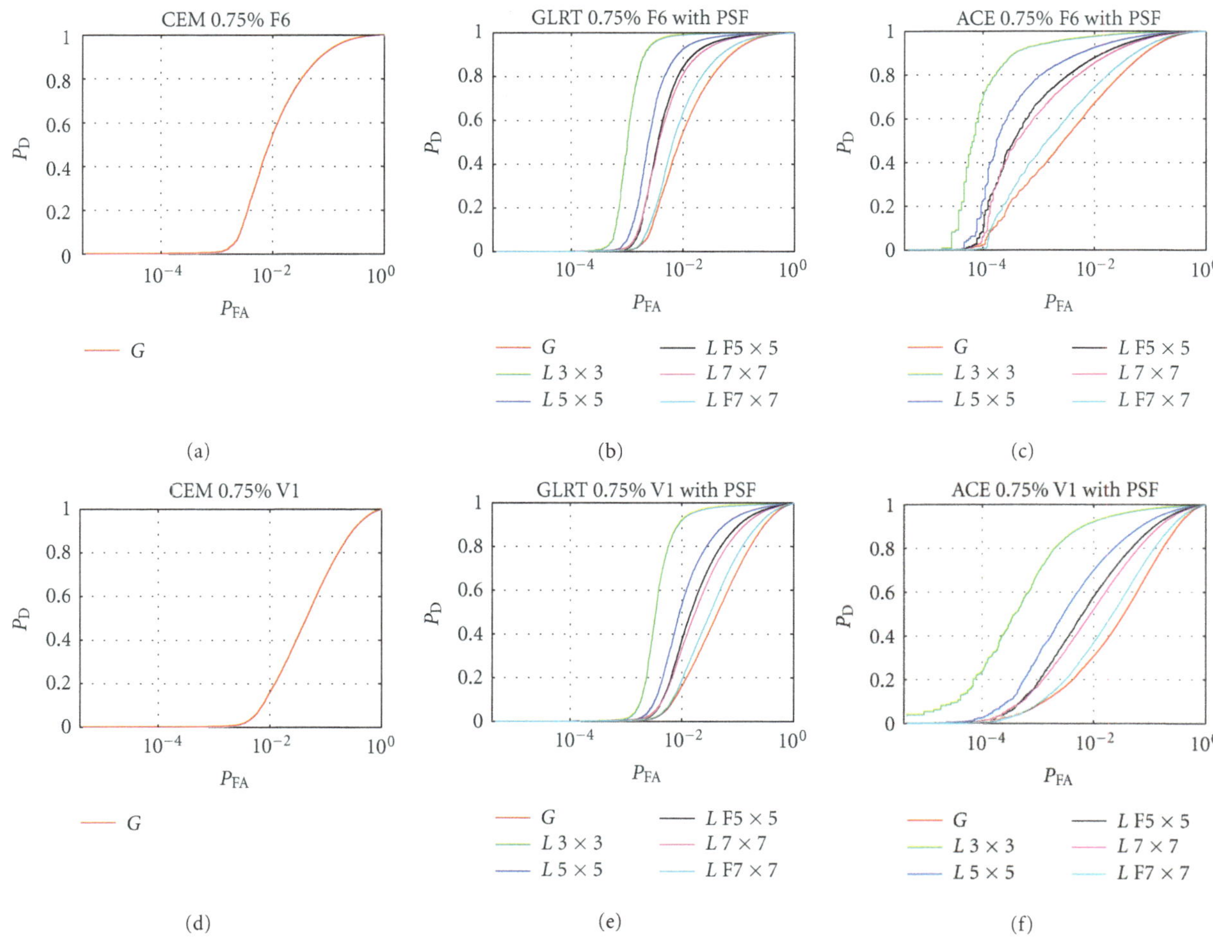

FIGURE 12: IRBTA results for different type of detectors.

detector scores that signified a significant advantage relative to other detectors. From our analysis, it is clear that for both the GLRT (Table 3) and ACE (Table 4) detectors, the optimal size of the local window for this size target is 3 × 3, a result that is in accordance with our estimation by IRBTA.

16. Benchmark Results

The RIT website provides a comparison between the results of different algorithms which have been submitted throughout the world. Table 5, which presents the web rank for the blind test, clearly shows that the performance of the local ACE 3×3 algorithm was superior relative to dozens of results that have been uploaded to the site. We achieved perfect detection for V1 (the nearest score was 15 for algorithms that work well only for V1) and reasonable detection for V2 (the nearest score was 196). There was no algorithm that achieved detection for V3 (the best score was 112). (The above conclusions are correct if we do not take into account one algorithm that was submitted to the RIT site, labeled "WTA", i.e., "Winner Take All." This algorithm exhibited perfect detection for all targets, including the vehicles, However, since the WTA algorithm has not been published, we cannot test it. In addition, there are objective reasons to believe that the algorithm does not actually exist; perfect detection of all targets with no false alarms ever would be an almost impossible result). While we included WTA in our scoring in Table 5, our comments following the Table do not consider this algorithm).

It is possible to compare the actual results obtained by the target detection algorithms on the RIT system to those obtained from the IRBTA simulation. In Tables 6 and 7, we show the results of several different algorithms when detecting the target V1; when we examine the results from the RIT test, we see that the ACE algorithm outperformed the GLRT algorithm, and that there was a definite preference for the 3×3 frame compared to the other frames (Table 6). These results are mirrored in our IRBTA results (Table 7). This (and results on the other targets) confirms our hypothesis that the RBTA can be used to simulate and predict target detection probabilities a priori in scenarios where actual targets are not yet present.

17. Conclusion

In this paper, we showed that there is no "best hyperspectral detection algorithm" for all images and targets. We noted the significant effect spatial distribution has on detector

Table 2: Results of global methods.

	Target ID		CEM	Global GLRT	ACE
Self-test	F1		15	13	*1*
	F2		*1*	*1*	*1*
	F3	1 m²	***1350***	***1366***	***1157***
		2 m²	30	28	**1**
	F4	1 m²	208	207	**118**
		2 m²	23	24	**1**
Blind test	F5	1 m²	163	156	**18**
		2 m²	19	19	**1**
	F6	1 m²	75	76	**7**
		2 m²	13	13	**1**
	F7	1 m²	***1204***	***1290***	***1792***
		2 m²	315	318	**2**
Self-test	V1		324	321	**34**
	V2		***2028***	***1852***	***1027***
	V3		***853***	***775***	***367***
Blind test	V1		422	428	**179**
	V2		***931***	***921***	***1365***
	V3		***3230***	***3154***	***2999***

Table 3: Results of local GLRT for different-sized windows.

	Target ID		GLRT local 3 × 3	5 × 5	F5 × 5	7 × 7	F7 × 7
Self-test	F1		**4**	5	6	6	8
	F2		*1*	*1*	*1*	*1*	*1*
	F3	1 m²	**188**	208	230	293	426
		2 m²	**9**	13	13	18	22
	F4	1 m²	**65**	101	121	120	152
		2 m²	**4**	8	10	12	18
Blind test	F5	1 m2	**15**	48	66	78	125
		2 m²	*1*	*1*	*1*	*1*	3
	F6	1 m²	**14**	24	34	33	54
		2 m²	**3**	5	5	5	6
	F7	1 m²	**92**	152	202	244	388
		2 m²	**81**	120	152	160	230
Self-test	V1		101	**74**	87	77	107
	V2		***6734***	***3456***	***3286***	***2812***	***2891***
	V3		***2506***	***875***	***741***	***706***	***712***
Blind test	V1		**37**	46	58	77	156
	V2		**283**	392	435	***518***	***694***
	V3		***11274***	***6455***	***5623***	***8259***	***10060***

Table 4: Results of local ACE for different-sized windows.

	Target ID		ACE local 3 × 3	5 × 5	F5 × 5	7 × 7	F7 × 7
Self-test	F1		*1*	*1*	*1*	*1*	*1*
	F2		*1*	*1*	*1*	*1*	*1*
	F3	1 m²	**16**	17	19	33	62
		2 m²	*1*	*1*	*1*	*1*	*1*
	F4	1 m²	**50**	68	75	72	75
		2 m²	*1*	*1*	*1*	*1*	*1*
Blind test	F5	1 m²	**11**	**11**	12	13	13
		2 m²	*1*	*1*	*1*	*1*	*1*
	F6	1 m²	6	5	5	5	5
		2 m²	*1*	*1*	*1*	*1*	*1*
	F7	1 m²	**5**	**5**	**5**	**5**	9
		2 m²	2	2	2	2	*1*
Self-test	V1		**3**	**3**	**3**	**3**	7
	V2		***3148***	***1524***	***1457***	***1171***	***1416***
	V3		***2387***	***1143***	***961***	***812***	***792***
Blind test	V1		**1**	5	8	15	21
	V2		**79**	106	119	198	359
	V3		***12513***	***6859***	***6327***	***10543***	***16328***

Table 5: Benchmark results.

	Target ID		Local ACE 3 × 3	Web rank
Blind test	F5	1 m²	**11**	12/148
		2 m²	*1*	1/148
	F6	1 m²	6	11/90
		2 m²	*1*	1/90
	F7	1 m²	**5**	3/82
		2 m²	**2**	5/82
Blind test		V1	**1**	1/50
		V2	**79**	3/82
		V3	***12513***	52/86

performances, and we showed that the RBTA can be used to select the proper detectors from among several detectors but without any need for ground truth. However, point targets can influence their neighboring pixels, due either to the PSF or to the target spreading across more than one pixel. To account for this potential source of inaccuracy, therefore, we introduced the improved RBTA (IRBTA), whose exact method of use depended on the target size. In addition, we showed that when detectors calculated the mean for estimating the pixel signature value, we did not need ground truth to find the best estimate. We tested our concept through the selection of the best detectors from among stochastic algorithms for target detection, that is, the constrained energy minimization (CEM), generalized likelihood ratio test (GLRT), and adaptive coherence estimator (ACE) algorithms, using the dataset and scoring methodology of the Rochester Institute of Technology (RIT) Target Detection Blind Test project. The results showed that our concepts predicted the best algorithms for the particular images and targets provided by the website.

Table 6: RIT results for the actual detection of V1 in RIT test image are shown in the first two lines. The percentage of implanted target was 0.75%. The GLRT and ACE algorithms were calculated as presented in the text. The size of the background was calculated for 3×3, 5×5 and 7×7 frames, excluding the center pixel. The "Only" 7×7 and 5×5 algorithms only used the outer ring of the window. The third and fourth lines represent the same results normalized by dividing by the values obtained by the 3×3 filter.

Website results						
Window	Only 7 × 7	7 × 7	Only 5 × 5	5 × 5	3 × 3	Global
V1_GLRT	156	77	58	46	37	428
V1_ACE	21	15	8	5	1	179
Normalize relative to 3 × 3						
V1_GLRT	**4.22**	**2.08**	**1.57**	**1.24**	**1**	**11.57**
V1_ACE	**21**	**15**	**8**	**5**	**1**	**179**

Table 7: The A_{th} $((A - th^2)/(th - th^2))$ where A is the area under the Pd-log (P_{fa}) curve and th is the threshold (i.e., the maximum false alarm rate) results of the IRBTA algorithm for V1 in the RIT test image are shown in the first two lines. The maximum false alarm rate is 10^{-3}; other parameters are as given in Table 6. The third and fourth lines represent the same results normalized by dividing the value of the 3×3 filter by the values of the other filters.

Ath by IRBTA-Threshold = 10^{-3}						
Window	Only 7 × 7	7 × 7	Only 5 × 5	5 × 5	3 × 3	Global
V1_GLRT	0.0007	0.0011	0.0010	0.0014	0.0032	0.0003
V1_ACE	0.0436	0.0908	0.0946	0.1657	0.4981	0.0420
Normalize relative to 3 × 3						
V1_GLRT	**4.81**	**2.97**	**3.35**	**2.36**	**1**	**11.83**
V1_ACE	**11.44**	**5.49**	**5.26**	**3.01**	**1**	**11.86**

Acknowledgment

The authors would like to acknowledge the contribution of the Rochester Institute of Technology (RIT) Digital Imaging and Remote Sensing (DIRS) Laboratory to this work by providing the dataset and an open forum to discuss hyperspectral target detection. The authors also acknowledge the Paul Ivanier Center for Robotics and Industrial Production, Beer-Sheva, Israel, for partial support of this work. (The work in this paper was performed and submitted before [21] came to our attention. There are several technical differences between the papers; together, they fortunately confirm the basic theses proposed in this paper).

References

[1] M. Bar-Tal and S. R. Rotman, "Performance measurement in point source target detection," *Infrared Physics and Technology*, vol. 37, no. 2, pp. 231–238, 1996.

[2] C. Caefer, J. Silvermana, O. Orthalb et al., "Improved covariance matrices for point target detection in hyperspectral data," *Optical Engineering*, vol. 47, no. 7, Article ID 076402, 2008.

[3] C. E. Caefer, M. S. Stefanou, E. D. Nielsen, A. P. Rizzuto, O. Raviv, and S. R. Rotman, "Analysis of false alarm distributions in the development and evaluation of hyperspectral point target detection algorithms," *Optical Engineering*, vol. 46, no. 7, Article ID 076402, 2007.

[4] C. E. Caefer and S. R. Rotman, "Local covariance matrices for improved target detection performance," in *Proceedings of the 1st Workshop on Hyperspectral Image and Signal Processing: Evolution in Remote Sensing (WHISPERS '09)*, August 2009.

[5] J. T. Casey and J. P. Kerekes, "Misregistration impacts on hyperspectral target detection," *Journal of Applied Remote Sensing*, vol. 3, no. 1, Article ID 033513, 2009.

[6] S. Matteoli, N. Acito, M. Diani, and G. Corsini, "An automatic approach to adaptive local background estimation and suppression in hyperspectral target detection," *IEEE Transactions on Geoscience and Remote Sensing*, vol. 49, no. 2, pp. 790–800, 2011.

[7] Y. Cohen, D. G. Blumberg, and S. R. Rotman, "Sub-pixel target detection using local spatial information in hyperspectral images," in *Proceedings of the Electro-Optical Remote Sensing, Photonic Technologies, and Applications V.*, Proceedings of SPIE, Prague, Czech Republic, 2011.

[8] D. Manolakis, R. Lockwood, T. Cooley et al., "Is there a best hyperspectral detection algorithm?" in *Proceedings of the Algorithms and Technologies for Multispectral, Hyperspectral, and Ultraspectral Imagery XV*, Proceedings of SPIE, Orlando, Fla, USA, 2009.

[9] J. Settle, "On constrained energy minimization and the partial unmixing of multispectral images," *IEEE Transactions on Geoscience and Remote Sensing*, vol. 40, no. 3, pp. 718–721, 2002.

[10] D. Manolakis, C. Siracusa, and G. Shaw, "Hyperspectral subpixel target detection using the linear mixing model," *IEEE Transactions on Geoscience and Remote Sensing*, vol. 39, no. 7, pp. 1392–1409, 2001.

[11] D. G. Manolakis and G. Shaw, "Detection algorithms for hyperspectral imaging applications," *IEEE Signal Processing Magazine*, vol. 19, no. 1, pp. 29–43, 2002.

[12] D. Manolakis, "Detection algorithms for hyperspectral imaging applications: a signal processing perspective," in *IEEE Workshop Advances in Techniques for Analysis of Remotely Sensed Data*, pp. 378–384, October 2003.

[13] D. Manolakis, "Hyperspectral signal models and implications to material detection algorithms," in *Proceedings of the IEEE International Conference on Acoustics, Speech, and Signal Processing (ICASSP '04)*, pp. III117–III120, May 2004.

[14] D. Manolakis, "Taxonomy of detection algorithms for hyperspectral imaging applications," *Optical Engineering*, vol. 44, no. 6, Article ID 066403, 2005.

[15] E. J. Kelly, "An adaptive detection algorithm," *IEEE Transactions on Aerospace and Electronic Systems*, vol. 22, no. 2, pp. 115–127, 1986.

[16] S. R. Rotman, J. Silverman, and C. E. Caefer, "Segmentation and analysis of hyperspectral data," in *Proceedings of the 22nd Convention of Electrical and Electronics Engineers in Israel*, 2002.

[17] J. Silverman, C. E. Caefer, J. M. Mooney, M. M. Weeks, and P. Yip, "Automated clustering/segmentation of hyperspectral images based on histogram thresholding," in *Proceedings of the Imaging Spectrometry VII*, vol. 4480 of *Proceedings of SPIE*, pp. 65–75, San Diego, Calif, USA, August 2001.

[18] J. Zhang, J. Chen, Y. Zhang, and B. Zou, "Hyperspectral image segmentation method based on spatial-spectral constrained

region active contour," in *Proceedings of the 30th IEEE International Geoscience and Remote Sensing Symposium (IGARSS '10)*, pp. 2214–2217, Honolulu, Hawaii, USA, July 2010.

[19] X. Yu, I. S. Reed, and A. D. Stocker, "Comparative performance analysis of adaptive multispectral detectors," *IEEE Transactions on Signal Processing*, vol. 41, no. 8, pp. 2639–2655, 1993.

[20] J. Schott, *Remote Sensing: The Image Chain Approach*, Oxford University Press, New York, NY, USA, 2007.

[21] W. F. Basener, E. Nance, and J. Kerekes, "The target implant method for predicting target difficulty and detector performance in hyperspectral imagery," in *Proceedings of the Algorithms and Technologies for Multispectral, Hyperspectral, and Ultraspectral Imagery XVII*, vol. 8048, 80481H of *Proceedings of SPIE*, Orlando, Fla, USA, 2011.

2

Parametric Yield-Driven Resource Binding in High-Level Synthesis with Multi-V_{th}/V_{dd} Library and Device Sizing

Yibo Chen,[1] Yu Wang,[2] Yuan Xie,[1] and Andres Takach[3]

[1] *Department of Computer Science and Engineering, The Pennsylvania State University, University Park, PA 16802, USA*
[2] *Department of Electronics Engineering, Tsinghua University, Beijing 100084, China*
[3] *Design Creation and Synthesis, Mentor Graphics Corporation, Wilsonville, OR 97070, USA*

Correspondence should be addressed to Yibo Chen, yxc236@cse.psu.edu

Academic Editor: Zhiru Zhang

The ever-increasing chip power dissipation in SoCs has imposed great challenges on today's circuit design. It has been shown that multiple threshold and supply voltages assignment (multi-V_{th}/V_{dd}) is an effective way to reduce power dissipation. However, most of the prior multi-V_{th}/V_{dd} optimizations are performed under deterministic conditions. With the increasing process variability that has significant impact on both the power dissipation and performance of circuit designs, it is necessary to employ statistical approaches in analysis and optimizations for low power. This paper studies the impact of process variations on the multi-V_{th}/V_{dd} technique at the behavioral synthesis level. A multi-V_{th}/V_{dd} resource library is characterized for delay and power variations at different voltage combinations. Meanwhile, device sizing is performed on the resources in the library to mitigate the impact of variation, and to enlarge the design space for better quality of the design choice. A parametric yield-driven resource binding algorithm is then proposed, which uses the characterized power and delay distributions and efficiently maximizes power yield under a timing yield constraint. During the resource binding process, voltage level converters are inserted between resources when required. Experimental results show that significant power reduction can be achieved with the proposed variation-aware framework, compared with traditional worstcase based deterministic approaches.

1. Introduction

Integrating billions of transistors on a single chip with nanoscale transistors has resulted in great challenges for chip designers. One of these challenges is that the pace of productivity gains has not kept up to address the increases in design complexity. Consequently, we have seen a recent trend of moving design abstraction to a higher level, with an emphasis on *electronic system level (ESL)* design methodologies. A very important component of ESL is raising the level of abstraction of hardware design. High-level synthesis (HLS) provides this component by providing automation to generate optimized hardware from a high-level description of the function or algorithm to be implemented in hardware. HLS generates a cycle-accurate specification at the register-transfer level (RTL) that is then used in existing ASIC or FPGA design methodologies. Commercial high-level synthesis tools [1] have recently gained a lot of attention as evidenced in recent conference HLS workshops (DATE2008, DAC2008, and ASPDAC2009), conference panels, and publications that track the industry. While high-level synthesis is able to quickly generate implementations of circuits, it is not intended to replace the existing low-level synthesis. The major benefit coming from high-level synthesis is the high design efficiency, the ability to perform fast prototyping, functional verification, and early-stage design space exploration, which in turn provide guidance on succeeding low-level design steps and help produce high-quality circuits.

Power consumption and process variability are among other critical design challenges as technology scales. While it is believed that tackling these issues at a higher level of the design hierarchy can lead to better design decisions, a lot of work has been done on low-power-high-level synthesis [2–4] as well as process-variation-aware-high-level synthesis [5–8]. These techniques have been successfully implemented but most of the existing work focuses on one side of the issues in isolation. Recently, Srivastava et al. [9] explore the multi-$V_{th}/V_{dd}/T_{ox}$ design space with the consideration of process variations at the gate level. Nevertheless, variation-aware-low power exploration for behavioral synthesis is still in its infancy.

Multiple threshold and supply voltages assignment (multi-V_{th}/V_{dd}) has been shown as an effective way to reduce circuit power dissipation [2, 3, 10, 11]. Existing approaches assign circuit components on critical paths to operate at a higher V_{dd} or lower V_{th}, and noncritical portions of the circuit are made to operate at lower V_{dd} or higher V_{th}, respectively. The total power consumption is thus reduced without degrading circuit performance. However, nowadays circuit performance is affected by process variations. If the variations are underestimated, for example, using nominal delays of circuit components to guide the design, non-critical components may turn to critical ones due to the variations, and circuit timing constraints may be violated. On the other hand, in existing corner-based worst-case analysis, variations are overestimated resulting in design specs that are hard to meet, and this consequently increases design effort and degrades circuit performance.

Device sizing is a well-studied technique for performance and power tuning at gate- or circuit-level [12]. To improve performance, upsizing of a high-V_{th} transistor, which increases switching power and die area, can be traded off against using a low-V_{th} transistor, which increases leakage power. Therefore, combining multi-V_{th} assignment and device sizing as integrated problem, can increase the design flexibility and further improve the design quality. Meanwhile, in terms of mitigating process variations, it is possible that increasing the size of transistors can reduce the randomness of the device parameters through averaging.

This paper presents a variation-aware power optimization framework in high-level synthesis using simultaneous multi-V_{th}/V_{dd} assignment and device sizing. Firstly, the impact of parameter variations on the delay and power of circuit components is explored at different operating points of threshold and supply voltages. Device sizing is then performed to mitigate the impact of variations and to enlarge the design space for better quality of the design choice. A variation-characterized resource library containing the parameters of delay and power distributions at different voltage "corners" and different device sizes, is built once for the given technology, so that it is available for high-level synthesis to query the delay/power characteristics of resources. The concept of parametric yield, which is defined as the probability that the design meets specified constraints such as delay or power constraints is then introduced to guide design space exploration. Statistical timing and power analysis on the data flow graph (DFG) is used to populate the delay and power distributions through the DFG and to estimate the overall performance and power yield of the entire design. A variation-aware resource binding algorithm is then proposed to maximize power yield under a timing yield constraint, by iteratively searching for the operations that have the maximum potential of performance/power yield improvement, and replacing them with better candidates in the multi-V_{th}/V_{dd} resource library. During the resource binding process, voltage level converters are inserted for chaining of resource units having different V_{dd} supplies.

The contribution of this paper can be summarized as follow:

(i) first, this is the first work to apply multi-V_{th}/V_{dd} techniques during high-level synthesis under the context of both delay and power variations. A flow for variation-aware power optimization in multi-V_{th}/V_{dd} HLS is proposed. This flow includes library characterization, statistical timing and power analysis methodologies for HLS, and resource binding optimization with variation-characterized multi-V_{th}/V_{dd} library;

(ii) combined multi-V_{th}/V_{dd} assignment and device sizing for high-level synthesis are performed at the granularity of function unit level, to improve the design quality and at the same time to reduce the design complexity;

(iii) voltage level conversion is explored during the resource binding in high-level synthesis, enabling the full utilization of multi-V_{dd} components for parametric yield maximization.

2. Related Work

Prior research work tightly related to this paper mainly falls into two categories: (1) gate level power minimization by simultaneous multi-V_{th} assignment and gate sizing; (2) low-power high-level synthesis using multi-V_{th} or multi-V_{dd}; (3) process variation aware high-level synthesis.

Several techniques were proposed to consider V_{th} allocation and transistor sizing as an integrated problem [13–16]. Wei et al. [14] presented simultaneous dual-V_{th} assignment and gate sizing to minimize the total power dissipation while maintaining high performance, while Karnik et al. [16] improved the simultaneous V_{th} allocation and device sizing using a Lagrangian Relaxation method. However, all of the reported techniques focus on tuning at transistor level or gate-level. While the fine granularity can yield optimal results, it also lead to high design complexity.

Shiue [2] proposed low-power scheduling schemes with multi-V_{dd} resources by maximizing the utilization of resources operating at reduced supply voltages. Khouri and Jha [3] performed high-level synthesis using a dual-V_{th} library for leakage power reduction. Tang et al. [4] formulated the synthesis problem using dual-V_{th} as a maximum weight-independent set (MWIS) problem, within which near-optimal leakage power reduction is achieved with greatly reduced run time. Very recently, Insup et al. explored optimal register allocation for high-level synthesis using dual supply voltages [17]. However, all of these techniques were applied under deterministic conditions without taking process variation into consideration.

Process variation-aware high-level synthesis has recently gained much attention. Jung and Kim [6] proposed a timing yield-aware HLS algorithm to improve resource sharing and reduce overall latency. Lucas et al. [8] integrated timing-driven floorplanning into the variation-aware high-level design. Mohanty and Kougianos's work [18] took into account the leakage power variations in low-power high-level synthesis; however, the major difference between [18] and our work is that, the delay variation of function units was not considered in [18], so the timing analysis during synthesis was still deterministic. Recently, Wang et al. [19] proposed a joint design-time optimization and postsilicon tuning framework that tackles both timing and power variations. Adaptive body biasing (ABB) was applied to function units to reduce leakage power and improve power yield.

3. Multi-V_{th}/V_{dd} Library Characterization under Process Variations

Scheduling and resource binding are key steps during the high-level synthesis process. The scheduler is in charge of determining the sequencing the operations of a control/data flow graph (CDFG) in control steps and within control steps (operator chaining) while obeying control and data dependencies and cycle constraints while optimizing for area/power/performance. The binding process binds operations to hardware units in the resource library to complete the mapping from abstracted descriptions of circuits into practical designs. This section presents the characterization of the variation-aware multi-V_{th}/V_{dd} resource library, including the delay and power characterization flow and the selection of dual threshold and supply voltages.

3.1. Variation-Aware Library Characterization Flow. In order to facilitate the design space exploration while considering process variations, the resource library of functional units for HLS has to be characterized for delay/power variations. As shown in Figure 1, under the influence of process variations, the delay and power of each component are no longer fixed values, but represented by probability density functions (PDFs). Consequently, the characterization of function units with delay and power variations requires statistical analysis methodologies.

Process variations come from a set of sources, including random doping fluctuation (RDF) [20] and geometric variations of the gate (primarily on channel length) [21]. Since both RDF and channel length variations manifest themselves as fluctuations on the effective threshold voltage of the transistor [22], their effects can be expressed by the variations of V_{th}. Since this work focuses on demonstrating the effectiveness of variation-aware synthesis, rather than a comprehensive modeling of all variation effects, we try to focus on V_{th} variations with a simplified assumption of normal distribution of V_{th} variations, rather than covering all physical-level variation factors with different distributions. The magnitude of V_{th} variations in real circuits can be obtained via on-chip sensing and measurement. In this work, we use NCSU FreePDK 45 nm technology library [23] for all the characterization and experiments. We set the standard deviation σ of V_{th} to be 50 mV, which is projected from the silicon measurement data in [24].

We then use a commercial gate-level statistical timing analysis tool, Synopsys PrimeTime *VX* to perform the characterization. This variation-aware tool increases the accuracy of timing analysis by considering the statistical distribution of device parameters such as threshold voltage and gate oxide thickness. Given the distribution of a device parameter *P*, PrimeTime *VX* calculates the distribution of gate delay continuously throughout the range of values, using linear interpolation and extrapolation at the library-defined functional operating points, as shown in Figure 2. Validation against SPICE Monte Carlo statistical analysis [25] shows that PrimeTime *VX* analysis holds the similar accuracy but significantly reduces the running time.

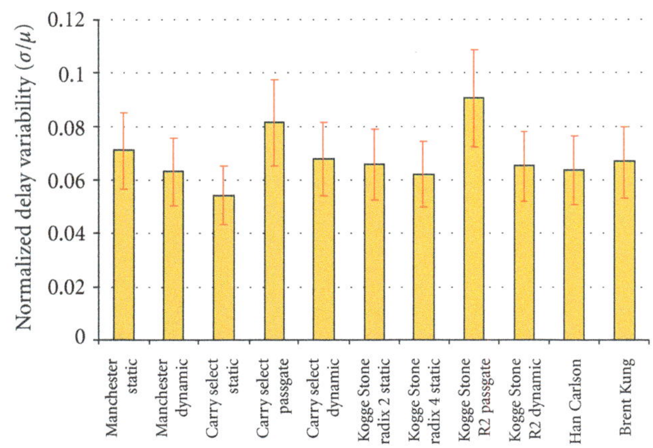

FIGURE 1: The delay variation for 16-bit adders in IBM Cu-08 technology (courtesy of IBM).

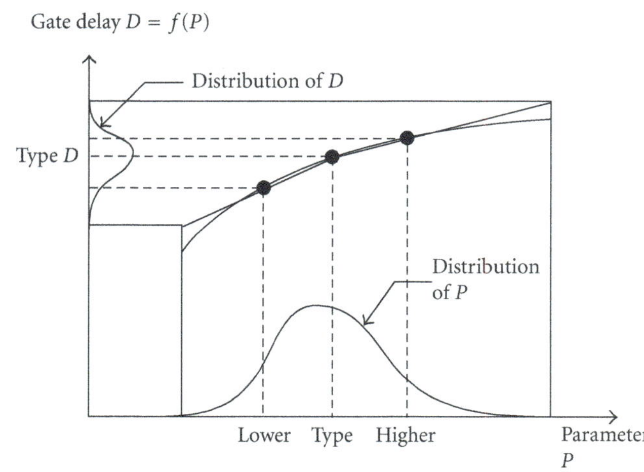

FIGURE 2: Calculating the delay distribution as a function of parameter *P*.

The characterization flow takes as input the statistical distributions of process parameters (V_{th} in this work) and generates the statistical distributions of delay and power for each resource in the library. To characterize the delay of function units under the impact of process variations, the following steps are performed:

(1) all the standard cells in a technology library are characterized using variation-aware analysis, and the

results including parameters of cell delay distributions are collected to build a variation-aware technology library;

(2) the function units used in HLS are then synthesized and linked to the variation-aware technology library;

(3) statistical timing analysis for the function units is performed using PrimeTime *VX*, and the parameters of delay distributions are reported.

Statistical power characterization for function units in the resource library can be done using Monte Carlo analysis in SPICE. The power consumption of function units consists of dynamic and leakage components. While dynamic power is relatively immune to process variation, leakage power is greatly affected and becomes dominant as technology continues scaling down [26]. Therefore, in this paper only leakage power is characterized using statistical analysis. However, this do not mean that considerations for dynamic power can be omitted. In fact, dynamic power optimization in high-level synthesis has been a well-explored topic [1]. Our variation-oriented work emphasizing leakage power optimization can be stacked on or integrated into existing power optimization approaches in high-level synthesis, to further reduce the total power consumption of circuits.

According to Berkeley short-channel BSIM4 model [27], higher threshold voltages can lead to exponential reduction in leakage power, which is given by:

$$I_{\text{leakage}} = I_0 \exp\left(\frac{V_{\text{gs}} - V_{\text{th}}}{nV_t}\right)\left(1 - \exp\left(\frac{-V_{\text{ds}}}{V_t}\right)\right), \quad (1)$$

$$I_0 = \mu_0 C_{\text{ox}} \frac{W}{L}(n-1)V_{\text{th}}^2, \quad (2)$$

where I_{leakage} is the gate leakage current, V_{gs} and V_{ds} are the gate voltages, V_t is the thermal voltage, and n is the subthreshold swing factor. Since we assume that the device parameter V_{th} follows normal distribution, I_{leakage} follow log-normal distribution. Therefore, the leakage power of a function units is the sum of a set of log-normal distributions, which describe the leakage power of each library cell. According to [28], the sum of several log-normal random variables can be approximated by another log-normal random variable, as shown in (3):

$$P_{\text{FU}} = P_1 + P_2 + \cdots + P_n = k_1 e^{V_1} + k_2 e^{V_2} + \cdots + k_n e^{V_n}, \quad (3)$$

where P_{FU} describes the leakage power distribution of the function unit; while P_n, k_n, and V_n are the corresponding variables for library cells that build up the function unit. The mean and deviation of P_{FU} can be estimated via iterative moment matching out of the leakage power distributions of library cells [28].

The power characterization flow is stated as follows. Process variations are set in the MOS model files, and 1000 runs of Monte Carlo iterations are performed for each library cell. After the characterization, the parameters of the leakage power distributions of library cells are extracted.

Note that in our work we only characterize subthreshold leakage, since it starts dominant for technology nodes of 45 nm and below. The gate leakage can also be characterized with similar methods.

3.2. Multi-V_{th}/V_{dd} Library Characterization. Previous implementations using multiple threshold and supply voltages in conjunction have shown a very effective reduction in both dynamic and leakage power [11]. Therefore, our approach considers the combination of dual threshold and dual supply voltages, and characterizations are performed at the four "corners" of voltage settings, namely $(V_{\text{th}}^L, V_{\text{dd}}^H)$, $(V_{\text{th}}^H, V_{\text{dd}}^H)$, $(V_{\text{th}}^L, V_{\text{dd}}^L)$, and $(V_{\text{th}}^H, V_{\text{dd}}^L)$, where $(V_{\text{th}}^L, V_{\text{dd}}^H)$ is the nominal case and the other three are low-power settings. Note that although only 4 voltage settings are discussed in this paper, it is natural to extend the approach presented here to deal with more voltage settings. To reduce the process technology cost, in this paper, the multi-$V_{\text{th}}/V_{\text{dd}}$ techniques are applied at the granularity of function units. That means, all the gates inside a function unit operate at the same threshold and supply voltages. Voltages only differ from function units to function units.

The selection of appropriate values of threshold and supply voltages for power minimization has been discussed under deterministic conditions [11]. Rules of thumb are derived for the second V_{dd} and V_{th} as functions of the original voltages [11]:

$$\begin{aligned} V_{\text{dd}}^L &= 0.43V_{\text{dd}}^H + 0.82V_{\text{th}}^H + \frac{0.72}{K} - \frac{0.55}{K^2} - 0.2, \\ V_{\text{th}}^L &= -0.024V_{\text{dd}}^H + 1.14V_{\text{th}}^H + \frac{0.72}{K} - \frac{0.49}{K^2} - 0.18, \end{aligned} \quad (4)$$

where K stands for the initial ratio between dynamic and static power. While the empirical models in [11] are validated on actual circuit benchmarks [29], they may not be accurate under the impact of process variations. A refined model taking into account the process variations is presented in [9]. As shown in Figure 3, the total power reduction with variation awareness is plotted under different combinations of Vth_2 (V_{th}^H) and Vdd_2 (V_{dd}^L), and this guides the optimal value selection in this work.

The characterization results (which will be further discussed in Section 6) show that, power reduction is always achieved at the cost of delay penalties. Moreover, larger delay variations are observed for slower units operating at high-V_{th} or low-V_{dd}, which means larger probability of timing violations when they are placed on the near-critical paths. This further demonstrates the necessity of statistical analysis and parametric yield-driven optimization approaches.

3.3. Device Sizing for the Resource Library. Conventionally, device sizing is an effective technique to optimize CMOS circuits for dynamic power dissipation and performance. In this work, we show that device sizing may also be utilized to mitigate the impact of process variations. As previously mentioned, the sources of process variations mainly consists of random doping fluctuation (RDF) [20] and geometric variations (GVs). GV affect the real V_{th} through the drain induced barrier lowering (DIBL) effect. Originally, both RDF and GV have almost the equal importance in determining the V_{th} variance. As we propose to use low-V_{dd} and high-V_{th} resource units in the design, the difference between supply voltage and threshold voltage diminishes, and this reduces DIBL effect. As a result, the uncertainty in V_{th} arising from GV rapidly falls as V_{dd}. On the other hand, the RDF-induced V_{th} variation is independent of V_{dd} changes and solely a function of channel area [30]. Therefore, V_{th} variation resulting from RDF becomes dominating as V_{dd} approaches V_{th}.

Due to the independent nature of RDF variations, it is possible to reduce their impact on circuit performance through averaging. Therefore, upsizing the device can be an effective way for variability mitigation with enlarged channel area. According to [31], V_{th} variance $\sigma_{V_{th}}$ resulting from RDF is roughly proportional to $(WL)^{-1/2}$, which means we can either increase the transistor width or channel length or both. Conventional sizing approaches focus on tuning the transistor width for performance. In terms of process variability mitigation, the measurement data of V_{th} variation for 4 different device sizes is plotted in Figure 4, which shows that increasing transistor width is a more effective way to reduce the V_{th} variance [24]. Although larger transistor width means larger leakage power, the fluctuations on leakage power are reduced, and the design space for resource binding is significantly enlarged, thus using resources with larger size in the design may still be able to improve the parametric yield.

In this work, we upsize all the function units in the resource library to generate alternatives for power tuning and variability mitigation. The sizing is performed on all the gates with two different settings: the basic size (1W1L) and the double-width size (2W1L). We then perform the variation characterization for the upsized function units under all the four voltage "corners" presented in the previous section. The characterization results will be presented in Section 6.

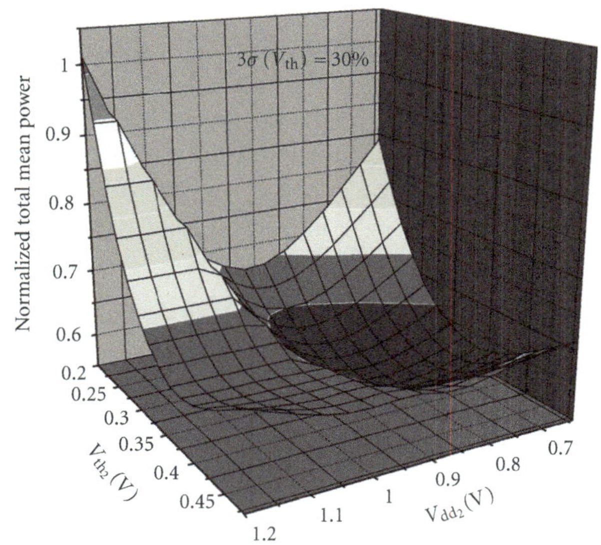

FIGURE 3: Optimal selection of dual threshold and supply voltages under process variation.

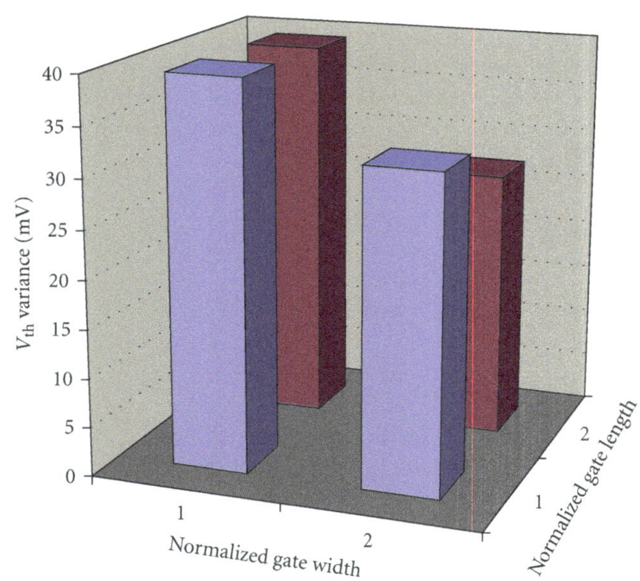

FIGURE 4: V_{th} variations with difference device sizes (normalized to minimal size).

4. Yield Analysis in Statistical High-Level Synthesis

In this section, a parametric yield analysis framework for statistical HLS is presented. We first show the necessity of statistical analysis by a simple motivational example and then demonstrate the statistical timing and power analysis for HLS as well as the modeling and integration of level converters for multi-V_{dd} HLS.

4.1. Parametric Yield. To bring the process-variation awareness to the high-level synthesis flow, we first introduce a new metric called parametric yield. The parametric yield is defined as the probability of the synthesized hardware meeting a specified constraint Yield $= P(Y \leq Y_{max})$, where Y can be delay or power.

Figure 5 shows a motivational example of yield-aware analysis. Three resource units $R1$, $R2$, and $R3$ have the same circuit implementation but operate at different supply or threshold voltages. Figure 5 shows the delay and power distributions for $R1$, $R2$, and $R3$. In this case the mean power follows up $\mu_P(R3) < \mu_P(R2) < \mu_P(R1)$, and the mean delay follows $\mu_D(R1) < \mu_D(R2) < \mu_D(R3)$, which is as expected since power reduction usually comes at the cost of increased delay. The clock cycle time T_{CLK} and the power consumption constraint P_{LMT} (e.g., the TDP (thermal design power) of most modern microprocessors) are also shown in the figure. If the variation is disregarded and nominal-case analysis is used, any of the resource units can be chosen since they all meet timing. In this case, $R3$ would be chosen as it has the lowest power consumption. However, under a statistical point of view, $R3$ has a low timing yield (approximately 50%) and is very likely to cause timing violations. In contrast, with corner-based worst-case analysis only $R1$ can be chosen under the clock cycle time constraint (the worst-case delay of

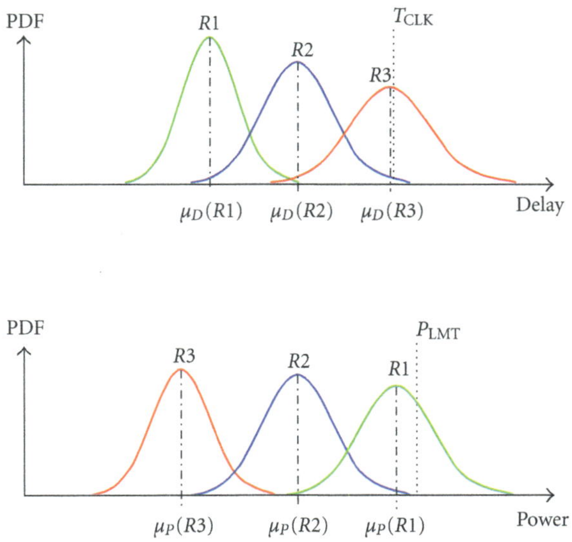

Figure 5: Motivating example of yield-driven synthesis.

$R2$ slightly violates the limit), whereas $R1$ has a poor power yield. In fact, if we set a timing yield constraint instead of enforcing the worst-case delay limitation, $R2$ can be chosen with a slight timing yield loss but a well-balanced delay and power tradeoff. Therefore, a yield-driven statistical approach is needed for exploring the design space to maximize one parametric yield under other parametric yield constraints.

4.2. Statistical Timing and Power Analysis for HLS. High-level synthesis (HLS) is the process of transforming a behavioral description into register level structure description. Operations such as additions and multiplications in the DFG are scheduled into control steps. During the resource allocation and binding stages, operations are bound to corresponding function units in the resource library meeting type and latency requirements.

Given the clock cycle time T_{CLK}, the timing yield of the entire DFG, Yield_T is defined as

$$\text{Yield}_T = P(T_1 \leq T_{CLK}, T_2 \leq T_{CLK}, \ldots, T_n \leq T_{CLK}), \tag{5}$$

where $P()$ is the probability function, $T_1, T_2, \ldots, T_n$ are the arrival time distributions at control step $1, 2, \ldots, n$, respectively.

The arriving time distribution of each clock cycle can be computed from the delay distributions of function units bound at that cycle. Two operations, sum and max, are used to compute the distributions:

(i) sum operation is used when two function units are chained in cascade within a clock cycle, as shown in CC1 and CC2 of Figure 6. The total delay can be computed as the "sum" of their delay distributions (normal distribution assumed);

(ii) max operation is used when the outputs of two or more units are fed to another function unit at the same clock cycle, as shown in CC1 of Figure 6. The "maximum" delay distribution can be computed out of the contributing distributions using tightness probability and moment matching [19].

With these two operations, the arriving time distribution of each clock cycle is computed, and the overall timing yield of the DFG is obtained using (5).

The total power consumption of a DFG can be computed as the sum of the power consumptions of all the function units used in the DFG. Given a power limitation P_{LMT}, the power yield of the DFG Yield_P is computed as the probability that total power P_{DFG} is less than the requirement, as expressed in (6):

$$\text{Yield}_P = P(P_{DFG} \leq P_{LMT}). \tag{6}$$

Since dynamic power is relatively immune to process variations, it is regarded as a constant portion which only affects the mean value of the total power consumption. Therefore, the total power is still normally distributed, although statistical analysis is only applied to the leakage power. As aforementioned in Section 3, our proposed yield-driven statistical framework can be stacked on existing approaches for dynamic power optimization, to further reduce the total power consumption of circuits.

4.3. Voltage Level Conversion in HLS. In designs using multi-V_{dd} resource units, voltage level convertors are required when a low-voltage resource unit is driving a high-voltage resource unit. Level conversion can be performed either synchronously or asynchronously. Synchronous level conversion is usually embedded in flip-flops and occurs at the active clock edge, while asynchronous level converters can be inserted anywhere within the combinational logic block.

When process variations are considered, asynchronous level converters are even more favorable, because they are not bounded by clock edges, and timing slacks can be passed through the converters. Therefore, time borrowing can happen between low-voltage and high-voltage resource units. As slow function units (due to variations) may get more time to finish execution, the timing yield can be improved, and the impact of process variations is consequently reduced.

While many fast and low-power level conversion circuits have been proposed recently, this paper uses the multi-V_{th} level converter presented in [32], taking the advantage that there is no extra process technology overhead for multi-V_{th} level converters, since multi-V_{th} is already deployed for function units. The proposed level converter is composed of two dual V_{th} cascaded inverters. Its delay and power are then characterized in HSPICE using the listed parameters [32].

The delay penalty of a level converter can be accounted by summing its delay with the delay of the function unit it is associated to. The power penalty can be addressed by counting the level converters used in the DFG and adding the corresponding power to the total power consumption.

5. Yield-Driven Power Optimization Algorithm

In this section, we propose our yield-driven power optimization framework based on the aforementioned statistical

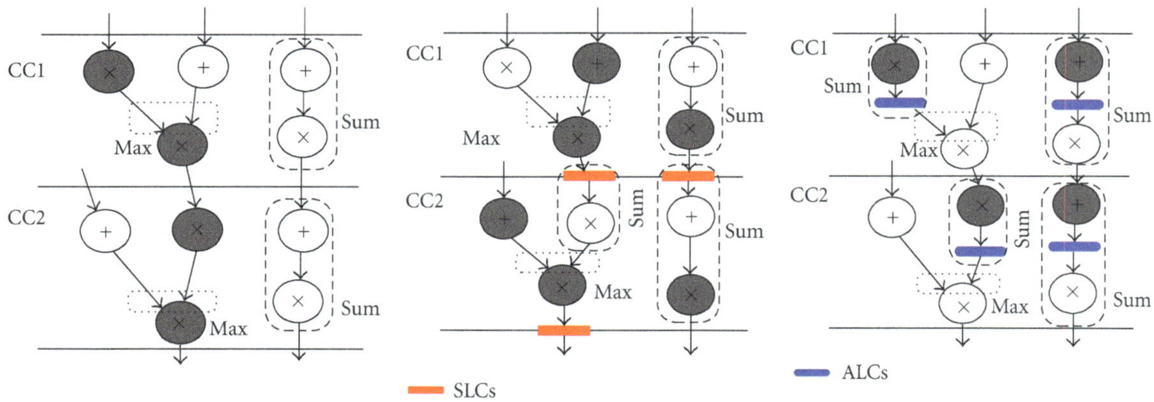

(a) No level conversion (b) synchronous conversion (c) synchronous conversion

FIGURE 6: Timing yield computation in multi-V_{dd} high-level synthesis with different level conversions. Shadowed operations are bound to function units with low-supply voltages, and the bars indicate the insertion of level converters.

timing and power yield analysis. During the high-level synthesis design loop, resource binding selects the optimal resource instances in the resource library and binds them to the scheduled operations at each control step. A variation-aware resource binding algorithm is then proposed to maximize power yield under a preset timing yield constraint, by iteratively searching for the operations with the maximum potential of timing/power yield improvement, and replacing them with better candidates in the multi-V_{th}/V_{dd} resource library.

5.1. Variation-Aware Resource Binding Algorithm Overview. Our variation-aware resource binding algorithm takes a search strategy called variable depth search [19, 33, 34] to iteratively improve the power yield under performance constraints. The outline of the algorithm is shown in Algorithm 1, where a DFG is initially scheduled and bound to resource library with nominal voltages (V_{th}^L, V_{dd}^H). A lower-bound constraint on the timing yield is set, so that the probability of the design can operate at a given clock frequency, will be larger than or equal to a preset threshold (e.g., 95%). In the algorithm, a move is defined as a local and incremental change on the resource bindings. As shown in the sub routine GENMOVE in Algorithm 1, the algorithm generates a set of moves and finds out a sequence of moves that maximizes the accumulated *gain*, which is defined as $\alpha * \Delta \text{Yield}_D + \Delta \text{Yield}_P$, where α is a weighting factor to balance the weights of timing and power yield improvements. The optimal sequence of moves is then applied to the DFG, and the timing and power yields of the DFG are updated before the next iteration. The iterative search ends when there is no yield improvement or the timing yield constraint is violated.

Note that our worst-case resource binding algorithm uses the same search strategy (variable depth search) [19, 33, 34] as the variation-aware resource binding algorithm. The key difference is that, instead of iteratively improving the power yield under performance constraints, the worst-case resource binding algorithm iteratively reduces the power consumption under specified performance constraints, where both the power consumption calculation and performance constraints are specified as deterministic numbers, rather than using the concept of power yield and performance yield.

5.2. Voltage Level Conversion Strategies. Moves during the iterative search may result in low-voltage resource units driving high-voltage resource units. Therefore, level conversion is needed during resource binding. However, if resources are selected and bound so that low-voltage resource units never drive high-voltage ones, level conversion will not be necessary, and the delay and power overheads brought by level converters can be avoided. This reduces the flexibility of resource binding for multivoltage module combinations, and may consequently decrease the attainable yield improvement. The tradeoff in this conversion-avoidance strategy, can be explored and evaluated within our proposed power optimization algorithm.

We also incorporate other two strategies of level conversions in the power optimization algorithm for comparison. All the three strategies are listed as follows:

(i) level conversion avoidance: resource binding is performed with the objective that low-voltage resources never drive high-voltage ones. As shown in Figure 6(a), no dark-to-light transition between operations is allowed (while dark operations are bound to low-V_{dd} units), so that level conversion is avoided. This is the most conservative strategy;

(ii) synchronous level conversion: voltage level conversion is done synchronously in the level-converting flip-flops (SLCs). As shown in Figure 6(b), the dark-to-light transition only happens at the beginning of each clock cycles. The flip-flop structure proposed in [35] is claimed to have smaller delay than the combination of an asynchronous converter and a conventional flip-flop. However, as discussed previously, synchronous level conversion may reduce the flexibility of resource binding as well as the possibility

VABinding(*DFG*, *ResLib*, *Constraints*, *LCStrategy*)
▷ Initialization
(1) Scheduling using ASAP strategy
(2) Initial Binding to $(V_{\text{th}}^{L}, V_{\text{dd}}^{H})$ resources
▷ Variation-aware resource binding
(3) **while** $\Delta\text{Yield}_P > 0$ AND $\text{Yield}_D \geq$ *Constraint*
(4) **do for** $i \leftarrow 1$ to *MAXMOVES*
(5) **do** $\text{Gain}_i \leftarrow$ GenMove(*DFG*, *ResLib*, *LCStrategy*)
(6) Append Gain_i to Gain_*List*;
(7) Find subsequence $\text{Gain}_1, \ldots, \text{Gain}_k$ in Gain_*List*
so that $G = \sum_{i=1}^{k} \text{Gain}_i$ is maximized
(8) **if** $G > 0$
(9) **do** Accept moves $1 \cdots k$
(10) Evaluate ΔYield_P and Yield_D
GenMove(*DEG*, *ResLib*, *LCStrategy*)
(1) MOVE: Choose a move using steepest descent heuristic [33]
(2) Check whether and where level conversion is needed
(3) **if** *LC Strategy* = *Avoidance* AND *NeedConversion*
(4) **do goto** MOVE
(5) **if** *LC Strategy* = *Synchronous* AND *NeedConversion*
▷Check whether conversion is synchronous or not
(6) **do if** Conversion is inside operation chaining
(7) **do goto** MOVE
(8) Count the overhead of level conversion
(9) Evaluate the Gain of this move
(10) Return Gain

Algorithm 1: Outline of the variation-aware resource binding algorithm.

of timing borrowing. The effectiveness of this strategy is to be explored by the optimization algorithm;

(iii) asynchronous level conversionL: asynchronous level converters (ALCs) are inserted wherever level conversion is needed, as dark-to-light transition can happen anywhere in Figure 6. This aggressive strategy provides the maximum flexibility for resource binding and timing borrowing. Although it brings in delay and power overhead, it still has great potential for timing yield improvement.

5.3. Moves Used in the Iterative Search. In order to fully explore the design space, three types of moves are used in the iterative search for resource binding;

(i) resource rebinding: in this move, an operation is assigned to a different function unit in the library with different timing and power characteristics. The key benefit of the multi-$V_{\text{th}}/V_{\text{dd}}$ techniques is that it provides an enlarged design space for exploration, and optimal improvements are more likely to be obtained;

(ii) resource sharing: in this move, two function units that are originally bound to different function units, are now merged to share the same function unit. The type of move reduces the resource usage and consequently improves the power yield;

(iii) resource splitting: in this move, the operation that originally shared function unit with other operations, is split from the shared function unit. This type of move might lead to other moves such as resource rebinding and resource sharing.

After each move, the algorithm checks where the low-supply voltage function units are used and decides whether to insert or remove the level converters, according to the predefined level conversion strategy. If a move is against the strategy, it is revoked, and new moves are generated until a qualifying move is found.

5.4. Algorithm Analysis. It has to be noted that, in the procedure GenMove shown in Algorithm 1, even though the returned Gain might be negative, it still could be accepted. Since the sequence of a cumulative positive gain is considered, the negative gains help the algorithm escape from local minima through hill climbing.

As for the computational complexity, it is generally not possible to give nontrivial upper bounds of run time for local search algorithms [33]. However, for variable depth search in general graph partitioning, Aarts and Lenstra [33] found a near-optimal growth rate of run time to be $O(n \log n)$, where n is the number of nodes in the graph. In our proposed algorithm, the timing and power yield evaluation, as well as the level converter insertion, are performed at each move. Since the yield can be updated using a gradient computation approach [19], the run time for each move is at most $O(n)$. Therefore, the overall run time for the proposed resource binding algorithm is $O(n^2 \log n)$.

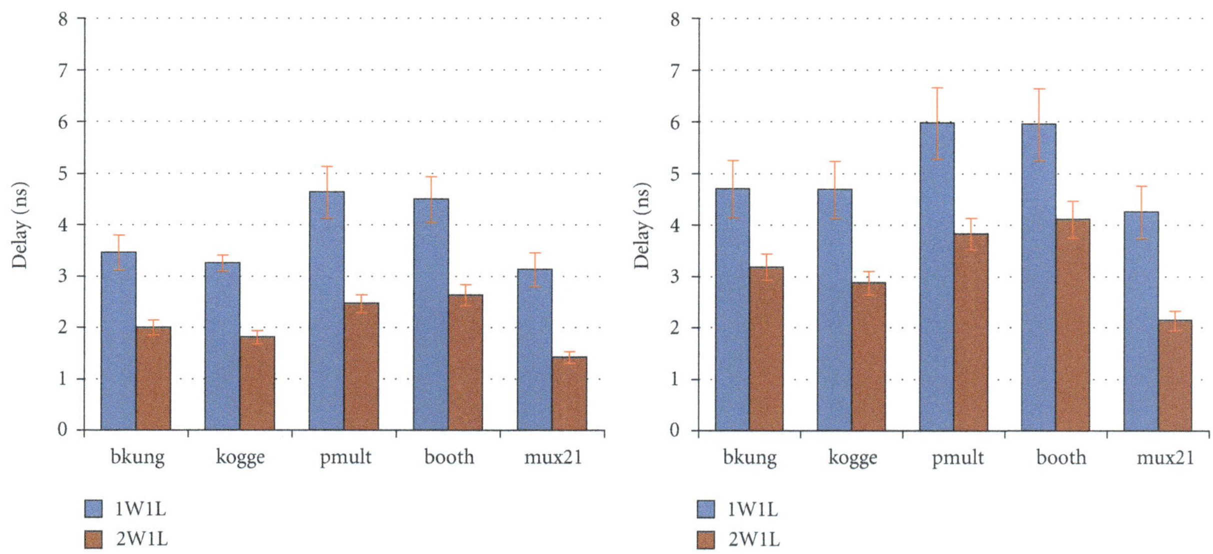

(a) $V_{th} = 0.37\,V$ (b) $V_{th} = 0.56\,V$

FIGURE 7: Delay characterization of function units with multi-V_{th}/V_{dd} and variation awareness.

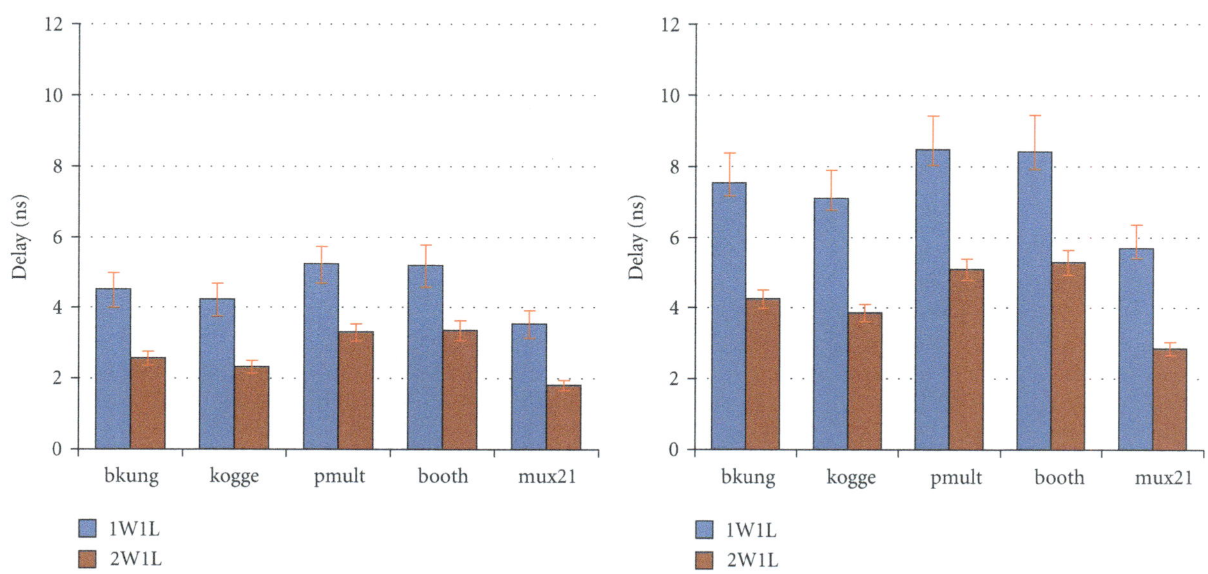

(a) $V_{th} = 0.37\,V$ (b) $V_{th} = 0.56\,V$

FIGURE 8: Delay characterization of function units with multi-V_{th}/V_{dd} and variation awareness.

6. Experimental Results

In this section, we present the experimental results of our variation-aware power optimization framework for high-level synthesis. The results show that our method can effectively improve the overall power yield of given designs and reduce the impact of process variations.

We first show the variation-aware delay and power characterization of function units. The characterization is based on NCSU FreePDK 45 nm technology library [23]. The voltage corners for the characterization are set as $V_{th}^{L} = 0.37\,V$, $V_{th}^{H} = 0.56\,V$, $V_{dd}^{L} = 0.9\,V$, and $V_{dd}^{H} = 1.1\,V$. The characterization results for five function units, including two 16-bit adders bkung and kogge, two 8-bit × 8-bit multipliers pmult and booth, and one 16-bit multiplexer *mux21*, are depicted in Figures 7, 8, 9, 10. In the figures, the color bars show the nominal case values, while the error bars show the deviations. It is clearly shown that with lower V_{dd} and/or higher V_{th}, significant power reductions are achieved at the cost of delay penalty. Meanwhile, up sizing the transistor can improve the circuit performance but also yield to larger power consumption. In terms of variability mitigation, both voltage scaling and device sizing have significant impact on the delay and leakage variations. We can explore this trend

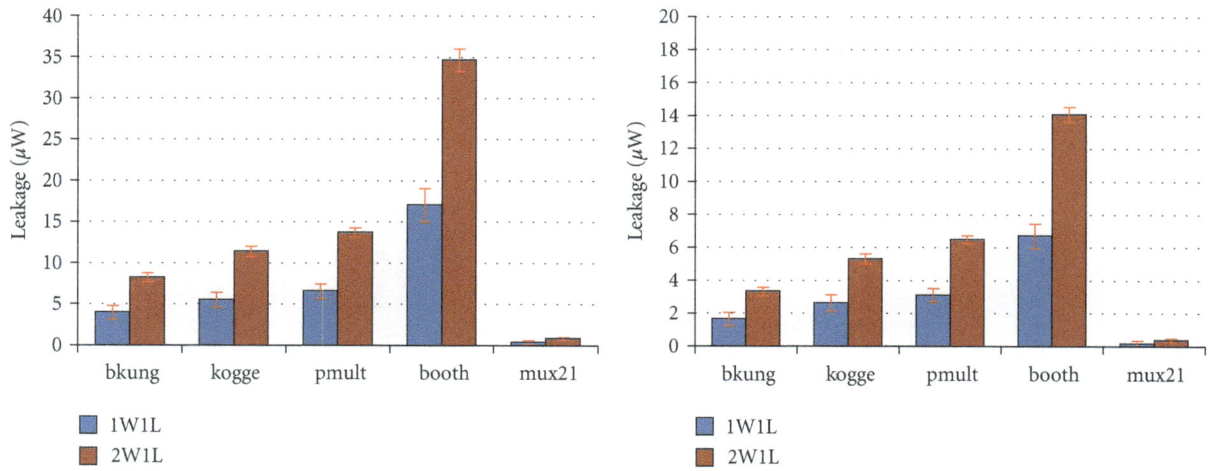

(a) $V_{th} = 0.37$ V (b) $V_{th} = 0.56$ V

FIGURE 9: Leakage power characterization of function units with multi-V_{th}/V_{dd} and variation awareness.

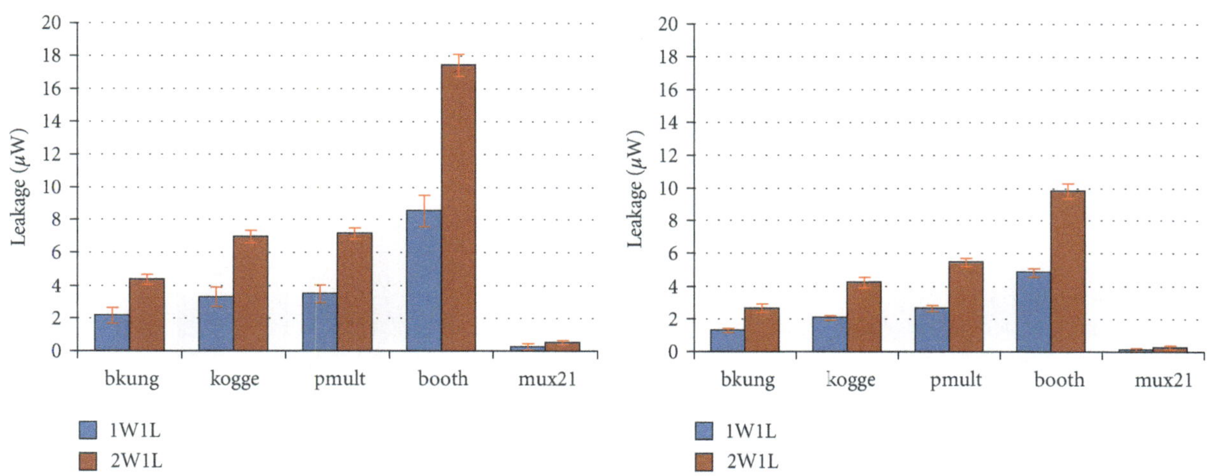

(a) $V_{th} = 0.37$ V (b) $V_{th} = 0.56$ V

FIGURE 10: Leakage power characterization of function units with multi-V_{th}/V_{dd} and variation awareness.

further in Figures 11 and 12, where the delay and power distributions of the function unit bkung is sampled at a third V_{th} of 0.45 V. The plotted curves show that the magnitude of delay variation increases for higher V_{th} units, which means larger probabilities of timing violations if these high V_{th} units are placed on near-critical paths. The figures also show that up-sizing the device can effectively reduce the delay and leakage variations, as depicted by the error bars in Figures 11 and 12.

With the variation-aware multi-V_{th}/V_{dd} resource library characterized, our proposed resource binding algorithm is applied on a set of industrial high-level synthesis benchmarks, which are listed in Table 1. A total power limitation P_{LMT} is set for each benchmark to evaluate the power yield improvement. The dynamic power consumption of function units is estimated by Synopsys Design Compiler with multi-V_{th}/V_{dd} technology libraries generated by Liberty NCX. In this work with FreePDK 45 nm technology, the dynamic power is about 2 times the mean leakage power. The power yield before and after the improvement is then computed using (6) in Section 4.2. The proposed resource binding algorithm is implemented in C++, and experiments are conducted on a Linux workstation with Intel Xeon 3.2 GHz processor and 2 GB RAM. All the experiments run in less than 60 s of CPU time.

We compare our variation-aware resource binding algorithm against the traditional deterministic approach, which uses the worst-case ($\mu + 3\sigma$) delay values of function units in the multi-V_{th}/V_{dd} library to guide the resource binding. For deterministic approach, we leverage a commercial HLS tool called Catapult-C to obtain the delay/area/power estimation. The worst-cased based approach will naturally lead to 100% timing yield; however, the power yield is poor as shown in the motivational example in Figure 5. In contrast, our yield-aware statistical optimization algorithm takes the delay and power distributions as inputs, explores the design space

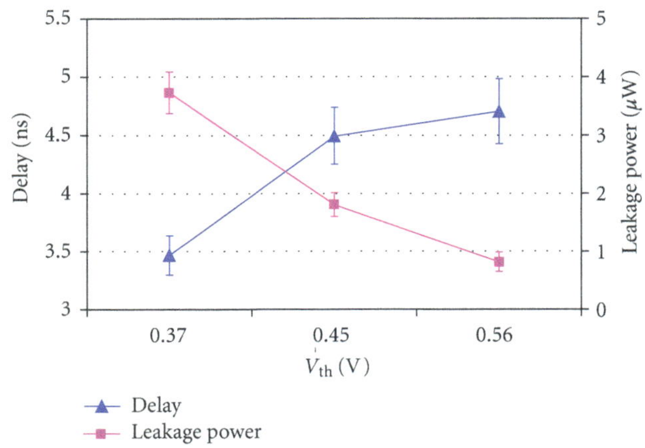

FIGURE 11: The delay and power tradeoff with increasing V_{th} for bkung with default device sizing (1W1L).

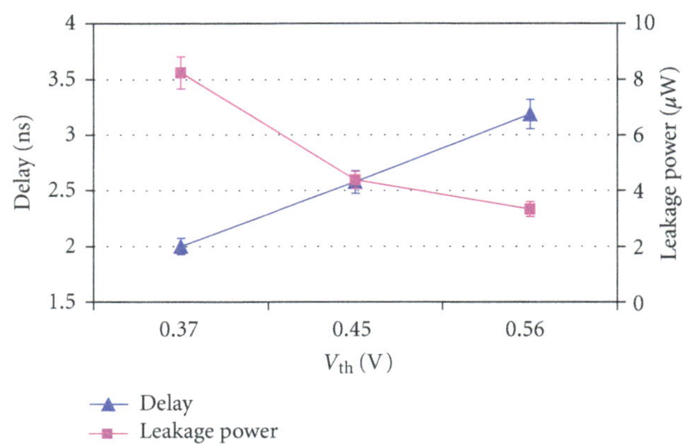

FIGURE 12: The delay and power tradeoff with increasing V_{th} for bkung with doubled device sizing (2W1L).

TABLE 1: The profile of test benchmarks.

Name	Nodes number	Edges number	Add number	Mult number
CHEM	33	51	15	18
EWF	34	49	20	12
PR	44	132	26	16
WANG	52	132	26	24
MCM	96	250	64	30
HONDA	99	212	45	52
DIR	150	312	84	64
STEAM	222	470	105	115

with the guidance of YieldGain, and iteratively improves the power yield under a slight timing yield loss. The comparison results are shown in Figures 13, 14, 15 and 16, respectively.

Figure 13 shows the power yield improvement against worst-case delay based approach, with different level conversion strategies. A fixed timing yield constraint of 95% is set for the proposed variation-aware algorithm, using the function units with default device sizes (1W1L). The overheads of the level converters used in this paper are listed in

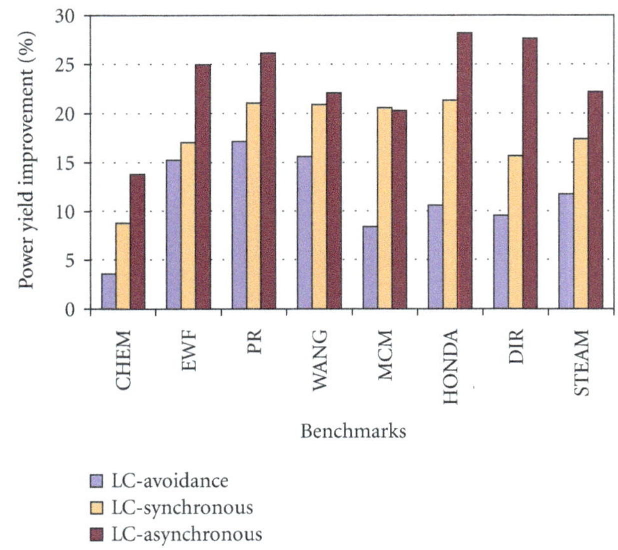

FIGURE 13: Power yield improvement against deterministic worst-case approach with different level conversion strategies and timing yield constraint of 95%.

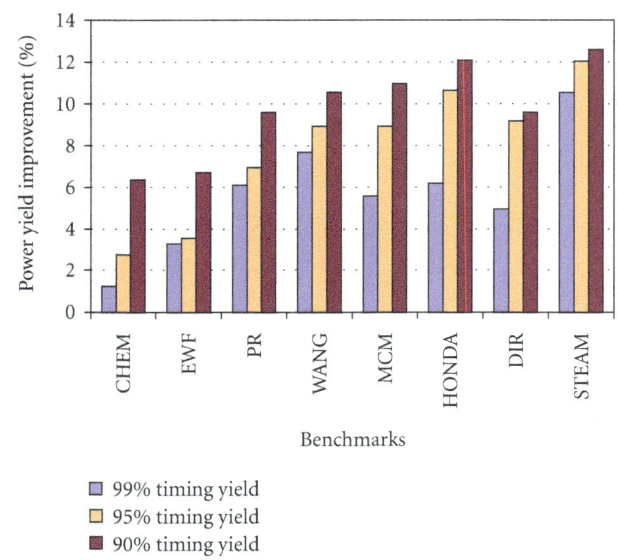

FIGURE 14: Power yield improvement against deterministic worst-case approach with multi-V_{th} only and different timing yield constraints.

TABLE 2: The Overheads of voltage level converters.

Type	Delay (ps)	Power (nW)
Synchronous converter	80	Negligible
Asynchronous converter	200	3790

Table 2. The usage of function units and level converters under the three listed conversion strategies (conversion avoidance, synchronous conversion and asynchronous conversion) is listed in Table 3, in which "Vdd-H FUs number" and "Vdd-L FUs number" show the numbers of function units with high/low supply voltages, respectively, and "LCs

TABLE 3: The usage of function units and level converters with different level conversion strategies.

Bench name	LC-avoidance Vdd-H	LC-avoidance Vdd-L	LC-synchronous Vdd-H	LC-synchronous Vdd-L	LC-synchronous LCs	LC-asynchronous Vdd-H	LC-asynchronous Vdd-L	LC-asynchronous LCs	Overhead
CHEM	5	1	3	3	1	2	4	2	4.2%
EWF	6	1	4	3	1	3	4	2	3.7%
PR	7	2	6	3	2	4	5	3	4.1%
WANG	9	2	8	3	2	5	6	3	3.4%
MCM	20	4	16	8	4	15	9	5	2.4%
HONDA	22	5	17	10	6	15	12	6	3.9%
DIR	28	4	20	12	8	14	18	12	4.8%
STEAM	34	8	25	19	11	21	23	13	5.0%

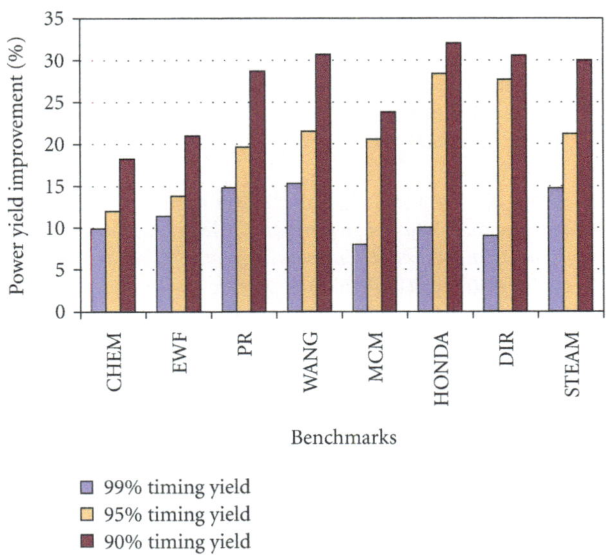

FIGURE 15: Power yield improvement against deterministic worst-case approach with asynchronous level conversion and different timing yield constraints.

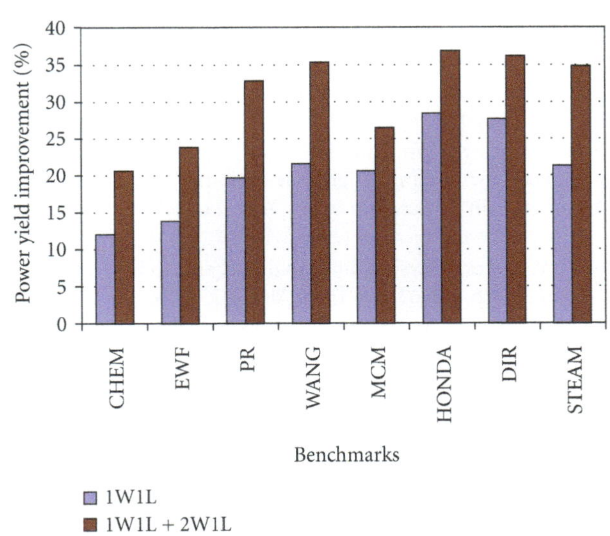

FIGURE 16: Power yield improvement against deterministic worst-case approach with asynchronous level conversion and different timing yield constraints.

number" counts the number of converters used in the design. The last column counts the total power overhead of the asynchronous level converters. The average power yield improvements for the three strategies are 11.7%, 17.9%, and 22.2%, respectively. From Figure 13 and Table 3 we can see that larger power yield improvements can be achieved when more low-Vdd function units are used in the design. The results also validate our claims in Section 4.3 and Section 5.2 that, asynchronous level conversion is more favorable in statistical optimization because it enables timing borrowing between function units and leads to the timing yield improvement that can compensate the overhead of the converters. Therefore, compared to the synchronous case, more asynchronous converters are used while yielding better results.

Figure 14 shows power yield improvement with multi-V_{th} technique only, which means only the resource units with nominal supply voltage V_{dd}^{H} can be selected. In this case, no level conversion is needed so there is no overhead for level converters. Only function units with default device sizes (1W1L) are used. The average power yield improvements against worst-case delay based approach, under timing yield constraints 99%, 95%, and 90% are 5.7%, 7.9%, and 9.8%, respectively. At timing yield 95%, the average power yield improvement (7.9%) is smaller than the LC-avoidance case (11.5%) in Figure 13, which shows that using multi-V_{dd} resource units can further improve the power yield.

Figure 15 shows the power yield improvement against worst-case delay based approach, under different timing yield constraints. Asynchronous level conversion is chosen in this series of experiments. Only function units with default device sizes (1W1L) are used. The average power yield improvements under timing yield constraints 99%, 95%, and 90% are 11.6%, 20.6%, and 26.9%, respectively. It is clearly shown that, the power yield improvement largely depends on how much timing yield loss is affordable for the design. This will further push forward the design space exploration for a well-balanced timing and power tradeoff.

Figure 16 shows the power yield improvement against worst-case delay-based approach, using function units with different device sizes. Asynchronous level conversion is chosen in this series of experiments, and a fixed timing yield constraint of 95% is set for the proposed variation-aware algorithm. Compared with the average 20.6% yield improvement in the case using default device size (1W1L) only, using both default-size (1W1L) and double-size (2W1L) resources can lead to an average power yield improvement of 30.9%. Obviously, upsized device with higher performance and smaller variability provide additional flexibility for design space exploration; however, this is achieved at the cost of larger silicon area.

7. Conclusions

In this paper, we investigate the impact of process variations on multi-$V_{\mathrm{th}}/V_{\mathrm{dd}}$ and device sizing techniques for low-power-high-level synthesis. We characterize delay and power variations of function units under different threshold and supply voltages, and feed the variation-characterized resource library to the HLS design loop. Statistical timing and power analysis for high-level synthesis is then introduced, to help our proposed resource binding algorithm explore the design space and maximize the power yield of designs under given timing yield constraints. Experimental results show that significant power reduction can be achieved with the proposed variation-aware framework, compared with traditional worst-case based deterministic approaches.

Acknowledgment

This work was supported in part by NSF 0643902,0903432, and 1017277, NSFC 60870001/61028006 and a grant from SRC.

References

[1] P. Coussy and A. Morawiec, *High-Level Synthesis: From Algorithm to Digital Circuit*, Springer, 2009.

[2] W. T. Shiue, "High level synthesis for peak power minimization using ILP," in *Proceedings of the IEEE International Conference on Application-Specific Systems, Architectures, and Processors*, pp. 103–112, July 2000.

[3] K. S. Khouri and N. K. Jha, "Leakage power analysis and reduction during behavioral synthesis," *IEEE Transactions on Very Large Scale Integration Systems*, vol. 10, no. 6, pp. 876–885, 2002.

[4] X. Tang, H. Zhou, and P. Banerje, "Leakage power optimization with dual-Vth library in high-level synthesis," in *Proceedings of the 42nd Design Automation Conference (DAC '05)*, pp. 202–207, June 2005.

[5] W. L. Hung, X. Wu, and Y. Xie, "Guaranteeing performance yield in high-level synthesis," in *Proceedings of the International Conference on Computer-Aided Design (ICCAD '06)*, pp. 303–309, November 2006.

[6] J. Jung and T. Kim, "Timing variation-aware high-level synthesis," in *Proceedings of the IEEE/ACM International Conference on Computer-Aided Design (ICCAD '07)*, pp. 424–428, November 2007.

[7] F. Wang, G. Sun, and Y. Xie, "A variation aware high level synthesis framework," in *Proceedings of the Design, Automation and Test in Europe (DATE '08)*, pp. 1063–1068, March 2008.

[8] G. Lucas, S. Cromar, and D. Chen, "FastYield: Variation-aware, layout-driven simultaneous binding and module selection for performance yield optimization," in *Proceedings of the Asia and South Pacific Design Automation Conference (ASP-DAC '09)*, pp. 61–66, January 2009.

[9] A. Srivastava, T. Kachru, and D. Sylvester, "Low-power-design space exploration considering process variation using robust optimization," *IEEE Transactions on Computer-Aided Design of Integrated Circuits and Systems*, vol. 26, no. 1, pp. 67–78, 2007.

[10] K. Usami and M. Igarashi, "Low-power design methodology and applications utilizing dual supply voltages," in *Proceedings of the Design Automation Conference (ASPDAC '00)*, pp. 123–128, Yokohama, Japan, 2000.

[11] A. Srivastava and D. Sylvester, "Minimizing total power by simultaneous Vdd/Vth assignment," *IEEE Transactions on Computer-Aided Design of Integrated Circuits and Systems*, vol. 23, no. 5, pp. 665–677, 2004.

[12] C. P. Chen, C. C. N. Chu, and D. F. Wong, "Fast and exact simultaneous gate and wire sizing by lagrangian relaxation," in *Proceedings of the IEEE/ACM International Conference on Computer-Aided Design (ICCAD '98)*, pp. 617–624, ACM, New York, NY, USA, 1998.

[13] S. Sirichotiyakul, T. Edwards, C. Oh et al., "Stand-by power minimization through simultaneous threshold voltage selection and circuit sizing," in *Proceedings of the 1999 36th Annual Design Automation Conference (DAC '99)*, pp. 436–441, June 1999.

[14] L. Wei, K. Roy, and C. K. Koh, "Power minimization by simultaneous dual-Vth assignment and gate-sizing," in *Proceedings of the 22nd Annual Custom Integrated Circuits Conference (CICC '00)*, pp. 413–416, May 2000.

[15] P. Pant, R. K. Roy, and A. Chatterjee, "Dual-threshold voltage assignment with transistor sizing for low power CMOS circuits," *IEEE Transactions on Very Large Scale Integration Systems*, vol. 9, no. 2, pp. 390–394, 2001.

[16] T. Karnik, Y. Ye, J. Tschanz et al., "Total power optimization by simultaneous dual-Vt allocation and device sizing in high performance microprocessors," in *Proceedings of the 39th Annual Design Automation Conference (DAC '02)*, pp. 486–491, June 2002.

[17] S. Insup, P. Seungwhun, and S. Youngsoo, "Register allocation for high-level synthesis using dual supply voltages," in *Proceedings of the 46th ACM/IEEE Design Automation Conference (DAC '09)*, pp. 937–942, July 2009.

[18] S. P. Mohanty and E. Kougianos, "Simultaneous power fluctuation and average power minimization during nano-CMOS behavioral synthesis," in *Proceedings of the 20th International Conference on VLSI Design held jointly with 6th International Conference on Embedded Systems (VLSID '07)*, pp. 577–582, January 2007.

[19] F. Wang, X. Wu, and Y. Xie, "Variability-driven module selection with joint design time optimization and post-silicon tuning," in *Proceedings of the Asia and South Pacific Design Automation Conference (ASP-DAC '08)*, pp. 2–9, March 2008.

[20] R. W. Keyes, "Physical limits in digital electronics," *Proceedings of the IEEE*, vol. 63, no. 5, pp. 740–767, 1975.

[21] D. S. Boning and S. Nassif, "Models of process variations in device and interconnect," in *Design of High Performance Microprocessor Circuits*, IEEE Press, 2000.

[22] B. Zhai, S. Hanson, D. Blaauw, and D. Sylvester, "Analysis and mitigation of variability in subthreshold design," in *Proceedings of the International Symposium on Low Power Electronics and Design*, pp. 20–25, August 2005.

[23] NCSU, "45 nm FreePDK," http://www.eda.ncsu.edu/wiki/FreePDK.

[24] M. Meterelliyoz, A. Goel, J. P. Kulkarni, and K. Roy, "Accurate characterization of random process variations using a robust

low-voltage high-sensitivity sensor featuring replica-bias circuit," in *Proceedings of the IEEE International Solid-State Circuits Conference (ISSCC '10)*, pp. 186–187, February 2010.

[25] C. Jacoboni and P. Lugli, *The Monte Carlo Method for Semiconductor Device Simulation*, Springer, 1990.

[26] N. S. Kim, T. Austin, D. Blaauw et al., "Leakage current: Moore's law meets static power," *Computer*, vol. 36, no. 12, pp. 68–64, 2003.

[27] K. M. Cao, W. C. Lee, W. Liu et al., "BSIM4 gate leakage model including source-drain partition," in *Proceedings of the IEEE International Electron Devices Meeting*, pp. 815–818, December 2000.

[28] N. C. Beaulieu, A. A. Abu-Dayya, and P. J. McLane, "Comparison of methods of computing lognormal sum distributions and outages for digital wireless applications," in *Proceedings of the IEEE International Conference on Communications*, pp. 1270–1275, May 1994.

[29] S. H. Kulkarni, A. N. Srivastava, and D. Sylvester, "A new algorithm for improved VDD assignment in low power dual VDD systems," in *Proceedings of the 2004 International Symposium on Lower Power Electronics and Design (ISLPED '04)*, pp. 200–205, August 2004.

[30] M. J. M. Pelgrom, A. C. J. Duinmaijer, and A. P. G. Welbers, "Matching properties of MOS transistors," *IEEE Journal of Solid-State Circuits*, vol. 24, no. 5, pp. 1433–1440, 1989.

[31] J. Kwong and A. P. Chandrakasan, "Variation-driven device sizing for minimum energy sub-threshold circuits," in *Proceedings of the 11th ACM/IEEE International Symposium on Low Power Electronics and Design (ISLPED '06)*, pp. 8–13, October 2006.

[32] S. A. Tawfik and V. Kursun, "Multi-Vth level conversion circuits for multi-VDD systems," in *Proceedings of the IEEE International Symposium on Circuits and Systems (ISCAS '07)*, pp. 1397–1400, May 2007.

[33] E. Aarts and J. K. Lenstra, *Local Search in Combinatorial Optimization*, Princeton University Press, 2003.

[34] A. Raghunathan and N. K. Jha, "Iterative improvement algorithm for low power data path synthesis," in *Proceedings of the IEEE/ACM International Conference on Computer-Aided Design*, pp. 597–602, November 1995.

[35] F. Ishihara, F. Sheikh, and B. Nikolić, "Level conversion for dual-supply systems," *IEEE Transactions on Very Large Scale Integration Systems*, vol. 12, no. 2, pp. 185–195, 2004.

3

Intelligent On/Off Dynamic Link Management for On-Chip Networks

Andreas G. Savva,[1] Theocharis Theocharides,[1] and Vassos Soteriou[2]

[1] *Department of Electrical and Computer Engineering, University of Cyprus, 1678 Nicosia, Cyprus*
[2] *Department of Electrical Engineering and Information Technology, Cyprus University of Technology, 3036 Limassol, Cyprus*

Correspondence should be addressed to Andreas G. Savva, eep7sa4@ucy.ac.cy

Academic Editor: Sao-Jie Chen

Networks-on-chips (NoCs) provide scalable on-chip communication and are expected to be the dominant interconnection architectures in multicore and manycore systems. Power consumption, however, is a major limitation in NoCs today, and researchers have been constantly working on reducing both dynamic and static power. Among the NoC components, links that connect the NoC routers are the most power-hungry components. Several attempts have been made to reduce the link power consumption at both the circuit level and the system level. Most past research efforts have proposed selective on/off link state switching based on system-level information based on link utilization levels. Most of these proposed algorithms focus on a pessimistic and simple static threshold mechanism which determines whether or not a link should be turned on/off. This paper presents an intelligent dynamic power management policy for NoCs with improved predictive abilities based on supervised online learning of the system status (i.e., expected future utilization link levels), where links are turned off and on via the use of a small and scalable neural network. Simulation results with various synthetic traffic models over various network topologies show that the proposed work can reach up to 13% power savings when compared to a trivial threshold computation, at very low (<4%) hardware overheads.

1. Introduction

Power management is a crucial element in modern-day on-chip interconnects. Significant efforts have been made in order to address power consumption in networks-on-chips (NoCs) [1–6]. One of the most power-hungry NoC components are the links connecting the routers to each other and the processing elements (PEs) of the on-chip interconnection network (NoC). Recent data from Intel's Teraflop NoC prototype [7] suggests that link power consumption could be as high as 17% of the network power and could be even more given the types of links used as well as the size and pipelining involved in designing the link structure. These links, which can be designed with differential signals and low-voltage swing hardware using level converters as circuit-based optimizations for low power consumption, are almost active all the time, even when not transmitting useful data thus spending energy when no inter-router communication exits. While such traditional hardware design techniques have contributed towards reducing the power of these links, a system-level technique becomes necessary for more efficient power reduction, as the number of links increases with the scaling and increasing sizes of NoCs, and as application-specific knowledge becomes available. For example, power-aware-encoding techniques [8] such as Gray coding cannot be efficiently used, as the hardware cost in the encoder/decoder increases drastically as the system scales to a higher number of network components. As such, recent research focuses on turning links off and on in order to reduce power consumption, and has been adopted by several works [1, 6, 9–11], as certain links in the system are severely underutilized during a specific operational time frame [1]. Techniques such as DVFS (dynamic voltage and frequency scaling) applied to the link hardware, [9, 11] have been used to vary the link frequency and power according to link utilization; however, even when not data is sent across a link, static power is still being consumed, especially in multipipelined links with pipeline buffers in place. In addition, CMOS technology scaling is pointing towards an increased portion of the allocated power budget being consumed as static energy instead of dynamic energy; hence, switching on/off links instead of just selectively reducing their frequency/voltage levels offers better power saving advantages as links still do burn power even at lower (i.e., nonzero) voltage-frequency settings [12]. The majority of these on/off link dynamic power-management works employ traditionally a statically-computed threshold value on the link utilization, and based on that threshold value, the link is turned off for an amount of time and then is turned back on when the algorithm decides so. This of course is a pessimistic approach by nature, and imposes harder performance constraints. Recently, the use of control theory for managing candidate links for turning off has been proposed as an idea in [10], with promising results when compared to the statically-based approaches.

Motivated by the findings in [10], this paper proposes the use of artificial neural networks (ANNs) as a dynamic link power consumption management mechanism, by utilizing application traffic information. Based on their ability to dynamically be trained by variable scenarios, ANNs can offer flexibility and high prediction capabilities [13]. An ANN-based mechanism can be used to intelligently compute dynamically the threshold value used to determine which links can be turned off and on during discrete time intervals. The ANN receives link utilization data in discrete time intervals, and predicts the links that should be turned off or on based on the computed threshold. ANNs can be dynamically trained to new application information, and have been proven that they can offer accurate prediction results in similar scenarios [14]. ANNs can be efficiently designed in hardware provided they remained relatively small, through efficient resource sharing and pipelining. Furthermore, by partitioning the NoC, individual small ANNs can be assigned to monitor each partition independently, and in parallel monitor the entire network. This work also introduces topology-based directed training as a pretraining scheme, using guided simulation, which helps to minimize the large training set and the ANN complexity. This work extends our initial idea presented in [15] by several new contributions: (a) the ANN architecture has been redesigned, making it flexible and smaller through trade-off simulations involving the size and structure of the ANN and the offered power savings, (b) extended discussion on the architecture and its hardware implementation, (c) extended discussion on the simulation platform, synthetic traffic benchmarks and power modeling, and (d) extended results related to the power savings versus the performance penalty and the associated hardware overheads.

The rest of this paper is organized as follows. Section 2 discusses background and related work. In Section 3, we introduce the ANN-based approach for managing link power in NoCs. Section 4 presents the simulation framework and simulation results and analysis through various topologies and synthetic traffic, and Section 5 concludes the paper giving brief future research directives.

2. Background and Related Work

2.1. ANN Background and Motivation. An ANN is an information-processing paradigm that is inspired by the way biological neurons systems process information. It is composed of a large number of highly interconnected processing elements (neurons) working in unison to solve specific problems. ANNs learn by training and are typically trained for specific applications such as pattern recognition or data classification. ANNs have been successfully used as prediction and forecasting mechanisms in several application areas, as they are able to determine hidden and strongly nonlinear dependencies, even when there is a significant noise in the data set [14]. ANNs have been used as branch prediction mechanisms in computer architecture, as forecasting mechanisms in stocks [14], and in several other prediction applications. The ANN operates in two stages: the training stage and the computational stage. A neuron takes a set of inputs and multiplies each input by a weight value, which is determined in training stage, accumulating the result until all the inputs are received. A threshold value is then substracted from the accumulated result, and this result is then used to compute the output of the neuron based on an activation function. The neuron output is then propagated to the neurons of the next layer which perform the same operation with the newly set of inputs and their own weights. This is repeated for all the layers of an ANN. A neural network can be realized in hardware by using interconnected neuron hardware models, each of which is composed by multiplier accumulator (MAC) units, and a look-up Table (LUT) as a shared activation function. Some memory is also required to hold the training weights.

2.2. Related Work in Link Dynamic Power Management. Recently published research practice surveys such as [16] which outline the design challenges and lay the roadmap in future NoC design have emphasized the critical need to conduct research in NoC power management due to concerns of battery life, cooling, environmental issues, and thermal management, as a means to safeguard the scalability of general-purpose multicore systems that employ NoCs as their communication backbone. Link dynamic power management has been given significant attention by NoC researchers, as circuit-based techniques such as differential signals and low-voltage swing hardware using level converters do not seem to adequately address the power management problem [17, 18]. As such, there is a significant shift towards high-level techniques such as selective turning of links on and off. The challenge involved in those techniques includes the computation of the decision on whether a certain link is to be turned off, and when it will be turned back on. These decisions typically rely on information from the system concerning link utilization, and, so far, have been taken using a threshold-based approach. There have been attempts in dynamic link frequency and dynamic link voltage (DVFS) management with most using these thresholds as well.

Among the proposed techniques, some approaches use software-based management techniques such as the one in

[17], which proposes the use of reducing energy consumption through compiler-directed channel voltage scaling. This technique uses proactive power management, where application code is analyzed during static compilation time to identify periods of network inactivity; power management calls are then inserted into the compiled application code to direct on/off link transitions to save link power. A similar approach was also taken in [19] for communication power management using dynamic voltage-scalable links. Both of these techniques, however, have been applied to highly predictive array-intensive applications, where precise idle and active periods can be extracted. Hence, run-time variability, applicable to NoCs found in general-purposed multicore chips, has not been examined. Further the work in [20] proposes software-hardware hybrid techniques that extend the flow of a parallelizing compiler in order to direct run-time network power reduction. In this paper, the parallelizing compiler orchestrates dynamic-voltage scaling of communication links, while the hardware part handles unpredicted online traffic variability in the underlying NoC to handle unexpected swings in link utilization that could not be captured by the compiler for improved power savings and performance attainability.

Low-level, hardware-based techniques that determine on/off periods and manage the voltage and frequency, exhibit however better energy savings as they can shorten the processing time required for a decision whether to turn a link off or on to be made. The most commonly used power management policies deal with adjusting processing frequency and voltage (dynamic voltage scaling—DVS). The works in [5, 18] present DVS techniques that feature a utilization threshold to adjust the voltage to the minimum value while maintaining the worst case execution time. In [21], the authors propose that the dynamic voltage scaling is performed based on the information concerning execution time variation within multimedia streams. The work in [22] proposes a power consumption scheme, in which variable-frequency links can track and adjust their voltage level to the minimum supply voltage as the link frequency is changed. Furthermore, [11] introduces a history-based DVS policy which adjusts the operating voltage and clock frequency of a link according to the utilization of the link/input buffer. Link and buffer utilization information are also used in [9], which proposes a DVS policy scheme that dynamically adapts its voltage scaling to achieve power savings with minimal impact on performance. Given the task graph of a periodic real-time application, the proposed algorithm in [9] assigns an appropriate communication speed to each link, which minimizes the energy consumption of the NoC while guaranteeing the timing constraints of real applications. Moreover, this algorithm turns off links statically when no communications are scheduled because the leakage power of an interconnection network is significant. In general on/off links have, in most cases, been more efficient than DVFS techniques, as links, even if operating at a lower voltage, still consume leakage and dynamic power [1, 6]. These works therefore present a threshold-based technique that turns links off when there is low utilization, using a statically computed threshold. Given that static computation by nature is pessimistic, dynamic policies have been proposed. Research work in [23] proposes a mechanism to reduce interconnect power consumption that combines dynamic on/off network link switching as a function of traffic while maintaining network connectivity, and dynamically reducing the available network bandwidth when traffic becomes low. This technique is also based on a threshold-based on/off decision policy. Next, the work in [24] considers a 3D torus network in a cluster design (off-chip interconnection network) to explore opportunities for link shutdown during collective communication operations. The scheme in [25] introduces the Skip-link architecture that dynamically reconfigures NoC topologies, in order to reduce the overall switching activity and hence associated energy consumption. The technique allows the creation of long-range Skip-links at run-time to reduce the logical distance between frequently communicating nodes. However, this is based on application communication behavior in order to extract such opportunities to save energy. Finally the related work in [26] explores how the power consumed by such on-chip networks may be reduced through the application of clock and signal-gating optimizations, shutting power to routers when they are inactive. This is applied at two levels: (1) at a granular level applied to individual router components and (2) globally at the entire router.

Run-time link power management has recently gained ground in research to address the leakage issues as well. As links become heavily pipelined to satisfy performance constraints, link buffers and pipeline buffers contribute significantly in leakage power consumption. As such, the problem becomes significant with the increased on-chip NoC sizes, impacting both the power consumption as well as the thermal stability of the chip. Dynamic link management techniques have therefore been proposed; the work in [2] proposes an adaptive low-power transmission scheme, where the energy required for reliable communications is minimized while satisfying a QoS constraint by varying dynamically the voltage on the links. The work in [27] introduces ideas of dynamic routing in the context of NoCs and focuses on how to deal with links or/and routers that become unavailable either temporarily or permanently. Such techniques are a little more complicated than a threshold-based approach, and inhere performance overheads during each dynamic computation. As such, the work in [10] introduces the idea of an intelligent method for dynamic (run-time) power management policy, utilizing control theory. A preliminary idea of a closed-loop power management system for NoCs is presented, where the estimator tracks changes in the NoC and estimates changes in service times, in arrival traffic patterns and other NoC parameters. The estimator then feeds any changes into the system model, and the controller sets the voltage and frequency of the processor for the newly estimated frequency rate. Motivated by the promising results presented in [10], and the potential performance benefits of dynamic threshold computation techniques, this work proposes a dynamic, intelligent, and flexible scheme based on ANNs for dynamic computation of the threshold that determines which links can be turned off or on.

3. ANN-Based Threshold Computation Methodology

3.1. Static Threshold Computation for On/Off Links. The first step in realizing the proposed ANN methodology is to establish a framework for comparing whether an intelligent management is comparable to the nonintelligent case, not only in terms of energy savings, but also in terms of throughput and hardware overheads. As such, a trivial case, where a simple threshold mechanism was used to determine whether or not a link would turn off or back on, was first implemented using an NoC simulation framework and the Orion power models [28] (explained later in Section 4.1). The mechanism chooses an appropriate threshold based on which the links turn on and off. This trivial algorithm takes as input the link utilizations of all the links in the experimental NoC system, and outputs control signals based on a statically defined threshold; based on this threshold, the algorithm then decides which links are turned off and then back on. The statically-defined threshold was computed based on simulation observations from different synthetic traffic models and based on the observed power savings and throughput reduction when compared to a system without the mechanism. Figure 1 shows the real-time power savings for four synthetic traffic models, observed over a 4×4 NoC.

This method was introduced in [1], and the results presented therein as well as the experiments with our framework indicate that such mechanisms can be quite effective. However, a run-time mechanism, which can potentially benefit from real-time information stemming from the network, can potentially outperform this method. Such mechanism is described next. Furthermore, [1] uses an open-loop mechanism, prone to oscillations that potentially can limit both the attainable performance and also the power savings, as power is still used during the transition [10].

3.2. Mechanism Overview. The ANN-based mechanism can be integrated as an independent processing element in the NoC (PE), potentially located in a central point in the network for easy access by the rest of the PEs, and each base ANN mechanism can be assigned to monitor an NoC partition. Such cases are shown in Figure 2(a). Each base ANN mechanism monitors all the average link utilization rates within its region. These values are processed by the ANN, which computes the threshold utilization value for each link within its region, during each interval. The threshold value is then used to turn off any links in the region that exhibit lower utilization. Links which have been turned off remain off for a certain period of time. Experiments in related work [1, 10] indicate that such time should be within a few hundred cycles, as longer periods tend to create a vast performance drop-off (as the network congestion increases due to lack of available paths), whereas shorter periods do not incur worthy power savings. The proposed ANN mechanism uses a 100-cycle interval, during which all new utilization rates are received. This interval was chosen based on existing experiments in [1], which shows that a 100-cycle interval incurs better performance to power savings. The interval, however, is a system parameter, which can also be

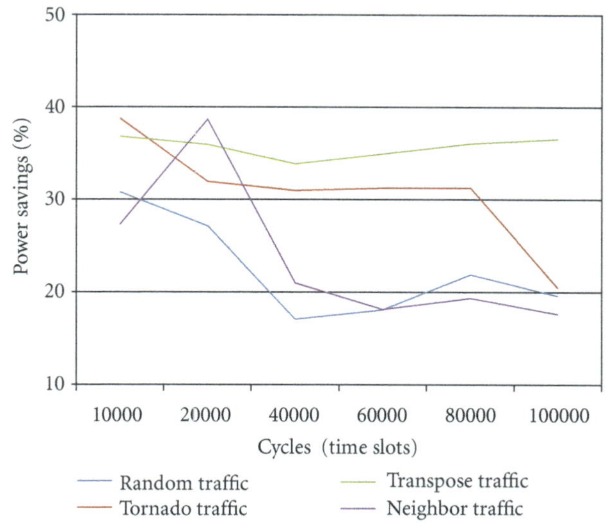

FIGURE 1: Power savings of a trivial threshold case compared to no on/off links case.

taken into consideration by the system training, and involves future work. During the interval span, the ANN computes and outputs the new threshold, which is then used by the link control mechanisms in each router to turn off underutilized links. The links remain off for another 100 cycles, and turn back on when a new threshold is computed. During the 100-cycle interval, links which are off, do not participate in the computation of the next threshold; instead, they are encoded with a sentinel value that represents them being fully utilized, so they are not kept off in two subsequent intervals. This reserves fair path allocation within the network.

Each ANN-based mechanism follows a fully connected multilayer perceptron model [13, 14], consisting of one hidden layer of internal neurons/nodes and a single output-layer neuron. The activation function used in this work is the hyperbolic tangent function, which is symmetric and asymptotic, henceforth easy to implement in hardware as a LUT [29]. Furthermore, the specific function has been extensively used in several ANNs and its accuracy has been very good [13]. The ANN system is shown in Figure 2(b). The number of internal neurons was chosen to be the half of the summation of the input and output neurons [13]. The input neurons depend on the number of links that the system receives as feedback. As such, the size of the ANN depends on the number of inputs to the system. The output neuron chooses the corresponding threshold that best matches the pattern observed through the hidden layer neurons and outputs the threshold value to the link controller.

The neuron computation involves computing the weighted sum of the link utilization inputs. An activation function is then applied to the weighted sum of the inputs of the neuron in order to produce the neuron output (i.e., *activate* the neuron). Equation (1) shows how the output of a neuron is calculated.

Function which calculates the output of a neuron:

$$f(x) = K\left(\sum_i w_i g_i(x)\right). \quad (1)$$

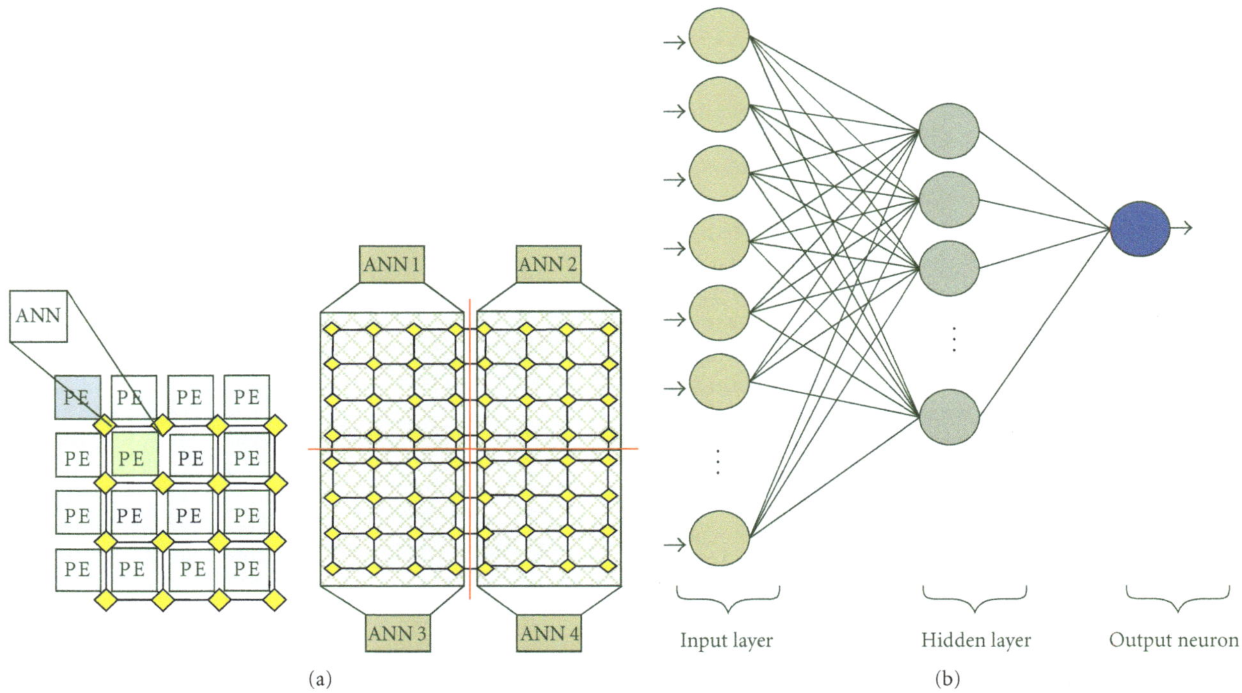

FIGURE 2: (a) ANN predictor with NoCs and an 8×8 network partition into four 4×4 networks with their ANNs; (b) Structure of the neural network.

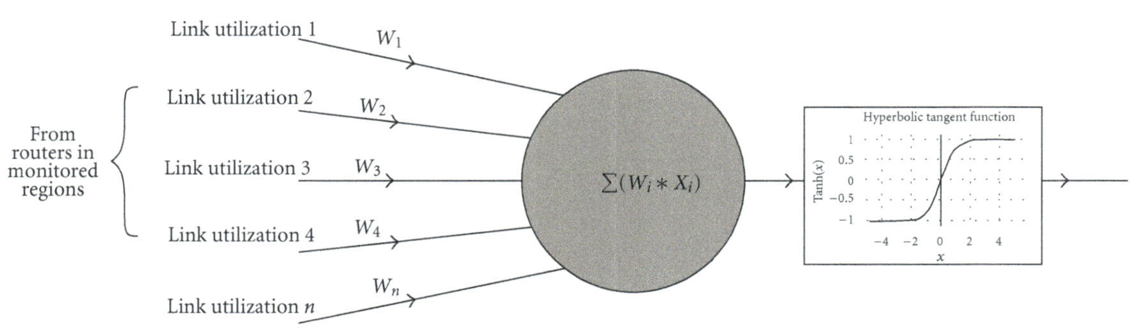

FIGURE 3: Neuron computations.

where K represents the activation function which is the hyperbolic tangent, w represents the weights which apply to the link utilization inputs which are represented by $g(x)$ input function. The overall procedure is shown in Figure 3.

3.3. ANN Training and Operation. The training stage can be performed off-line, that is, when the NoC is not used, and the training weights can be stored in SRAM-based LUTs for fast and on-line reconfiguration of the network. The network is trained using application traffic patterns, off-line, using the back-propagation ANN training algorithm [14]. In our experiments, we used synthetic traffic patterns and the Matlab ANN toolbox; the weight values were then fed to the simulator as inputs, where the actual prediction was then implemented and simulated. The operation of the ANN can be potentially improved, by categorizing the applications that a system will practically run. As such, for each application category (and subsequently traffic patterns with certain common characteristics), the ANN can be trained with the corresponding weights. Each training set can then be dynamically loaded during long operation intervals, where the system migrates to a new application behavior.

3.4. Intelligent Threshold Computation—ANN Size and NoC Scalability Issues. While ANNs are heavily efficient in predicting scenarios based on learning algorithms, they require careful hardware design considerations, as their size and complexity depend on the number of inputs received as well as the number of different output predictions (classes) that they have to do. NoCs consist of a large number of links which grow exponentially as the size of the NoC grows. Therefore, receiving link utilization and having to determine the threshold that controls which links are candidates for turning off and on would require an exponentially scalable ANN. As such, we devise a preprocessing technique, which identifies, based on simulation and observations, the set of candidate links for turning off and on, eliminating links which are almost always utilized. This depends obviously on the chosen network topology (e.g., in a 2D mesh topology, links that are likely to be less busy include links which are at the edges of the mesh, whereas central links are usually more active and can be left on all the time), so that the ANN mechanism can handle the output decision in a more manageable way. This can be aided by intelligent floor planning and placement of processing elements inside the NoC as well, but is beyond the scope of this paper. Through various synthetic traffic simulations, for each given NoC topology, the average utilization values for each link through various phases in the simulation are computed, and the links with the highest utilization values are always assumed that they will be on. Obviously this step reduces a little the effectiveness of the ANN, but it is necessary to minimize the size and overheads of the ANN both in terms of performance and in terms of hardware resources. This step has to be done for a given topology, prior to the ANN training. However, both steps (determining the links that the ANN will use, as well as the ANN training) can be done off-line, during the NoC design stage. The ANN training can also be done repeatedly whenever new application knowledge becomes available that might alter the on-chip network traffic behavior. This particular property of ANNs provides a comparative advantage against a statically computed threshold, making the NoC flexible under any application that it is required to facilitate. It must be stated that the number of links that will be considered as likely candidates for on/off activity (i.e., the ones which do tend to have low utilization during the pretraining stage) impact both the size of the ANN itself and the overall size of the mechanism (which involves logic that sends the appropriate control signals). Through the two steps, pretraining and training, each ANN can be trained and configured independently to satisfy its targeted NoC structure (topology and number of monitored links).

Furthermore, large NoCs can be partitioned into smaller regions. As such, a base ANN architecture can be assigned to monitor each region, and all the link utilizations of the routers of the NoC partition arrive at the ANN which is responsible for that region. The size of this NoC region, however, depends on two major factors; the incurred power savings that the corresponding base ANN offers, which depend on its ability to process and evaluate the input information, and the resulting ANN size and hardware overheads (and subsequently power consumed within the ANN) which grow exponentially as the size of the NoC region grows. Choosing a small NoC region will likely result in a small ANN, but will result in smaller savings since the ANN will not have enough information to compute a good threshold value. On the other hand, a large NoC region will provide the ANN with much more information and potentially result in a much better threshold value, but its size and overheads would reduce the power savings making the ANN ineffective. As such, we experimented with several NoC regions and base ANNs, comparing their hardware overheads (a product of the ANN power consumption and the gate count required to implement each ANN in hardware) and responding savings incurred with the computed threshold.

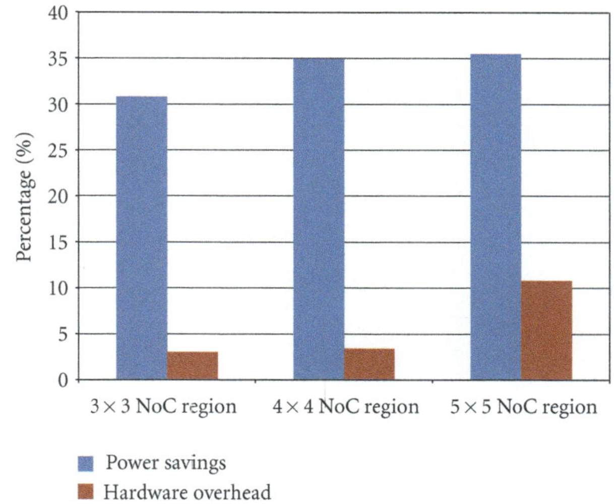

FIGURE 4: Power savings versus CMOS hardware overheads corresponding to various sizes of ANN monitoring regions in an NoC.

Figure 4 shows a comparison between hardware overheads (power × gate count) and power savings in the cases of 3×3, 4×4, and 5×5 ANN sizes for monitoring regions in an NoC. Results show that computation over a 4×4 NoC region offers satisfactory power savings and significantly less ANN overheads when compared to a 5×5 NoC region. A 3×3 NoC region does not provide enough information to the ANN in order to make accurate predictions. Based on these observations, we designed the base ANN system to monitor 4×4 NoC regions.

3.5. Base (4 × 4) Artificial Neural Network Operation. The ANN mechanism is responsible to compute for all the link utilizations the minimum values during each interval. Based on these values, the ANN calculates an optimal threshold. Figure 5 shows the procedure of the ANN mechanism for a 4×4 NoC partition. The ANN mechanism receives all the average link utilizations from all the links of the 4×4 NoC partition. These values are fed to the ANN in order to calculate an optimal threshold. Each router contains a control hardware monitor that measures the average link utilization for each of the four links in each router, and this value is sent to the ANN every n cycle (where n is the size of the time interval). If a router fails to transmit the values at a single interval, its value is set to sentinel value, which shows that its buffers are fully utilized. This mechanism acts also as a congestion information mechanism because links which are heavily active are not candidates to be turned off. The ANN uses the utilization values to find the threshold which will determine if a link is going to be turned off or on for the next n-cycle interval. As said earlier, we used 100-cycle intervals [1] (i.e., $n = 100$) in our simulations.

3.6. Base (4 × 4) Artificial Neural Network Hardware Architecture. One of the main advantages of ANNs is their simple hardware implementation when the number of neurons remains small and the activation function remains simple

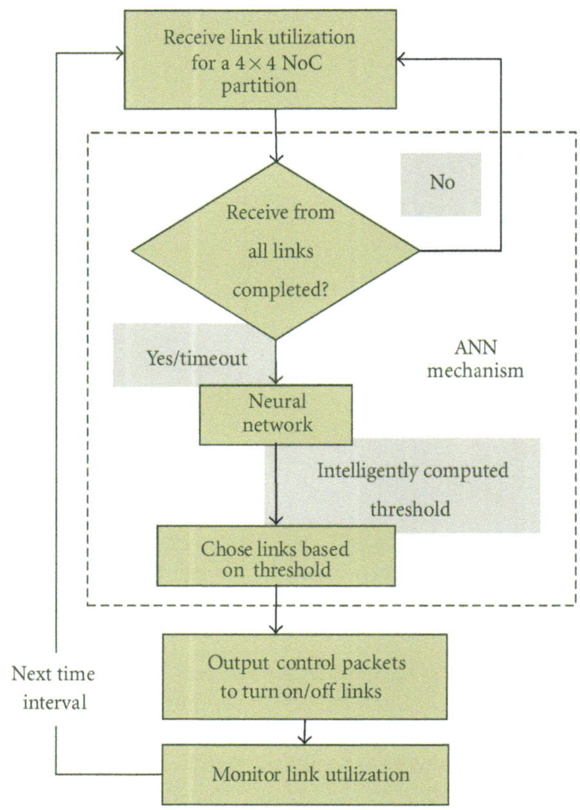

Figure 5: Main steps of a 4 × 4 ANN predictor.

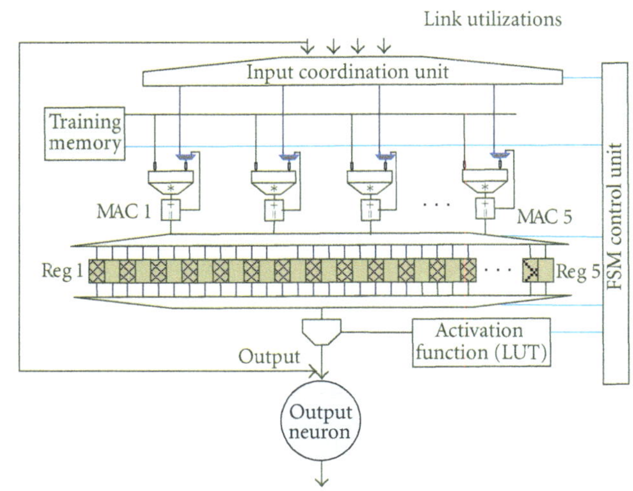

Figure 6: ANN hardware architecture and its hardware realizations.

[13]. The neuron operation can be designed efficiently in hardware since it can be modeled as a multiply-accumulate operation. The ANN hardware implementation depends on the number of hidden layer neurons; each neuron is implemented as a multiplier-accumulator (MAC) unit, with the accumulators being multiplexed for each neuron, so that the number of multipliers is minimized. The base ANN hardware architecture is shown in Figure 6. Utilization values for each link arrive and sorted through an input coordination unit, which distributes the values to each of the appropriate multipliers. The multipliers receive these values and through a shared weights memory, the corresponding weight. The weights and inputs product is then accumulated in the corresponding accumulator, with the entire process controlled via a finite-state machine controller. Each neuron has an assigned storage register, to enable data reuse; when one layer of neurons is computed, their outputs are stored inside a corresponding register. As such, the same hardware is reused for computing the next layer (i.e., from input layer to hidden layer and from hidden layer to output layer). When each neuron finishes its MAC computation, the result is then computed through the activation function LUT and propagates to the output neuron.

An ANN monitoring a 4×4 region in a torus topology, for example, receives 64 different inputs; if we are to assume that each router transmits a packet with its own link utilization during each interval, and if we also assume one packet per cycle delivered to the ANN during each interval, then, during each cycle, the ANN will receive at most 4 input values. Hence, if we use pipelined multipliers, we need only 4 multipliers for each ANN to achieve maximum throughput. The ANN therefore remains small and flexible, regardless of the size of the network it monitors. Furthermore, an ANN monitoring a 4 × 4 NoC partition receives 16 packets (one for each router); as such, it requires $16m$ cycles (where m is the cycle delay of each multiplier), plus 16 cycles for each accumulator, plus one cycle for the activation function plus one cycle for the output neuron, to output the new threshold (total of $16m + 18$ cycles). The overall data flow and architecture is shown in Figure 6.

3.7. ANN Hardware Optimization and Trade-Offs. In order to make the ANN architecture simpler and smaller we studied how the number of neurons of the hidden layer affect the total power savings of the system. Given that the 4 × 4 ANN monitors 16 routers, we need at least 8 input neurons [14]. Having eight neurons at the input layer of the ANN means that the hidden layer should have five neurons (based on the rule of thumb that a satisfactory number of the hidden layer neurons equals to half the number of input neurons plus one neuron) [14]. Three different ANNs were developed with five, four, and three neurons at the hidden layer, respectively. Figure 7 shows the power savings for these ANNs under the use of four different traffic patterns (Random, Tornado, Transpose, and Neighbor). Using four neurons therefore (instead of five), in the hidden layer exhibits the best power savings for all the traffic patterns. In addition, we studied how the bit representation of the training weights affects the threshold computation and subsequently the total power savings. Figure 8 shows how the bits used in representing the training weights influence the power savings of the system. As we can see 24, 16, 8, and 6 bits show similar power savings, but these savings are significantly reduced when 4 bits are used, due to reduced training accuracy. Based on the above, we selected the weight bit representation from 6 bits, which made the multiplier-accumulation hardware very small, requiring a 6-bit port for each weight and a 5-bit port for the utilization values.

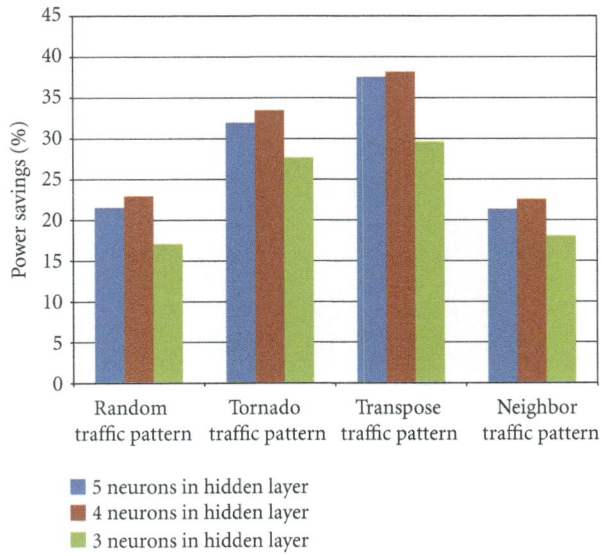

FIGURE 7: Power savings for five, four and three neurons in the hidden layer of the ANN.

4. Simulation and Results

4.1. Experimental Setup. In order to evaluate the ANN-based on/off link prediction mechanism, we developed a simulation framework based on the Java-based cycle-accurate *gpNoCsim* simulator (general-purpose simulator for network-on-chip architectures) [30]. The framework enables simulation of multiple topologies, utilizing dimension-ordered XY routing algorithm with virtual channel support and 4-stage pipelined router operation. The simulated router supports 64-bits flit width, 2 virtual channels per link and two buffers per virtual channel. The routers used are all the same, and we assume wormhole flow control. The framework supports various synthetic and user-defined traffic models. We experimented with a 4×4 mesh topology, and mesh and torus 8×8 topologies. Simulations are done over a range of 200,000 cycles, with a warm-up period of 100,000 cycles. In the 8×8 topologies, we partitioned the NoC into four regions of 4×4 routers/links, where each ANN-based model was assigned as responsible for monitoring. The ANN-based models monitored all links in their corresponding partition, all links were candidates for off/on, and all ANN results related to the size and operation of the ANN are given based on these architectural details.

Time was divided into 100-cycle intervals [1]; at the end of each interval, all routers in the NoC partition transmit their average utilization data for that span (computed via a counter and LUT-based multiplication with the reciprocal of the interval). A time-out mechanism equal to the expected delay of each router towards the ANN mechanism is imposed to maintain reasonable delays. The ANN receives one packet from each router with four utilization values, one for each port. The ANN then proceeds to compute the new threshold which is transmitted to each router through a control packet. Each router then turns off each link, depending whether or not its utilization value is above or below the new threshold. The router continues operation until the end of the new interval. It must be repeated that when a link is turned off or on, an extra 100-cycle penalty is inserted into the simulation to indicate the impact on the network throughput.

While the savings could significantly be improved using a more intelligent routing algorithm than the trivial XY dimension-ordered algorithm (DOR), we experimented with the XY algorithm and induced blocking when a link was off and the output buffer of that link fully utilized. In the case that a packet arrives in a router, and its destination link is off, the packet resides in the buffer until the link is turned back on. While this incurs a performance penalty, it is necessary to maintain correctness; this is a pessimistic approach obviously—a better routing algorithm would likely yield much better throughput and power savings, but is beyond the scope of this paper.

In order to study the power savings and the throughput of the dynamic ANN-based prediction algorithm for turning links on/off, we compare this to a static threshold-based algorithm and to a system without any on/off mechanism. Prior to discussing the simulation results, we first explain the power modeling followed in the experiments.

4.2. Power Modeling. We adopted the Orion power models for the dynamic power consumption of each router [28]. Router and link hardware were designed and synthesized in Verilog and Synopsys Design Compiler in order to obtain the leakage power values. We used a commercial CMOS 65 nm library, and a sequence of random input vectors, for several thousand cycles, and measured the leakage power of each router and link, through all computation cycles and combinations of events. The leakage values are then fed into the simulator, along with the Orion models for active power, and the overall power is computed. In addition, we take into consideration the start-up power consumed when we turn a link back on.

4.3. Simulation Results and Discussion. Using synthetic traffic patterns with varied injection rates (Random, Tornado, Transpose, and Neighbor) [31, 32], we first evaluated the power savings of the ANN-based mechanism when compared to the same system without any on/off link capability, and when compared to a system that employs a statically determined threshold. The traffic patterns, for which we experimented, are a superset of the patterns used to train the ANN; we measured power savings and the impact of the throughput on all the traffic patterns. However, in order to compute the power savings in the torus network, we follow the guided-training approach as described in Section 3.3, and we measure link utilizations in all possible partitions of the torus network to compensate for the toroidal links. The link utilizations with the least values (from all the link utilizations, from all the partitions of the torus network) are then passed through the ANNs. Figure 9 shows the comparison when targeting 8×8 mesh and torus NoCs. The power savings of the ANN-based mechanism are better than the savings in the cases of statically determined threshold

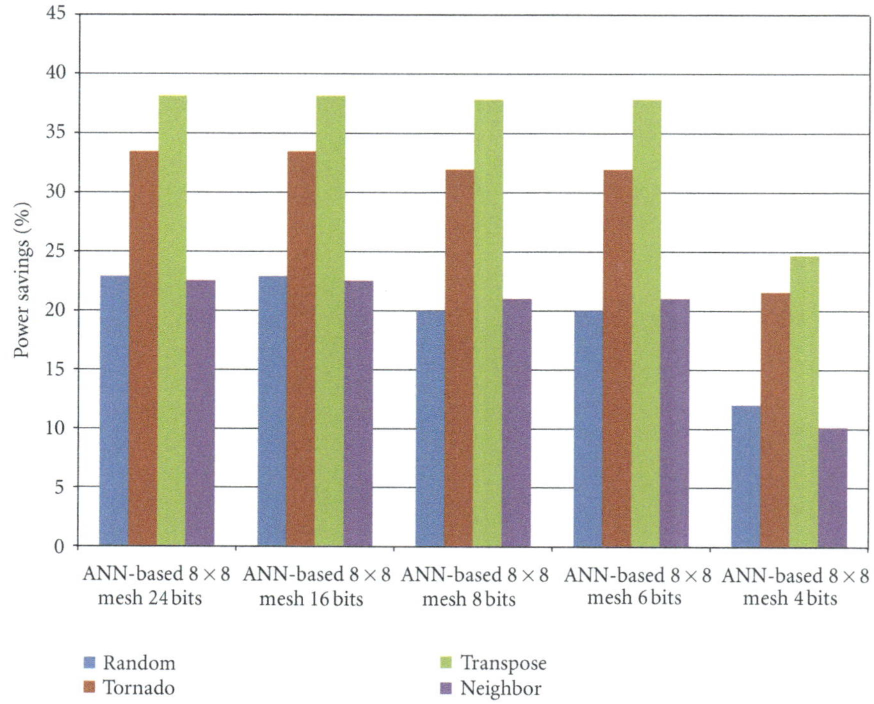

FIGURE 8: Power savings for different training weight bit representations.

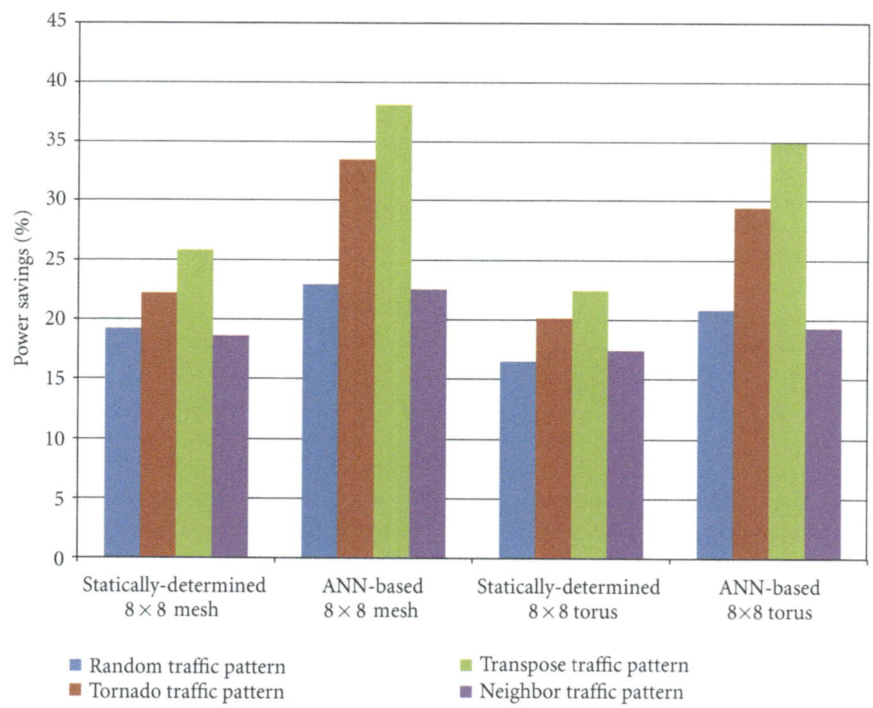

FIGURE 9: Power Savings for 8 × 8 mesh and 8 × 8 torus networks for the ANN-based technique, static threshold technique and no on/off technique.

and the case without any on/off links. The ANN-based mechanism can identify a significant amount of future behavior in the observed traffic patterns; therefore, it can intelligently select the threshold necessary for the next timing interval.

Next, we measure the impact of the throughput in each mechanism; while having no on/off mechanism obviously yields a higher throughput, the ANN-based technique shows better throughput results compared to statically determined threshold techniques. Figure 10 shows the throughput

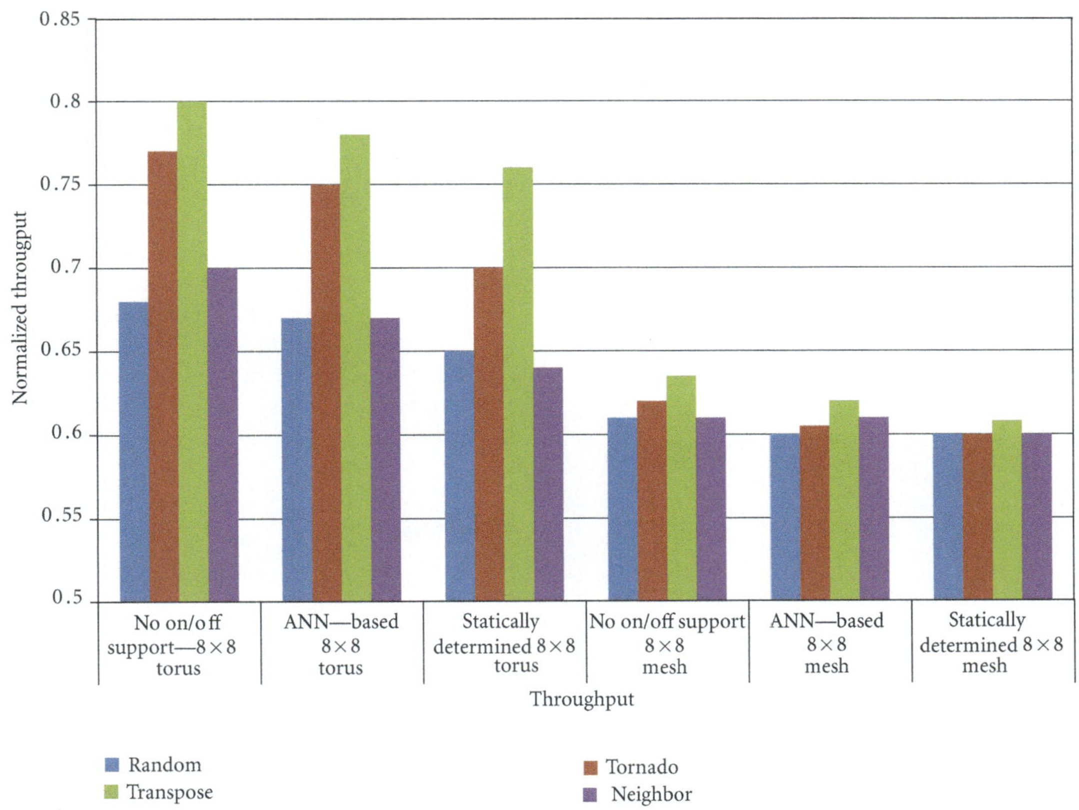

FIGURE 10: Average network throughput comparisons for 8 × 8 mesh and torus networks.

TABLE 1: Power savings/hardware overhead comparisons.

Related work	Characteristics	Power savings comparing to no algorithm	Hardware overhead
[1]	8 × 8 2D mesh topology, Uniform traffic	~37,5%—turning on/off 2 links per router	N/A
[11]	8 × 8 2D mesh topology, Pareto distribution—0.5 packet injection rate	~30%	500 equivalent logic gates per router port Delay ignored
Proposed ANN-based technique	8 × 8 2D mesh topology and 8 × 8 torus topology,uniform traffic	Up to ~40%—turning on/off links based on ANN prediction	4% of the NoC hardware for a complete 4 × 4 Mesh NoC

comparisons for an 8 × 8 mesh and an 8 × 8 torus network. The throughput values are normalized based on the number of the simulation cycles.

Figure 11 represents the normalized energy consumed in a 8 × 8 mesh network. We observe that the energy consumed using the ANN mechanism is less than the cases of statically-computed threshold and without on/off link management algorithm. The ANN exhibits a reduction in the overall energy, because of a balanced performance-to-power savings ratio, when compared to not having on/off links or when compared to static threshold computation.

Figure 12 presents the average packet delay in packets per cycle for the 8 × 8 mesh, when the ANN-based mechanism is used compared to the cases where no on/off mechanism is used and the statically computed threshold case. The ANN-based mechanism incurs more delay, but we believe that the delay penalty is acceptable when compared to the associated power savings.

4.4. ANN Hardware Overheads: Synthesis Results. To compute the hardware overheads of the proposed scheme, the ANN-based mechanism for one 4 × 4 NoC region was synthesized and implemented targeting a commercial 65 nm CMOS technology. The ensuing synthesized ANN-based controller and the associated hardware overheads in each router consume approximately 4 K logic gates (for comparison purposes, an NoC router similar to the one used in our simulation [29] consumes roughly 21 K gates),

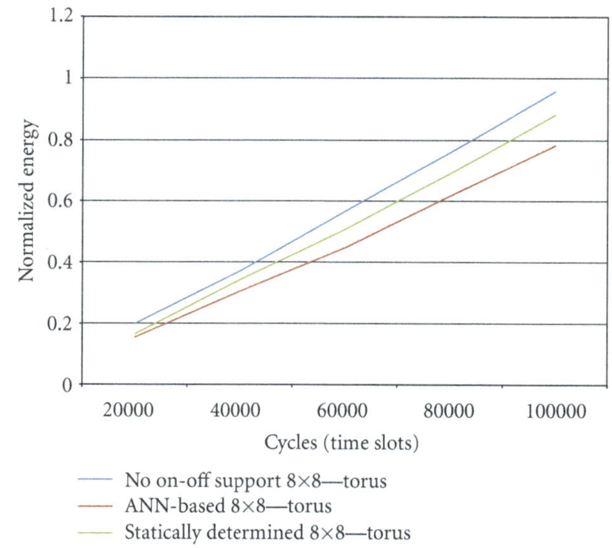

FIGURE 11: Energy consumption for a 8 × 8 mesh network.

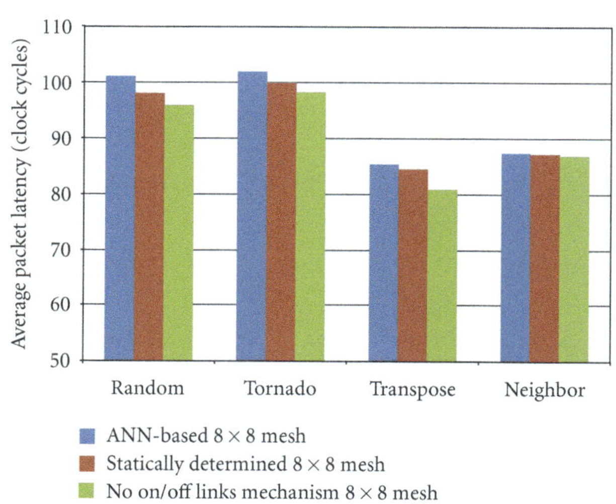

FIGURE 12: Average packet latency for the cases where ANN-based mechanism is used, when trivial case is used and when there is no on/off mechanism.

bringing the estimated hardware overhead for an 4 × 4 mesh network to roughly 4% of the NoC hardware.

4.5. Comparison with Related Works. Lastly, we briefly give a comparison with relevant related works that follow dynamic threshold techniques in Table 1. When compared to both [1, 11], the ANN-based prediction yields better power savings than having no prediction mechanism, while still maintaining lower hardware overheads. We must note that while [10] was the motivating idea behind our paper, it presented only a preliminary implementation of the idea, without enough information about hardware overheads and power savings in order to make an informed comparison.

5. Conclusions

This paper presented how an ANN-based mechanism can be used to dynamically compute a utilization threshold, that can be in turn used to select candidate links for turning on or off, in an effort to achieve power savings in an NoC. The ANN-based model utilizes very low hardware resources, and can be integrated in large mesh and torus NoCs, exhibiting significant power savings. Simulation results indicate approximately 13% additional power savings when compared to a statically determined threshold methodology under synthetic traffic models. We hope to expand the results of this paper to further explore dynamic reduction of power consumption in NoCs using ANNs and other intelligent methods.

References

[1] V. Soteriou and L.-S. Peh, "Dynamic power management for power optimization of interconnection networks using on/off links," in *Proceedings of the 11th Symposium on High Performance Interconnects*, pp. 15–20, 2003.

[2] F. Worm, P. Thiran, P. Ienne, and G. De Micheli, "An adaptive low-power transmission scheme for on-chip networks," in *Proceedings of the 15th International Symposium on System Synthesis*, pp. 92–100, October 2002.

[3] S. Kumar, A. Jantsch, J.-P Soininen et al., "A network on chip architecture and design methodology," in *Proceedings of the IEEE Computer Society Annual Symposium on VLSI*, pp. 117–124, 2002.

[4] L. Benini and G. De Micheli, "Networks on chips: a new SoC paradigm," *Computer*, vol. 35, no. 1, pp. 70–78, 2002.

[5] K. Govil, E. Chan, and H. Wasserman, "Comparing algorithms for dynamic speed-setting of a low-power CPU," in *Proceedings of the 1st Annual International Conference on Mobile Computing and Networking*, pp. 13–25, November 1995.

[6] V. Soteriou and L. S. Peh, "Exploring the design space of self-regulating power-aware on/off interconnection networks," *IEEE Transactions on Parallel and Distributed Systems*, vol. 18, no. 3, pp. 393–408, 2007.

[7] Y. Hoskote, S. Vangal, S. Dighe et al., "Teraflops Prototype Processor with 80 Cores," Microprocessor Technology Labs, Intel Corporation.

[8] W.-C. Cheng and M. Pedram, "Low power techniques for address encoding and memory allocation," in *Proceedings of the Asia and South Pacific Design Automation Conference (ASP-DAC '01)*, pp. 245–250, 2001.

[9] D. Shin and J. Kim, "Power-aware communication optimization for networks-on-chips with voltage scalable links," in *Proceedings of the 2nd IEEE/ACM/IFIP International Conference on Hardware/Software Codesign and System Synthesis (CODES+ISSS '04)*, pp. 170–175, September 2004.

[10] T. Simunic and S. Boyd, "Managing power consumption in networks on chips," in *Proceedings of the Design, Automation and Test in Europe*, pp. 110–116, 2002.

[11] L. Shang, L.-S. Peh, and N. K. Jha, "Dynamic voltage scaling with links for power optimization of interconnection networks," in *Proceedings of the International Symposium on High Performance Computer Architecture*, pp. 91–102, 2003.

[12] Semiconductor Industry Association, "International Technology Roadmap for Semiconductors," 2009, http://www.itrs.net/Links/2009ITRS/Home2009.htm.

[13] A. K. Jain, J. Mao, and K. M. Mohiuddin, "Artificial neural networks: a tutorial," *Computer*, vol. 29, no. 3, pp. 31–44, 1996.
[14] R. Schalkoff, *Artificial Neural Networks*, McGrow-Hill, 1997.
[15] A. Savva, T. Theocharides, and V. Soteriou, "Intelligent on/off link management for on-chipnetworks," in *Proceedings of the IEEE Annual Symposium on VLSI*, pp. 343–344, 2011.
[16] R. Marculescu, U. Y. Ogras, L. S. Peh, N. E. Jerger, and Y. Hoskote, "Outstanding research problems in NoC design: system, microarchitecture, and circuit perspectives," *IEEE Transactions on Computer-Aided Design of Integrated Circuits and Systems*, vol. 28, no. 1, pp. 3–21, 2009.
[17] G. Chen, F. Li, M. Kandemir, and M. J. Irwin, "Reducing NoC energy consumption through compiler-directed channel voltage scaling," in *Proceedings of the ACM SIGPLAN Conference on Programming Language Design and Implementation (PLDI '06)*, pp. 193–203, June 2006.
[18] T. Pering, T. Burd, and R. Brodersen, "Voltage scheduling in the IpARM microprocessor system," in *Proceedings of the 2000 Symposium on Low Power Electronics and Design (ISLPED '00)*, pp. 96–101, July 2000.
[19] F. Li, G. Chen, and M. Kandemir, "Compiler-directed voltage scaling on communication links for reducing power consumption," in *Proceedings of the IEEE/ACM International Conference on Computer-Aided Design (ICCAD '05)*, pp. 455–459, November 2005.
[20] V. Soteriou, N. Eisley, and L.-S. Peh, "Software-directed power-aware interconnection networks," *ACM Transactions on Architecture and Code Optimization*, vol. 4, no. 1, pp. 274–285, 2007.
[21] E. Y. Chung, L. Benini, and G. De Micheli, "Contents provider-assisted dynamic voltage scaling for low energy multimedia applications," in *Proceedings of the International Symposium on Low Power Electronics and Design*, pp. 42–47, August 2002.
[22] J. Kim and M. Horowitz, "Adaptive supply serial links with sub—1V operation and per-pin clock recovery," in *Proceedings of the International Solid State Circuits Conference (ISSCC '02)*, pp. 1403–1413, 2002.
[23] M. Alonso, S. Coll, J. M. Martínez, V. Santonja, P. López, and J. Duato, "Power saving in regular interconnection networks," *Parallel Computing*, vol. 36, no. 12, pp. 696–712, 2010.
[24] S. Conner, S. Akioka, M. J. Irwin, and P. Raghavan, "Link shutdown opportunities during collective communications in 3-D torus nets," in *Proceedings of the 21st International Parallel and Distributed Processing Symposium (IPDPS '07)*, pp. 1–8, March 2007.
[25] C. Jackson and S. J. Hollis, "Skip-links: a dynamically reconfiguring topology for energy-efficient NoCs," in *Proceedings of the 12th International Symposium on System-on-Chip 2010 (SoC '10)*, pp. 49–54, September 2010.
[26] R. Mullins, "Minimising dynamic power consumption in on-chip networks," in *Proceedings of the International Symposium on System-on-Chip (SoC '06)*, pp. 1–4, November 2006.
[27] M. Ali, M. Welzl, and S. Hellebrand, "A dynamic routing mechanism for network on chip," in *Proceedings of the 23rd NORCHIP Conference*, pp. 70–73, November 2005.
[28] H.-S. Wang, X. Zhu, L.-S. Peh, and S. Malik, "Orion: a power-performance simulator for interconnection networks," in *Proceedings of the International Symposium on Microarchitecture*, pp. 294–305, 2002.
[29] E. Kakoullit, V. Soteriou, and T. Theocharides, "An artificial neural network-based hotspot prediction mechanism for NoCs," in *Proceedings of the IEEE Computer Society Annual Symposium on VLSI (ISVLSI '10)*, pp. 339–344, July 2010.
[30] H. Hossain, M. Ahmed, A. Al-Nayeem, T. Z. Islam, and M. M. Akbar, "gpNoCsim—a general purpose simulator for network-on-chip," in *Proceedings of the International Conference on Information and Communication Technology (ICICT '07)*, pp. 254–257, March 2007.
[31] W. J. Dally and B. Towles, *Principles and Practices of Interconnection Networks*, Morgan Kaufmann, San Francisco, Calif, USA, 2004.
[32] J. Duato, S. Yalamanchili, and L. M. Ni, *Interconnection Networks: An Engineering Approach*, Morgan Kaufmann, San Francisco, Calif, USA, 2003.

FPGA Implementation of Gaussian Mixture Model Algorithm for 47fps Segmentation of 1080p Video

Mariangela Genovese, Ettore Napoli, Davide De Caro, Nicola Petra, and Antonio G. M. Strollo

Department of Biomedical, Electronic and Telecommunications Engineering, University of Napoli Federico II, 80125 Napoli, Italy

Correspondence should be addressed to Mariangela Genovese; mariangela.genovese@unina.it

Academic Editor: Ashkan Ashrafi

Circuits and systems able to process high quality video in real time are fundamental in nowadays imaging systems. The circuit proposed in the paper, aimed at the robust identification of the background in video streams, implements the improved formulation of the Gaussian Mixture Model (GMM) algorithm that is included in the OpenCV library. An innovative, hardware oriented, formulation of the GMM equations, the use of truncated binary multipliers, and ROM compression techniques allow reduced hardware complexity and increased processing capability. The proposed circuit has been designed having commercial FPGA devices as target and provides speed and logic resources occupation that overcome previously proposed implementations. The circuit, when implemented on Virtex6 or StratixIV, processes more than 45 frame per second in 1080p format and uses few percent of FPGA logic resources.

1. Introduction

Real time background identification in high definition video streams is an important task in smart video systems [1–4]. Various algorithms that address the problem have been proposed in the scientific literature throughout the years [5–16].

Foreground detection algorithms based on temporal difference techniques [5–7] allow the detection of relevant objects (Foreground, Fg) by comparing two or more consecutive frames but are unreliable in presence of illumination changes.

More robust algorithms, named background subtraction algorithms [8–15] are based on the creation of a model that includes the static parts of the scene (Background, Bg). The Fg is detected by analyzing the difference between a frame and the model of the Bg. The Bg subtraction algorithms regularly update the Bg model and are resilient to slow changes of the scene. They differ for the considered model of the Bg and for the algorithm that takes care of model updating.

References [13–15] propose Bg subtraction algorithms based on statistical models.

In [14] the Bg is represented with a statistical model based on a single Gaussian distribution per pixel. This allows a pixel by pixel model of the Bg with larger variance for pixels that experience wide changes of lighting.

The Gaussian Mixture Model (GMM) algorithm has been proposed in [15]. It uses a statistical model in which the probability distribution of the luminosity of each pixel is modeled with a mixture of Gaussian distributions. The GMM algorithm provides good performances in both presence of illumination changes and multimodal Bg. A multimodal Bg is characterized by objects showing repetitive motions, for example, waves, moving leaves, or flickering light. When a pixel lies in a region where a repetitive motion occurs, its brightness oscillates between two or more values. This results in false Fg detection in most algorithms. On the contrary, when the GMM algorithm is used, the intensity distribution of the pixel is modeled using two or more Gaussian distributions and the problem of false Fg detection is mitigated.

The OpenCV (Open source Computer Vision software library developed by Intel [16]) provides a common base of computer vision instruments that is becoming the de facto standard for video/image processing. The Bg detection algorithm included in the OpenCV is an optimized version of the GMM algorithm of [15]. It changes the initial learning phase for the Bg model, that is improved with respect to [15], and calculates the learning rate in a simplified yet efficient way. As a consequence, the OpenCV version of the GMM algorithm is both widely used and provides good performances.

1.1. Previous Hardware Implementations. In order to generate the updated Bg model, the GMM algorithm processes the video stream by computing a great number of parameters for each pixel of each frame, with a computational burden unfeasible for computers in real time applications. As an example, in [15], only a frame rates of 11–13 frame per second (fps) is obtained for frame size 160 × 120 on an SGI O2 workstation.

Real time video applications with larger frame size require a dedicated hardware architecture. Hardware processors have been proposed in [17–22]. Reference [17] proposes a circuit able to process video sequences with frame size 1024 × 1024 at 38 fps when implemented on a VirtexII FPGA platform. Processing capabilities of [17] are 39.8 Mega pixel per second (Mps). In [18] the design of an automated digital surveillance system running in real time on an embedded platform is presented. The segmentation unit of the circuit proposed in [18] is able to run at 83 MHz on VirtexII xc2pro30. In [19], the research group of [18] improves the memory throughput with respect to [18] by employing a memory reduction scheme. The improvement in memory throughput is paid with a reduction of the processing capability that is limited to 7.68 Mps.

Reference [20] proposes an OpenCV GMM algorithm implementation able to process 22 HD (1920 × 1080 pixels) frames per second when implemented on Virtex5 xc5vlx50 FPGA. The circuit of [20] is used in [21] in conjunction with a denoising block that implements erosion, dilation, opening, and closing.

A circuit that implements the OpenCV GMM algorithm and provides increased performances has been presented in [22]. The improvement in circuit performance derives from the optimization of the bit width of the signals and from a better formulation of the GMM equations.

1.2. Proposed Circuit. In this paper a hardware implementation of the GMM algorithm that allows real time processing of HD videos and comply with the OpenCV algorithm is proposed. To the best of author's knowledge, the proposed circuit is the one with highest performance proposed to date. The circuit processes gray scale videos. When color videos need to be processed, the circuit, as reported in [19], can be fed with the luminance channel of the YCrCb color space.

The proposed circuit is optimized for an FPGA implementation and, with respect to the best performing previous work, [22], increases the maximum working frequency (+77.4%) and reduces the area utilization (−3.8% of Slices and 0 versus 10 DSP).

The improved performances are obtained by

(i) replacing full-width binary multipliers with truncated binary multipliers [23–26];

(ii) approximating nonlinear functions with piecewise linear approximations [27–31];

(iii) exploiting an innovative formulation of the updating equations of the GMM algorithm.

It is worth noting that the implemented design techniques (mainly the use of truncated multipliers and the approximation of nonlinear functions with piecewise linear functions) while allowing improved performances in terms of circuit speed and logic resources occupation are a modification of the reference OpenCV GMM algorithm and introduce a computation error that could result in a worsening of the quality of the processed video. The tests presented in Section 5, however, demonstrate that the proposed circuit is able to obtain very high performances while providing a good quality of the processed video sequences.

2. GMM Algorithm

The GMM algorithm has been proposed by Stauffer and Grimson, [15], with the target of efficiently dealing with multimodal Bg by using a statistical model composed by a mixture of Gaussian distributions. The GMM algorithm has been modified and included in the OpenCV libraries. A short description of the OpenCV GMM algorithm is given in the following. The differences with respect to the algorithm of [15] are indicated in the text.

2.1. Statistical Model. The statistical model for each pixel of the video sequence is composed by a mixture of K Gaussian distributions, each one represented by four parameters: weight (w), mean (μ), variance (σ^2), and *matchsum*, a counter introduced in the OpenCV algorithm.

Gaussian parameters differ for each Gaussian of each pixel and change for every frame of the video sequence. They are therefore defined by three indexes (p, k, t), where "p" is the index for the pixel, "k" is the index for the Gaussian distribution, and "t" is for the frame. In the following the pixel index is omitted since the same operations are repeated for every pixel.

2.2. Parameters Update. When a frame is acquired, for each pixel, the K Gaussian distributions are sorted in decreasing order of a parameter named Fitness ($F_{k,t}$):

$$F_{k,t} = \frac{w_{k,t}}{\sigma_{k,t}}. \tag{1}$$

A match condition is checked with the K Gaussian distributions that model the pixel. The match condition is

$$M_k = 1 \quad \text{if } \left|(\text{pixel} - \mu_{k,t})\right| < 2.5\sigma_{k,t}. \tag{2}$$

Equation (2) establishes if the pixel can be considered part of the Bg. A pixel can verify (2) for more than one Gaussian. The Gaussian that matches with the pixel ($M_k = 1$) and has the highest $F_{k,t}$ value is considered as the "matched distribution" and its parameters are updated as follows:

$$\mu_{k,t+1} = \mu_{k,t} + \alpha_{k,t} \cdot (\text{pixel} - \mu_{k,t}),$$
$$\sigma^2_{k,t+1} = \sigma^2_{k,t} + \alpha_{k,t} \cdot \left[(\text{pixel} - \mu_{k,t})^2 - \sigma^2_{k,t} \right],$$
$$w_{k,t+1} = w_{k,t} - \alpha_w \cdot w_{k,t} + \alpha_w,$$
$$\text{matchsum}_{k,t+1} = \text{matchsum}_{k,t} + 1. \quad (3)$$

The parameter α_w is the learning rate for the weight while $\alpha_{k,t}$ is the learning rate for mean and variance. It is derived from α_w as follows:

$$\alpha_{k,t} = \frac{\alpha_w}{w_{k,t}}. \quad (4)$$

Equation (4) is characteristic of the OpenCV algorithm and differs from what is proposed in [15] where $\alpha_{k,t}$ is calculated, being η the Gaussian probability density function as follows:

$$\alpha_{k,t} = \alpha_w \cdot \eta(\text{pixel}, \mu_{k,t}, \sigma_{k,t}). \quad (5)$$

The approach of [15] results, as shown in [32], in a slow convergence if compared with (4).

For the unmatched Gaussian distributions mean and variance are unchanged while the weights are updated as follows:

$$w_{k,t+1} = w_{k,t} - \alpha_w \cdots w_{k,t}. \quad (6)$$

When the pixel does not match any Gaussians, a specific "no_match" procedure is executed and the Gaussian distribution with smallest $F_{k,t}$ is updated as follows:

$$\mu_{k,t+1} = \text{pixel} \qquad \text{matchsum}_{k,t+1} = 1,$$
$$\sigma^2_{k,t+1} = \text{vinit} \qquad w_{k,t+1} = \frac{1}{\text{msumtot}}, \quad (7)$$

where vinit is a fixed initialization value and msumtot is the sum of the values of the matchsum of the $K-1$ Gaussians with highest $F_{k,t}$. The weights of the $K-1$ Gaussians with highest $F_{k,t}$ are decremented as in (6) while their means and variances are unchanged.

2.3. Background Identification. The Bg identification uses the following algorithm:

$$B = \arg\min_b \left(\sum_{k=1}^{b} w_{k,t} > T \right). \quad (8)$$

Equation (8) adds in succession the weights of the first b Gaussian distributions, sorted beginning from the one with highest $F_{k,t}$ value, until their sum is greater than T, a fixed threshold belonging to the $[0, 1]$ interval. The set of Gaussian distributions that verify (8) represents the Bg and a pixel that matches one of these Gaussians is classified as a Bg pixel. The algorithm entails that if the "no_match" condition occurs, the pixel is classified as Fg.

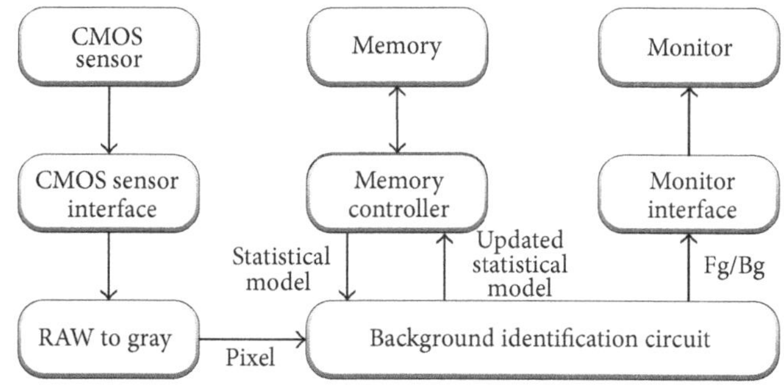

FIGURE 1: Conceptual overview of a background identification system.

3. Background Identification Circuit

Figure 1 shows a conceptual overview of a Bg identification system. A CMOS sensor captures each frame of the video sequence and a CMOS interface gives the pixel values to the Bg identification circuit. The Bg identification circuit processes the luminance value of the input "Pixel," and the "Statistical Model" of the pixel for the given frame. The output data are the "Updated Statistical Model" and the "Fg/Bg" tag. For each pixel, the Gaussian parameters are read from an external memory and the updated parameters are stored in the memory. For each frame of the input video sequence, the Fg/Bg tags produce a binary image that can be displayed on a monitor.

In this paper, the "background identification circuit" of Figure 1 has been implemented on FPGA devices. Figure 2 shows the block diagram of the proposed circuit. The circuit is described in VHDL code and implements the GMM algorithm using, as suggested in [15], three Gaussian distributions for each pixel ($k = [1, 2, 3]$). As a consequence, the proposed circuit processes 13 parameters per pixel (the luminance value and 12 Gaussians parameters for the statistical model of the pixel).

The software algorithm of the OpenCV library represents the Gaussians parameters as double precision (64 bit) floating point numbers. A double precision hardware implementation would be impractically large and slow and would require a very high bandwidth towards the memory. The proposed circuit represents the signals as fixed point numbers on a limited number of bits. The word length and the representation of the Gaussians parameters are determined by using the word length optimization algorithm proposed in [22] that minimizes the word length while constraining the Peak Signal-to-Noise Ratio (PSNR) of a set of test video streams. The input bits of the resulting circuit are the 99 bits for the parameters of the statistical model and the 8 bits of the luminance of the input pixel. The output bits are the updated statistical model, that requires 99 bits, and the Fg/Bg tag.

The optimized fixed point representation, an innovative formulation of the GMM algorithm equations (described in Section 4), the use of truncated multipliers, and the application of ROM optimization techniques (details in Section 5),

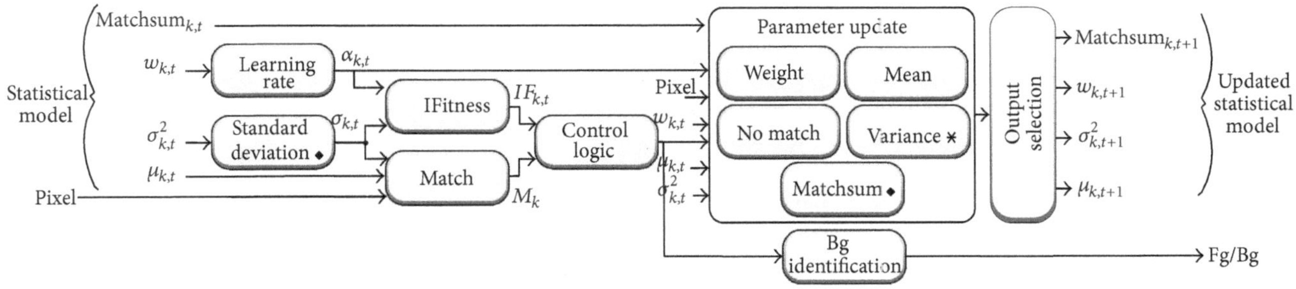

FIGURE 2: Block diagram of the proposed FPGA implementation. The functionality of the circuit is described in Section 5. The star indicates the circuital block that has been improved by using the truncated binary multipliers. The diamond indicates the circuital blocks that have been improved by using piecewise linear approximation of nonlinear functions.

allow the design of a Bg identification circuit with high working frequency and low logic resources utilization.

4. Hardware Oriented Formulation of the GMM Equations

The equations of the GMM algorithm, detailed in Section 2, if straightforwardly implemented, would be too complex for the realization of a fast and compact circuit. The proposed circuit is therefore based on the manipulation and simplification of the above-cited equations. A first step towards the simplification is the use of an optimized fixed point representation for circuit signals [22]. A second step is the smart implementation of the nonlinear operations as will be shown in Section 5.

In addition to the above-cited techniques, in this paper, in order to reduce the logic occupation, a new order parameter for Gaussian distributions is introduced. The parameter is named *IFitness* (indicated in the following as $IF_{k,t}$) and is defined as the inverse of the $F_{k,t}$ factor:

$$IF_{k,t} = \frac{1}{F_{k,t}}. \tag{9}$$

With simple manipulations, taking into account (1) and (4), $IF_{k,t}$ can be written as

$$IF_{k,t} = \frac{\sigma_{k,t}}{w_{k,t}} = \frac{\sigma_{k,t}}{w_{k,t}} \cdot \frac{\alpha_{k,t}}{\alpha_{k,t}} = \sigma_{k,t} \cdot \frac{\alpha_{k,t}}{\alpha_w}. \tag{10}$$

Equation (10) shows that the $IF_{k,t}$ factor can be computed as a function of the learning rate $\alpha_{k,t}$.

Moreover, in the proposed implementation, the learning rates α_w and $\alpha_{k,t}$ are quantized as power of two values:

$$\alpha_w = 2^{n_w} \qquad \alpha_{k,t} = 2^{n_{k,t}}. \tag{11}$$

As a consequence, the $IF_{k,t}$ factor can be computed as follows:

$$IF_{k,t} = \sigma_{k,t} \cdot 2^{(n_{k,t}-n_w)} \tag{12}$$

and the multiplication between $2^{(n_{k,t}-n_w)}$ and $\sigma_{k,t}$ can be performed using a shifter instead of a binary multiplier.

It is worth noting that ordering the Gaussian distributions as a function of $F_{k,t}$, a process needed in the GMM algorithm in order to determine the Bg model according to (8), is equivalent to ordering versus the $IF_{k,t}$ value.

Equations (4) and (12) can be implemented using only a shifter and a square root circuit that calculates the standard deviation $\sigma_{k,t}$. The implementation does not require the multiplier that would be needed to implement (1). The removal of the multiplier is not only beneficial for the implied reduction of hardware usage but, being the multiplier on the critical path of the Bg identification circuit, is also effective in increasing overall circuit speed. More details regarding the HW implementation of the above-described equations are given in the Section 5.

5. FPGA Implementation of the Background Identification Circuit

The schematic of the FPGA implementation of the GMM algorithm is shown in Figure 2. A detailed explanation of the functionalities and of the hardware implementation of the circuital blocks of Figure 2 is given in the following.

5.1. Learning Rate. Computes $\alpha_{k,t}$ following (4).

The "learning rate" unit is implemented with three ROM, each of them stores the power of two values that better approximates $\alpha_{k,t}$ as a function of $w_{k,t}$ ((4), (11)). The equation implemented by the ROM is

$$n_{k,t} = \text{round}\left[\log_2\left(\alpha_{k,t}\right)\right] \tag{13}$$

in which the function "round(·)" is the approximation to the nearest integer value. The $n_{k,t}$ values are then used by the "*IFitness*" unit to shift the $\sigma_{k,t}$ in order to calculate the $IF_{k,t}$ factor according to (12).

5.2. Standard Deviation. The formulation of the GMM equations described in Section 4 requires the standard deviation values, $\sigma_{k,t}$, for the three Gaussians while the input data for the circuit include the variance values, $\sigma^2_{k,t}$.

The "Standard Deviation" block determines the variance values by implementing a square root circuit.

In the proposed implementation the $\sigma^2_{k,t}$ signals are represented on 11 bits and range in [8 : 16376]. Due to the reduced number of bits of the variance signal, a ROM implementation for the square root circuit could be feasible but not optimal. In this paper the square root function has instead been implemented by using a piecewise linear approximation of the nonlinear function, a technique that has been previously proposed as an effective method to reduce hardware complexity [27–31].

The interval [8 : 16376] is divided, as shown in Figure 3, into four nonuniform intervals, $I_1, \ldots, I_4$. In each interval, a linear function y_i approximates $\sigma_{k,t}$ as follows:

$$y_i = q_i - m_i \cdot \sigma^2_{k,t} \quad \sigma^2_{k,t} \in I_i \tag{14}$$

with $i = 1, \ldots, 4$ and $q_i, m_i > 0$. In order to simplify the hardware, the m_i coefficients are quantized as powers of two and the multiplication between m_i and $\sigma^2_{k,t}$ is performed by using a shifter resulting in lower area utilization and higher working frequency.

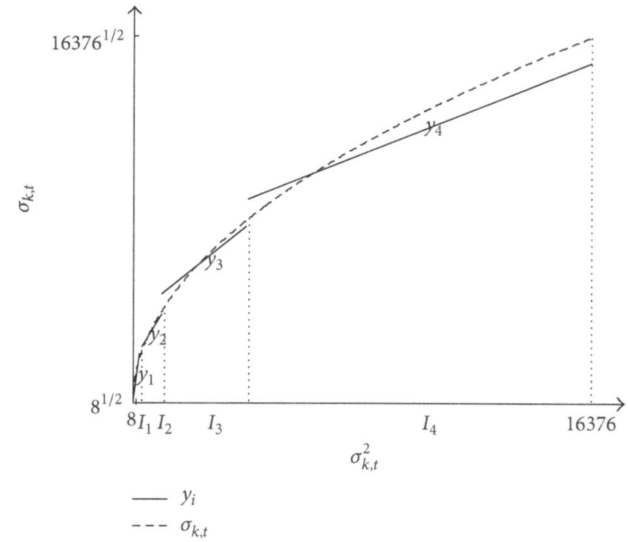

Figure 3: Piecewise linear polynomial approximation of the $\sigma_{k,t}$ function implemented in the "Standard Deviation" block of Figure 2.

5.3. IFitness. It computes $IF_{k,t}$ according to (12).

The circuit is actually a combinatorial shifter that shifts $\sigma^2_{k,t}$ as a function of the $n_{k,t}$ value and obtains the $IF_{k,t}$ factor.

The proposed formulation of the GMM equations, that use (4) and (12) computed by the "learning rate" and "*IFitness*" block, implemented on Virtex5 vlx50 FPGA requires 64 Slices. This result can be compared with the implementation of (4) and (1) (the technique used in [22]) that, on the same FPGA, requires 123 Slices and 3 DSP. This demonstrates that the new formulation of the GMM algorithm equations is effective in reducing circuit complexity.

5.4. Match. It verifies the match condition for the three Gaussian distributions and is composed by three identical units (one for each Gaussian). Each unit implements (15) and employs a subtractor and one comparator as follows:

$$M_k = 1 \quad \text{if } |\text{pixel} - \mu_{k,t}| < 2.5 \cdot \sigma_{k,t}. \tag{15}$$

5.5. Control Logic. It sorts the Gaussians in increasing $IF_{k,t}$ order and establishes which Gaussian has to be updated as in (3), (6), and (7).

5.6. Parameter Update. This circuital block is composed by five subunits ("Weight", "Mean", "Variance" "No Match", and "Matchsum") that implement the parameters update equations (3), (6), and (7) detailed in Section 2.

5.6.1. "Weight" and "Mean". If the match condition is verified, these blocks update the parameters according to (3) and (6). The quantization of the learning rates as power of two allows to implement (3) by using shifters instead of multipliers.

5.6.2. "Variance". The left hand of the variance update equation of (3) requires the computation of a square. In the proposed circuit the square has been implemented by using a truncated multiplier [26].

Truncated binary multipliers, for which the optimality in terms of mean square error is analytically demonstrated, have been proposed in [23–26]. Truncated multipliers have been used with good results in a variety of applications [33–35]. A full-width digital $n \times n$ bits multiplier computes the $2n$ bits output as a weighted sum of partial products. A truncated multiplier is an $n \times n$ multiplier with an $m < 2n$ bits output.

In the proposed circuit the inputs of the multipliers are represented on 12 bits. The output of a 12×12 full-width multiplier is hence on 24 bits. In order to reduce HW complexity and improve circuit speed, the proposed circuit implements a truncated binary multiplier, based on the architecture proposed in [24], in which the output is directly produced on 15 bits (9 less-significant bits are discarded with an almost 40% reduction of the required HW for the multiplier).

5.6.3. "No Match". When no Gaussians match the pixel, the "No Match" block updates the parameters of the Gaussian with highest *IFitness* value according to (7). Equation (7) requires the calculation of the inverse of the msumtot signal. In [22] a ROM-based approach is used to implement the required function. Since the msumtot signal is represented on 6 bits and its inverse is on 8 bits, the ROM size is equal to $2^6 \times 8$.

In the proposed paper the ROM is eliminated by using a piecewise linear approximation technique similar to those used for the calculation of the $\sigma_{k,t}$ signals. The msumtot signal range [1 : 63] is divided into four nonuniform intervals J_i and in each interval, a function z_i approximates 1/msumtot as follows:

$$z_i = \lambda_i - \eta_i \cdot \text{msumtot}, \quad \text{msumtot} \in J_i, \tag{16}$$

with $i = 1, \ldots, 4$, $\lambda_i, \eta_i > 0$, and η_i quantized as power of two so that the multiplication between η_i and msumtot is performed using a shifter.

TABLE 1: Performances of the circuit of Figure 2 implemented on different FPGA and comparison with previous art.

Target FPGA	Circuit	LUT/ALUT	Flip Flop	Slice/LAB	DSP/MULT	Frequency (MHz)	HD fps
Virtex6 (xc6vlx75t)	proposed circuit	922/46560	0/93120	265/11640	0/288	97.19	46
Virtex5 (xc5vlx50)	proposed circuit	844/28800	0/28800	301/7200	0/48	91.03	43
	Reference [22]	705/28800	0/28800	313/7200	10/48	51.30	24
StratixIV (SGX230HF35C2)	proposed circuit	1150/182400	0/182400	81/9120	0/1288	98.12	47
Spartan3 (xc3s1000)	proposed circuit	1797/15360	0/15360	1010/7680	0/24	35.30	17

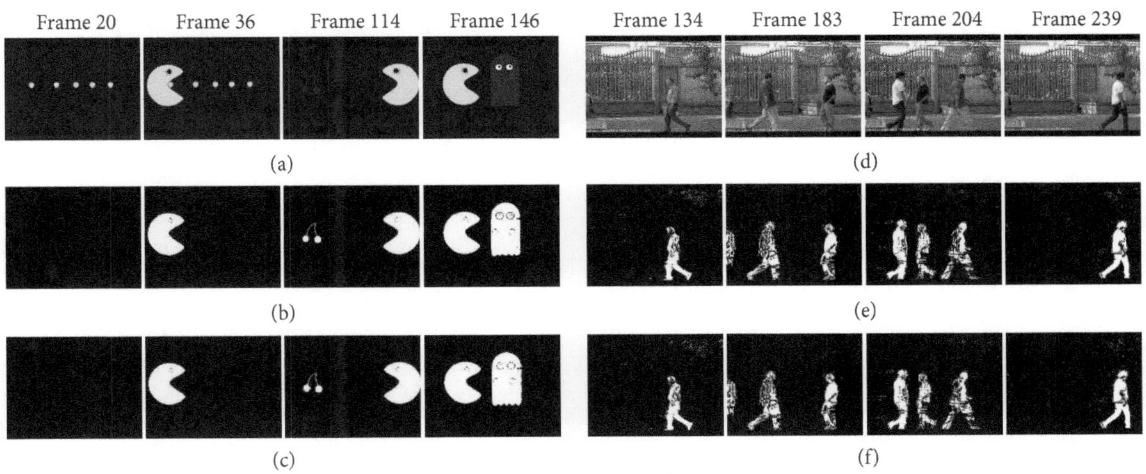

FIGURE 4: Comparison between the Bg masks obtained with the circuit of [22] and with the proposed implementation. (a), (d) unprocessed video frames; (b), (e) output of the proposed circuit; (c), (f) output of the circuit proposed in [22].

5.6.4. "Matchsum". This is the circuit that updates the *matchsum* signal according to (3) and (7). The matchsum signal is a counter, associated to every Gaussian, introduced in the OpenCV with the aim to improve the initial learning phase of the GMM algorithm (as demonstrated in [22]). The matchsum signal counts the number of times that a given Gaussian is matched according to (2). The introduction of the matchsum entails the synthesis of three counters that are not on the critical path. The only drawback is an increase of circuit area.

5.7. Output Selection. It establishes the values of the updated parameters depending on whether the match condition is verified or not.

5.8. Bg Identification. It verifies the background identification condition shown in (8) determining, also as a function of the "No Match" condition, if the input pixel belongs to the background or not.

6. Results

6.1. CAD Tools for Design and Test. The proposed FPGA implementation of the OpenCV GMM algorithm is synthesized and implemented on Virtex6 (xc6vlx75t), Virtex5 (xc5vlx50), and Spartan3 (xc3s1000) Xilinx FPGA and on Stratix IV (SGX230HF35C2) Altera FPGA. The performances of the proposed circuit on FPGA devices and the comparison with previous art are shown in Table 1.

For Virtex6, Virtex5, and Spartan3 the synthesis has been carried out using XST and fitting and Place&Route have been carried out using ISE. The implementation on StratixIV has been conducted with Quartus II. ModelSim PE has been used to perform the circuit simulations.

6.2. Quality of the Processed Video Streams. The implementation of (12), the utilization of truncated multipliers, and the implementation of ROM approximation techniques could worsen the accuracy of the processed video streams. Figure 4 shows a selection of frames with resolution 320×240 extracted from two video sequences. The original frames of Figures 4(a) and 4(d) have been processed by using the proposed implementation (Figures 4(b) and 4(e)) or by using the circuit of [22] that uses full-width multipliers and no ROM compression techniques (Figures 4(c) and 4(f)). The Bg masks of Figures 4(b), 4(e), and 4(c), 4(f) differ in few pixels (in both the video sequences less than the 1% of the total number of the pixels is differently classified by the two circuits). This demonstrates that the proposed implementation provides negligible effect on the quality of the processed video streams.

6.3. Circuit Performances. Table 1 shows that the circuit of Figure 2, implemented on high-performances FPGA devices, is able to process more than 43 HD fps by using less than the 5% of the FPGA resources. Implemented on Virtex6 processes 46 HD fps by using the 2% of the total number of Slice of the FPGA. On Virtex5 the frame rate is 43 HD fps while the logic utilization is equal to 301 of 7200 Slices, when implemented on StratixIV processes 47 HD fps by using less than the 1% of Logic Adaptive Blocks (LAB).

The performances analysis has been conducted by including input and output registers that synchronize the circuits and provide timing performances that are not dependent on the I/O pads. The reported frame rate values are obtained by considering the maximum work frequency of the circuits (Frequency in Table 1), obtained by imposing a constraint on the minimum period of the clock signal. For each implementation of Table 1, the synthesis of the circuit has been carried out by repeatedly reducing the constraint. The process has been stopped when a 0.1 ns reduction made the implementation impossible. The last synthesized clock period has determined the maximum work frequency of the circuit. The timing analysis of the obtained implementations has allowed to identify the critical path of the circuit that includes "Match," "Control Logic," "Variance," and "Output Selection" units of Figure 2. However, the very closed constraint on the minimum clock period entails that all the paths between inputs and outputs of the circuit have delay values almost equal to the maximum delay.

The previous art of the background identification circuits is represented by [22] that provides improved performances if compared against [17–21].

The proposed FPGA implementation, as shown in Table 1 outperforms the circuit proposed in [22]. Since that the proposed circuit improves many blocks that in [22] was on the critical path, when the proposed circuit is implemented on Virtex5 xc5vlx50, the maximum working frequency almost doubles (+77.4%), while the area utilization is strongly reduced (−3.8% in number of used Slices and zero DSP used versus 10 DSP).

Table 1 also shows the performances of the circuit on low-cost Spartan3 FPGA. Implemented on Spartan3, the proposed circuit is able to run at 35.30 MHz by using the 13% of the total number of Slice of the FPGA.

7. Conclusion

An OpenCV compatible hardware implementation of the GMM algorithm for Bg identification is presented in the paper. The circuit has been designed having commercial FPGA devices as a target and is able to process 1080p video at 47 frames per second when implemented on Statix IV FPGA.

The circuit exploits an innovative formulation of the equations of the GMM algorithm, approximates some non-linear functions with piecewise linear polynomials, and uses truncated binary multipliers. The circuit is implemented on both Altera and Xilinx FPGA and, when compared with previously proposed Bg identification circuits, provides improved speed and reduced logic utilization.

References

[1] I. Haritaoglu, D. Harwood, and L. S. Davis, "A fast background scene modeling and maintenance for outdoor surveillance," in *Proceedings of the 15th International Conference on Pattern Recognition*, no. 4, pp. 179–183, Barcelona, Spain, 2000.

[2] P. KaewTrakulPong and R. Bowden, "A real time adaptive visual surveillance system for tracking low-resolution colour targets in dynamically changing scenes," *Image and Vision Computing*, vol. 21, no. 10, pp. 913–929, 2003.

[3] B. Gloyer, H. K. Aghajan, K. Y. S. Siu, and T. Kailath, "Video-based freeway-monitoring system using recursive vehicle tracking," in *Image and Video Processing III*, vol. 2421 of *Proceedings of SPIE*, pp. 173–180, San Jose, Calif, USA, February 1995.

[4] H. Jiemnez and J. Salas, "Temporal templates for detecting the trajectories of moving vehicles," in *Advanced Concepts for Intelligent Vision Systems*, vol. 5807 of *Lecture Notes in Computer Science*, pp. 485–493, 2009.

[5] C. Zhan, X. Duan, S. Xu, Z. Song, and M. Luo, "An improved moving object detection algorithm based on frame difference and edge detection," in *4th International Conference on Image and Graphics, ICIG 2007*, pp. 519–523, Chengdu, China, August 2007.

[6] M. Ren and H. Sun, "A practical method for moving target detection under complex background," *Computer Engineering*, vol. 31, no. 20, pp. 33–40, 2005.

[7] B. Xiao, C. Lu, H. Chen, Y. Yu, and R. Chen, "Moving object detection and recognition based on the frame difference algorithm and moment invariant features," in *Proceedings of the 27th Chinese Control Conference (CCC '08)*, pp. 578–581, July 2008.

[8] S. Y. Chien, S. Y. Ma, and L. G. Chen, "Efficient moving object segmentation algorithm using background registration technique," *IEEE Transactions on Circuits and Systems for Video Technology*, vol. 12, no. 7, pp. 577–586, 2002.

[9] C. Ridder, O. Munkelt, and H. Kirchner, "Adaptive background estimation and foreground detection using Kalman filtering," in *Proceedings of the 10th International Congress for Applied Mineralogy (ICAM '95)*, pp. 193–199, 1995.

[10] K. P. Karmann and A. Brandt, "Moving object recognition using an adaptive background memory," in *Time-Varying Image Processing and Moving Object Recognition*, V. Capellini, Ed., pp. 289–307, Elsevier, 2nd edition, 1990.

[11] K. Toyama, J. Krumm, B. Brumitt, and B. Meyers, "Wallflower: principles and practice of background maintenance," in *Proceedings of the 1999 7th IEEE International Conference on Computer Vision (ICCV '99)*, pp. 255–261, September 1999.

[12] J. C. S. Jacques, C. R. Jung, and S. R. Musse, "Background subtraction and shadow detection in grayscale video sequences," in *Proceedings of the 18th Brazilian Symposium on Computer Graphics and Image Processing (SIBGRAPI '05)*, pp. 189–196, October 2005.

[13] C. C. Chiu, M. Y. Ku, and L. W. Liang, "A robust object segmentation system using a probability-based background extraction algorithm," *IEEE Transactions on Circuits and Systems for Video Technology*, vol. 20, no. 4, pp. 518–528, 2010.

[14] C. R. Wren, A. Azarbayejani, T. Darrell, and A. P. Pentland, "Pfinder: real-time tracking of the human body," *IEEE Transactions on Pattern Analysis and Machine Intelligence*, vol. 19, no. 7, pp. 780–785, 1997.

[15] C. Stauffer and W. E. L. Grimson, "Adaptive background mixture models for real-time tracking," in *Proceedings of the IEEE Computer Society Conference on Computer Vision and Pattern Recognition (CVPR '99)*, pp. 246–252, Fort Collins, Colo, USA, June 1999.

[16] OpenCV library on source forge, http://sourceforge.net/projects/opencvlibrary/.

[17] H. Jiang, H. Ardo, and V. Öwall, "Hardware accelerator design for video segmentation with multi-modal background modelling," in *Proceedings of the IEEE International Symposium on Circuits and Systems (ISCAS '05)*, vol. 2, pp. 1142–1145, 2005.

[18] F. Kristensen, H. Hedberg, H. Jiang, P. Nilsson, and V. Öwall, "An embedded real-time surveillance system: implementation and evaluation," *Journal of Signal Processing Systems*, vol. 52, no. 1, pp. 75–94, 2008.

[19] H. Jiang, H. Ardö, and V. Öwall, "A hardware architecture for real-time video segmentation utilizing memory reduction techniques," *IEEE Transactions on Circuits and Systems for Video Technology*, vol. 19, no. 2, pp. 226–236, 2009.

[20] M. Genovese, E. Napoli, and N. Petra, "OpenCV compatible real time processor for background foreground identification," in *Proceedings of the International Conference on Microelectronics (ICM '10)*, pp. 467–470, December 2010.

[21] M. Genovese and E. Napoli, "FPGA-based architecture for real time segmentation and denoising of HD video," *Journal of Real-Time Image Processing*, 2011.

[22] M. Genovese and E. Napoli, "FPGA implementation of OpenCV compatible background identification circuit," in *Proceedings of the 3rd International Symposium on Computational Modelling of Objects Represented in Images: Fundamentals, Methods and Applications (CompIMAGE '12)*, pp. 75–80, Rome, Italy, September 2012.

[23] A. G. M. Strollo, N. Petra, and D. De Caro, "Dual-tree error compensation for high performance fixed-width multipliers," *IEEE Transactions on Circuits and Systems II*, vol. 52, no. 8, pp. 501–507, 2005.

[24] N. Petra, D. De Caro, V. Garofalo, E. Napoli, and A. G. M. Strollo, "Truncated binary multipliers with variable correction and minimum mean square error," *IEEE Transactions on Circuits and Systems I*, vol. 57, no. 6, pp. 1312–1325, 2010.

[25] N. Petra, D. De Caro, V. Garofalo, E. Napoli, and A. G. M. Strollo, "Designof fixed-width multipliers with linear compensation function," *IEEE Transactions on Circuits and Systems I*, vol. 58, no. 5, pp. 947–960, 2011.

[26] V. Garofalo, N. Petra, and E. Napoli, "Analytical calculation of the maximum error for a family of truncated multipliers providing minimum mean square error," *IEEE Transactions on Computers*, vol. 60, no. 9, pp. 1366–1371, 2011.

[27] D. De Caro, E. Napoli, and A. G. M. Strollo, "Direct digital frequency synthesizers using high-order polynomial approximation," in *Proceedings of the IEEE International Solid-State Circuits Conference*, vol. 1, pp. 134–135, San Francisco, Calif, USA, February 2002.

[28] D. De Caro, E. Napoli, and A. G. M. Strollo, "Direct digital frequency synthesizers with polynomial hyperfolding technique," *IEEE Transactions on Circuits and Systems II*, vol. 51, no. 7, pp. 337–344, 2004.

[29] D. De Caro and A. G. M. Strollo, "High-performance direct digital frequency synthesizers in 0.25 μm CMOS using dual-slope approximation," *IEEE Journal of Solid-State Circuits*, vol. 40, no. 11, pp. 2220–2227, 2005.

[30] A. G. M. Strollo, D. De Caro, and N. Petra, "Elementary functions hardware implementation using constrained piecewise-polynomial approximations," *IEEE Transactions on Computers*, vol. 60, no. 3, pp. 418–432, 2011.

[31] D. De Caro, N. Petra, and A. G. M. Strollo, "High-performance special function unit for programmable 3-D graphics processors," *IEEE Transactions on Circuits and Systems I*, vol. 56, no. 9, pp. 1968–1978, 2009.

[32] P. KaewTraKulPong and R. Bowden, "An improved adaptive background mixture model for real-time tracking with shadow detection," in *Proceedings of the 2nd European Workshop Advanced Video Based Surveillance Systems*, 2001.

[33] V. Garofalo, N. Petra, D. De Caro, A. G. M. Strollo, and E. Napoli, "Low error truncated multipliers for DSP applications," in *Proceedings of the 15th IEEE International Conference on Electronics, Circuits and Systems (ICECS '08)*, pp. 29–32, September 2008.

[34] S. S. Kidambi and P. El-Guibaly, "Area-efficient multipliers for digital signal processing applications," *IEEE Transactions on Circuits and Systems II*, vol. 43, no. 2, pp. 90–95, 1996.

[35] V. Garofalo, "Fixed-width multipliers for the implementation of efficient digital FIR filters," *Microelectronics Journal*, vol. 39, no. 12, pp. 1491–1498, 2008.

5

Compensation of Linear Multiscale Doppler for OFDM-Based Underwater Acoustic Communication Systems

A. E. Abdelkareem, B. S. Sharif, C. C. Tsimenidis, and J. A. Neasham

School of Electrical and Electronic Engineering, Newcastle University, Newcastle upon Tyne NE1 7RU, UK

Correspondence should be addressed to A. E. Abdelkareem, a.e.abdelkareem@ncl.ac.uk

Academic Editor: Yuriy Zakharov

In particular cases, such as acceleration, it is required to design a receiver structure that is capable of accomplishing time varying Doppler compensation. In this paper, two approaches are taken into consideration in order to estimate the symbol timing offset parameter. The first method employed to achieve an estimate of this particular parameter is based upon centroid localization and this prediction is reinforced by a second technique which utilises linear prediction, based on the assumption that the speed changes linearly during the OFDM symbol time. Subsequently, the two estimations of the symbol timing offset parameter are smoothed in order to obtain a fine tuned approximation of the Doppler scale. Additionally, the effects of weighting coefficients on smoothing the Doppler scale and on the performance of the receiver are also investigated. The proposed receiver is investigated, incorporating an improvement that includes fine tuning of the coarse timing synchronization in order to accommodate the time-varying Doppler. Based on this fine-tuned timing synchronization, an extension to the improved receiver is presented to assess the performance of two point correlations. The proposed algorithms' performances were investigated using real data obtained from an experiment that took place in the North Sea in 2009.

1. Introduction

Several time-domain receivers that adopt coherent modulation with an emphasis on channel equalization in order to increase communication reliability have been suggested. However, the time-varying doubly-spread characteristic of the underwater acoustic (UWA) channel requires a highly complex equalizer. Recently, an alternative low-complexity, high-speed multicarrier communication system, in the form of OFDM, has attracted considerable interest in the field of underwater acoustic communication (UWAC). This is mainly due to its simplicity of operation by means of the fast Fourier transform (FFT) for the purpose of modulation/demodulation. This system deals with the frequency selectivity of the channel by dividing the broadband data into parallel narrowband channels. Additionally, in a delay-dispersive environment, adopting a CP of a length greater than the maximum delay spread provides an excellent way to assure the orthogonality of the carriers. However, propagation is considered to be time selective due to the Doppler shift in which one subcarrier may introduce ICI to the adjacent subcarriers [1]. The Doppler shift sensitivity is inversely proportional to the OFDM symbol duration; therefore, even slightly moving platforms can cause serious impairments as far as synchronization is concerned.

Previous studies in the field of UWAC have addressed several approaches for synchronization in the presence of Doppler distortion. For single carrier transmission, a block-based approach [2] has been used to estimate and compensate the Doppler shift. In this approach, two LFMs are used for a coarse estimation of the time scaling factor and then an equalizer is employed for residual Doppler-shift compensation. This method is particularly well suited for constant speed. An alternative approach utilizes an adaptive Doppler compensation technique, as suggested by [3], in order to accommodate autonomous vehicles (AV). This closed-loop Doppler correction method demands a high degree of complexity when it is applied to OFDM systems because there is a requirement for the demodulation of each interpolation factor.

As far as multicarrier transmission is concerned, the authors in [4] employed the principle in [2] and null subcarriers for the purpose of resampling factor estimation and residual Doppler compensation, respectively. Although these algorithms do attain precise estimation by adopting preamble and postamble, the bandwidth utilization factor is compromised. A point estimate of the Doppler shift is adopted in [5]; therefore, it is suitable for situations where the Doppler shift stays constant or varies slowly during the packet time. The concept in [6] was extended to work in UWAC by [7] with an iterative cyclic prefix correlation. To estimate the Doppler shift, the author employed the symmetry of the guard interval with its replica. This parameter is estimated iteratively, depending on the peak location and its phase with respect to the new sampling interval; therefore it is a computationally expensive search. In [8], the authors deal with the different Doppler distortion in the multiple-input multiple-output (MIMO) OFDM by adopting multiple resampling followed by FFT.

Although resampling the signal removes the Doppler shift, a major problem with its residual or carrier frequency offset (CFO) is the destruction of the orthogonality of the subcarriers due to the resulting ICI. A considerable amount of literature on combating ICI has been published. These studies [1, 9–11] have presented conclusions that mitigating ICI will result in successful communication.

All the aforementioned papers assume that the Doppler shift is constant during the symbol period and all paths have equal Doppler shift; therefore, resampling the signal with a unique time scaling factor is valid and a symbol by symbol approach works effectively. A recent study by [12] has highlighted the need to estimate the optimal time scaling factor in a multipath channel of different Doppler shift in each path. However, in our proposed method, it is assumed that the channel variation is mainly caused by the motion of both transmitter and receiver, leading to a significant time-varying Doppler shift. Consequently, this will create acceleration that may exceed 1 m/s^2 due to speed alterations; therefore, ignoring this effect yields a significant ICI. In this paper, the acceleration is considered and the Doppler shift is assumed to be changing linearly during the symbol time, but the same for all paths. This variation is dealt with by measuring the time expansion/compression frequently within a fraction of a sample period and then compensating the Doppler by means of an efficient 4th order Lagrange interpolation.

In this paper, an algorithm which combines the centroid-based localization and linear equation in estimating the time-varying Doppler shift is proposed. Furthermore, this paper introduces two improvements in order to increase the confidence of the Doppler shift estimation and consequently a low bit error rate (BER) is obtained. These improvements include:

(1) fine tune the coarse timing estimation; then,

(2) combining two correlation lags and fine tuning using the weighting coefficients [13]. The first lag is estimated based on the centroid localization [14], and the second lag is estimated based on the correlation of the cyclic prefix (CP) with its replica.

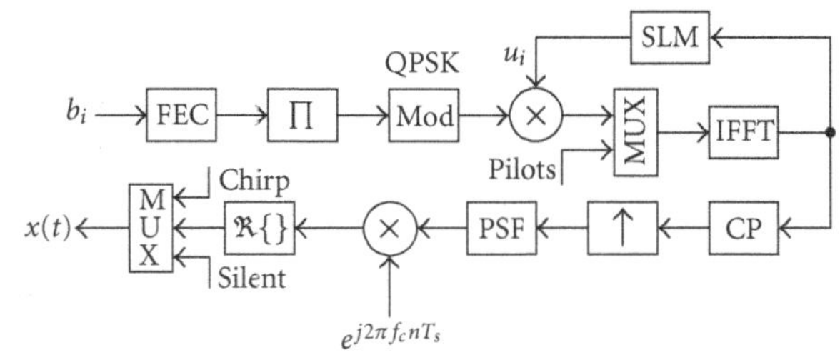

Figure 1: Proposed transmitter structure, where the operator $\Re$ represents the real part of the signal.

Additionally, for the case of low acceleration, these improvements result in a precise estimation of the Doppler shift and no CFO estimation is required.

The rest of this paper is organized as follows. In Section 2, the OFDM system model is introduced. The signal processing in the proposed receiver is presented in Section 3. In Section 4, Doppler extraction and channel estimation are demonstrated. In Section 5, the performance of the proposed algorithms are presented by means of experimental results. Finally, Section 6 draws the conclusions.

2. System and Channel Model

The proposed system contains the transmitter depicted in Figure 1. At each instant i, the encoder receives a vector of information bits b_i of length K_d at its input to produce a binary code of length $K_c = K_d/R_c$ encoded bits, where $R_c \in (0, 1]$ is the coding rate of the nonsystematic convolutional (NSC) code. The coded bits are permuted by a random interleaver, then converted, in groups of m successive bits, into alphabet symbols of constellation size $M = 2^m$. This mapping operation generates a sequence of $N_d = K_c/m$: $\mathbf{s} = \{s_0, \ldots, s_{N_d-1}\}$, where $s_i \in \mathbb{C}$ and $\mathbb{C}$ denotes the set of complex symbols. Subsequently, in the OFDM symbol to be constructed, pilot symbols of phase shift keying (PSK) with unit amplitude are embedded with the data symbols in a comb method, where all pilots and data are transmitted simultaneously on all symbols. These pilot symbols are used for the purpose of estimating channel response at the receiver. A peak-to-average power reduction (PAPR) is introduced using the selective mapping (SLM) technique [15]. To implement this technique, a sequence of phases $\mathbf{U}$ is added in the transmitted signal to be multiplied by the input data sequences, and the symbol sequence of minimum PAPR is selected for transmission. The resulting OFDM symbol, containing N_p pilots and N_d data-bearing subcarriers, where $N_d \cup N_p = N_c$, is then modulated by an IFFT of size N_c, and the last samples are copied and prefaced to the symbol to form the CP-OFDM frame. The guard interval of length N_g is chosen to be longer than the channel dispersion time in order to minimize the intersymbol interference (ISI). The resulting frame is pulse shaped, using a pulse shape filter (PSF), and then up converted using carrier modulation. Let T_d denote the OFDM symbol duration and T_g the guard interval. The total OFDM frame duration is $T = T_d + T_g$. Let $f_n = f_c + n\delta f$, being the carrier frequency corresponding to each of the subcarriers of the OFDM spectrum, where $\delta f = 1/T_d$ is the frequency separation between alternate subcarriers, and f_c is the carrier frequency, so the bandwidth is $B = N_c\delta f$. The time-domain representation of the ith OFDM symbol is given by

$$x_i(t) = \frac{1}{\sqrt{N_c}} \sum_{n\in\mathcal{I}} d_i(n) u_i^{\text{opt}}(n) e^{j2\pi(n/T_d)(t-T_g-iT)} p_{\text{rc}}(t-iT), \quad \text{for } iT \le t < (i+1)T, \tag{1}$$

where $d_i(n)$ is the symbol transmitted over the nth subcarrier, $\mathbf{U}^{\text{opt}}$ is the optimum phase set $[u_i(1), u_i(2), \ldots, u_i(n)]$ for lower PAPR with $u_i(n) = e^{j\varphi_n}$, $\varphi_n \in [0, 2\pi]$, $\mathcal{I}$ denotes the set of modulated subcarriers, and $p_{\text{rc}}(t-iT)$ is the pulse shaping filter, which is realized as an up-sampled raised cosine FIR filter. An equivalent passband model of (1) is

$$\begin{aligned} x(t) &= \Re\left\{ e^{j2\pi f_c t} \sum_{i=0}^{\infty} \frac{1}{\sqrt{N_c}} \times \sum_{n\in\mathcal{I}} d_i(n) u_i^{\text{opt}}(n) e^{j2\pi(n/T_d)(t-T_g-iT)} p_{\text{rc}}(t-iT) \right\} \\ &= \Re\left\{ \sum_{i=0}^{\infty} \frac{1}{\sqrt{N_c}} \times \sum_{n\in\mathcal{I}} d_i(n) u_i^{\text{opt}}(n) e^{j2\pi f_n(t-T_g-iT)} p_{\text{rc}}(t-iT) \right\}. \end{aligned} \tag{2}$$

It is assumed that the signal is transmitted over a multipath fading channel as follows:

$$h(\tau, t) = \sum_{l=0}^{L-1} h_l(t)\delta[\tau - \tau_l(t)], \tag{3}$$

where $\{h_l(t)\}$ are the path amplitudes, $\{\tau_l(t)\}$ are the time-varying path delays, and L is the total number of paths. As in [16], we assume the path delay τ_l and the gains h_l are constant over the frame duration T. For perfect OFDM synchronization, and providing that the maximum delay spread is within the guard interval, the received passband signal can be written as

$$\tilde{r}(t) = \Re\left\{ \frac{1}{\sqrt{N_c}} \sum_{n\in\mathcal{I}} d_i(n) u_i^{\text{opt}}(n) e^{j2\pi f_n t} \times \sum_{l=0}^{L-1} h_l p_{\text{rc}}(t-\tau_l) e^{-j2\pi f_n \tau_l} \right\} + \tilde{w}_i(t), \tag{4}$$

where $w_i(t)$ is a white Gaussian noise with variance σ^2.

When the Doppler shift is present, a transmitted signal is received as

$$r(t) = x\left[\left(1 \pm \frac{v}{c}\right)t - \tau_l\right], \tag{5}$$

where v denotes the induced speed due to the mobility of the transmitter and/or receiver, and c is the acoustic propagation speed of 1500 m/s. The (+) sign indicates an expansion of the signal since the distance is increased and vice versa. It is assumed that all paths have a similar Δ; therefore, the received signal in (4) can be rewritten as

$$\tilde{r}(t) = \Re\left\{ \frac{1}{\sqrt{N_c}} \sum_{n\in\mathcal{I}} d_i(n) u_i^{\text{opt}}(n) e^{j2\pi f_n(1+\Delta)t} \cdot \sum_{l=0}^{L-1} h_l p_{\text{rc}}((1+\Delta)t - \tau_l) e^{-j2\pi f_n \tau_l} \right\} + \tilde{w}_i(t). \tag{6}$$

Based on the assumption that the speed of the motion changes linearly during the ith OFDM symbol interval $t \in [iT, T(i+1))$, the Doppler shift is varied with time. Therefore the constant $\Delta = v/c$ does not hold to accommodate this variation and it should be replaced by $\Delta(t)$. Thus, the time varying Doppler shift can be modelled as

$$\Delta(t) = \frac{v(t)}{c}, \tag{7}$$

where $v(t)$ represents the speed variation during the symbol time. Therefore, the received passband signal in (6) can be rewritten as

$$\tilde{r}(t) = \Re\left\{ \frac{1}{\sqrt{N_c}} \sum_{n\in\mathcal{I}} d_i(n) u_i^{\text{opt}}(n) e^{j2\pi f_n(1+\Delta(t))t} \cdot \sum_{l=0}^{L-1} h_l p_{\text{rc}}[(1+\Delta(t))t - \tau_l] e^{-j2\pi f_n \tau_l} \right\} + \tilde{w}_i(t), \tag{8}$$

and its corresponding complex baseband signal model can be written as

$$r(t) = \sum_{i=0}^{\infty} \sum_{n\in\mathcal{I}} H_i(n) d_i(n) u_i^{\text{opt}}(n) e^{j2\pi n\delta f t} e^{j2\pi\Delta(t) f_n t} + w_i(t), \tag{9}$$

where $H_i(n)$ is the channel transfer function of the ith symbol at nth subcarrier with a time varying Doppler-shift that can be written as

$$H_i(n) = \sum_{l=0}^{L-1} h_l e^{-j2\pi f_n \tau_l} p_{\text{rc}}[1 + \Delta(t)t - \tau_l]. \tag{10}$$

As referred to in [17], it is obvious in (9) that the effect of the Doppler shift on the received signal is twofold. Firstly, it scales the received OFDM frame duration T by a factor of $1 + \Delta(t)$, yielding sampling frequency errors that result in a symbol timing error [18]. Secondly, there is a time-varying CFO.

2.1. Sampling Frequency Errors. In discrete time, the sampled transmitted signal $x[kT_s]$ in (5) is equivalent to a scaling of the sampling period (interpolation or decimation).

$$\tilde{r}[kT_s] = x[k(1 \mp \Delta(t))T_s - \tau_l], \tag{11}$$

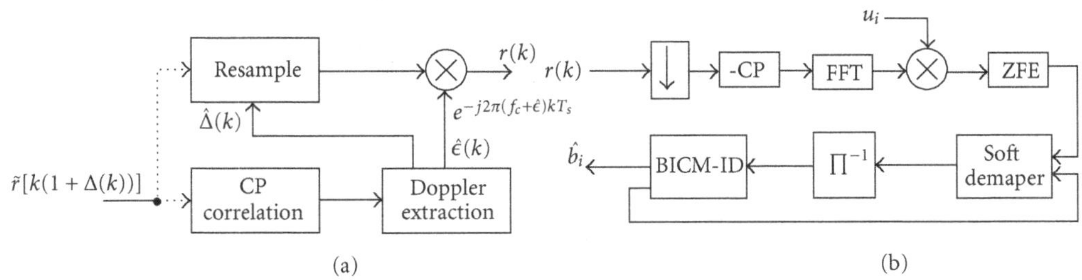

FIGURE 2: Proposed receiver structure.

where k is an integer, and T_s and $\tilde{r}(kT_s)$ are the sampling period and Doppler shifted received sampled signals respectively.

The bidirectional effect of the Doppler shift causes symbol timing errors, which are increased or decreased proportionally to $\Delta(t)$. To align the symbol within its period, samples should be removed if $(\Delta > 0)$ or added if $(\Delta < 0)$ at regular intervals [19].

Let $\pm\phi$ be the deviation of samples of the received sequence for each OFDM symbol due to the speed change; the sampling period results in expansion or compression of the samples' length, hence the Doppler-shifted received frame's length is modelled by

$$L'_f = \left(L_f \mp \phi\right), \tag{12}$$

where $L_f = (N_c/B \cdot T_s)$ represents the transmitted passband samples' length. It is apparent that L_f is only affected by T_s and any expansion/compression in the timescale will result in ϕ. Therefore, (12) is implicitly equivalent to (11). To remove both CFO and symbol shift, an inverse time scaling of the received (compressed/expanded) signal should be achieved, providing that the amount of Doppler shift $\Delta(t)$ is known. This is equivalent to changing the sampling rate of the passband signal by $1+\Delta(t)$ in discrete-time processing. From (12), it can be inferred that increasing or decreasing the length of samples is equivalent to adjusting the sampling frequency f_s by the same Doppler shift $1 + \Delta(t)$; thus, (11) is rewritten as

$$\tilde{r}[k] = x\left[\frac{k(1 \mp \Delta(t))}{f'_s} - \tau_l\right], \tag{13}$$

where $f'_s = f_s(1 \mp \Delta(t))$. By substituting f'_s in (13), $\tilde{r}[k] = x[k]$, the signal received is then in conformity with the transmitted signal.

2.2. Carrier Frequency Offset Errors. The factor $e^{j2\pi\Delta(t)f_n t}$ in the received signal in (9) represents a time varying CFO, where $\Delta(t)f_n = \Delta(t)f_c + \Delta(t)n\delta f$. The CFO (ϵ) is due to the residual Doppler shift. It is destructive because it deviates the subcarrier spacing δf and introduces ICI, which must be removed prior to the FFT to design an optimum receiver [18]. The resampling process removes the Doppler shift and converts the wideband system into narrowband. However, the residual Doppler shift produced by the fractional part of the time expansion/compression degrades the receiver.

3. Signal Processing in the Proposed Receiver

To utilize the available bandwidth efficiently, the algorithm employs a low-complexity blind technique to estimate the Doppler shift based on estimating the coarse timing metric for each OFDM symbol by exploiting the inherent periodicity of the CP. Centroid-based localization has been used to refine the maximum amplitude of the timing metric, that is, the timing offset, as explained in [14]. Using this coarse timing metric, the Doppler shift and its residual are frequently estimated by deriving a tracking step in the Doppler extraction unit (DEU). This unit comprises linear expectation of the timing offset, fine tuning of the estimated parameters, tracking the Doppler shift, and CFO estimation. In this technique, the fractional deviation of the subcarrier spacing, which is the source of ICI, is estimated by exploiting the fractional part of the normalized sampling frequency offset, whereas the integer part of this offset is used to estimate the integer Doppler shift.

3.1. Coarse Timing Metric Estimation. The receiver structure of the proposed system is depicted in Figure 2. The received signal $\tilde{r}(t)$ in (8) is fed through the transducer, preamplifier, and analogue-to-digital converter, and then filtered in the frequency band $[f_c - B/2, f_c + B/2]$. The resultant Doppler shifted passband signal $\tilde{r}[k(1 + \Delta(k))]$ is correlated with the Doppler tolerant training (chirp) to detect the start of the packet ζ that contains several OFDM symbols. Based on the existing guard interval, the drift in the received Doppler-shifted signal $\tilde{r}[k(1 + \Delta(k))]$ is measured by correlating the guard samples $(N_g \cdot N_s)$ with an anticipated observation window in order to estimate the coarse timing metric for each OFDM symbol within the packet, as in [14]. In the case of time-varying Doppler shift, that is, multi time scaling factor, the resulting timing metric is affected by the velocity perturbation. Consequently, there is a demand on estimating this timing metric of the same OFDM symbol, but using an alternative approach to increase the accuracy of the Doppler shift estimation. Therefore, in Figure 2(a), linear prediction is adopted to extract the Doppler shift for the purpose of reinforcing the symbol timing offset parameter that was estimated using CP correlation.

3.2. Time-Varying Doppler Shift Estimation. Thus far, the timing metric has only been considered for the case of a common Doppler shift during the OFDM symbol time. A worst case scenario may occur when there is a velocity that

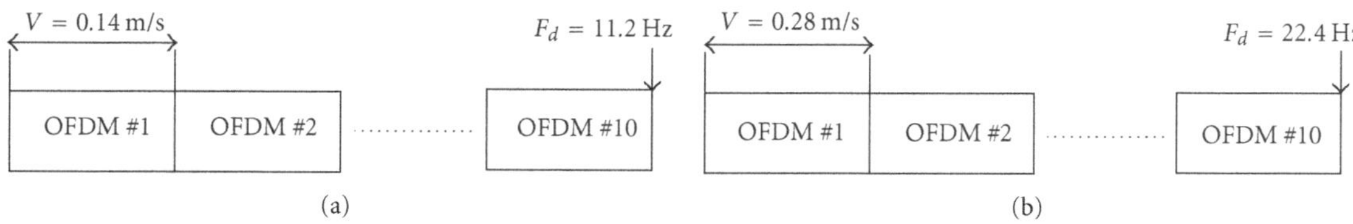

FIGURE 3: Acceleration effect over Doppler frequency change during each symbol time at $f_c = 12$ kHz. (a) Acc $= 0.5\,\mathrm{m/s^2}$, (b) Acc $= 1\,\mathrm{m/s^2}$.

accelerates or decelerates within the symbol period. This situation can be explained in Figure 3. This figure shows that the start of the OFDM symbol undergoes a different speed relative to the speed at the end of the symbol due to the acceleration, in which the speed is changing linearly with time. As a result, a linear multi Doppler shift during the OFDM symbol period is produced. In addition, the acceleration is a useful indication of how fast the change is, where in Figure 3(a) the Doppler frequency shift is 1.12 Hz at OFDM symbol 1 and it increases to 11.2 at OFDM symbol 10. The same case is demonstrated in Figure 3(b), where the acceleration is $1\,\mathrm{m/s^2}$ and the Doppler frequency at OFDM symbol 10 is 22.4 Hz, in terms of time-selectivity measurement which is given as

$$T_d F_d > 1. \tag{14}$$

This rapid change within the symbol duration gives an indicator of the amount of distortion caused by the channel on the signal.

Alternatively, frequent estimation of the Doppler shift within the OFDM symbol or reducing the frame length are viable solutions. However, in OFDM signal design, there is a tradeoff between the number of subcarriers, Doppler estimation resolution, and sensitivity to the CFO. Hence, frequent estimation of the interpolation factor is more feasible than shortening the OFDM symbol length.

When the channel has a velocity that accelerates or decelerates in both directions (up or down) within the symbol period, the following example is considered. If T_d is 256 ms and the maximum acceleration $1\,\mathrm{m/s^2}$ starting from initial speed v_0, then the symbol needs approximately $4T_d$ to attain the maximum speed $v_0 + 1$ m/s. From this assumption, it can be inferred that the maximum speed change in each OFDM symbol is approximately 0.25 m/s.

For a system of 12 kHz carrier frequency, 48 kHz sampling frequency, and a symbol time of 0.256 seconds, such speed variation causes a Doppler frequency shift F_d to increase by 2 Hz within each symbol up to 20 Hz by symbol number 10. In such circumstances, estimating a common timing metric may not hold to attain acceptable performance. Alternatively, a better solution and more accurate Doppler compensation can be realized by adopting a frequent estimation of the Doppler shift within the OFDM symbol.

4. Doppler Extraction and Channel Estimation

The Doppler extraction unit in Figure 2(a) comprises linear prediction of the symbol timing offset, fine symbol timing offset, tracking the Doppler shift and CFO or residual Doppler shift estimation.

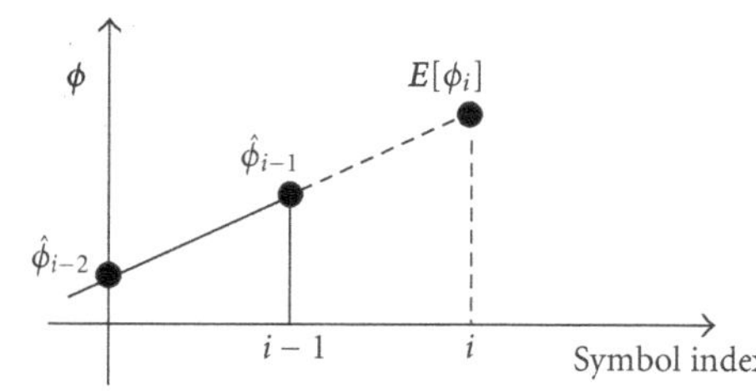

FIGURE 4: Estimation of timing offset during the packet time.

4.1. Linear Prediction of the Symbol Timing Offset. As the transmission structure contains multiple OFDM frames within a packet, the synchronization between these frames is paramount to reduce both the ISI and ICI on the receiver side. In the proposed technique, an improvement is obtained by involving the estimated timing offset at time $i - 1$ in predicting the timing offset at time i. To accomplish this, it is assumed that due to the first order Doppler shift, the OFDM frame could be expanded towards the leading edge or compressed towards the trailing edge in the range $[T(1+\Delta(t)) + \tau_{\max}, T(1-\Delta(t)) + \tau_{\max}]$, respectively. Therefore, the linear part of the speed variation can be formulated by the following first order equation:

$$y = m x_i + b, \tag{15}$$

where $m = (\hat{\phi}_i - \hat{\phi}_{i-1})/(x_i - x_{i-1})$ denotes the slope, and x_i is the OFDM symbol at index i, as shown in Figure 4. Accordingly, the gradient will vary gradually in accordance with the speed change and, subsequently, the output value y_i is obtained. The slope here is determined based on the previous two OFDM symbols estimated in (17) and subsequently used to predict the timing offset ϕ for the next OFDM symbol. Therefore, the first order predicted timing offset of the current OFDM symbol can be formulated as

$$E[\phi_i] = 2\hat{\phi}_{i-1} - \hat{\phi}_{i-2}. \tag{16}$$

4.2. Fine Symbol Timing Offset and Synchronization. Thus far, two estimations of the same parameter $\hat{\phi}$ have been obtained. It should be stressed that attaining accurate timing offset estimation may be difficult in the presence of noise and/or ISI, especially with a short observation window. Therefore, for the purpose of increasing the reliability of

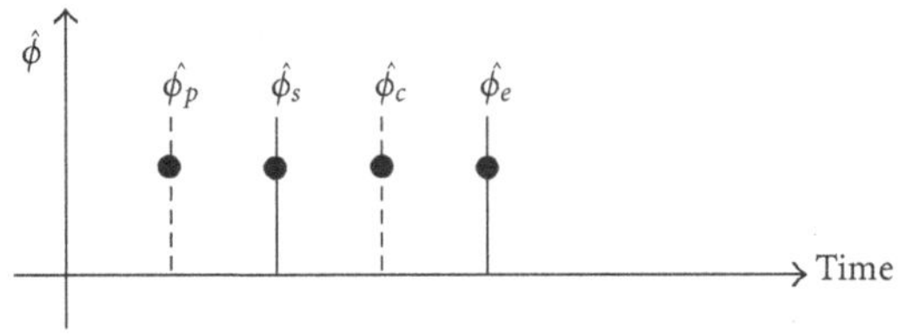

FIGURE 5: Tracking the Doppler within the OFDM symbol.

estimation, smoothing the timing offset is adopted. This yields the following fine tuned estimated timing offset:

$$\widehat{\phi}_i = \widehat{\phi}_i W_1 + E[\phi_i] W_2, \quad (17)$$

where the coefficients W_1 and W_2 are empirically obtained and satisfy the condition of $0 < W_1 + W_2 \leq 1$. These coefficients are designed to attain a tradeoff between estimation accuracy and tracking capabilities. It is crucial to mention that these coefficients have an effect on adapting the slope variation, where $W_1 = 1, W_2 = 0$ indicates fast slop variation and the linear expectation does not hold. At the same time, $W_1 = 0, W_2 = 1$ accommodate a constant gradient between symbols. The estimated fine timing offset $\widehat{\phi}$ in (17) still represents the average. Assuming the change in the time scale is linear within the OFDM symbol, the change in the speed is considered unidirectional. This will enable tracking of the Doppler shift caused by speed variation within the OFDM symbol time. Performing such tracking demands knowledge of the timing offset at both edges of the symbol in order to determine the tracking step. By involving previous estimation of fine symbol timing offset $\widehat{\phi}_p$ and current fine symbol timing offset $\widehat{\phi}_c$, the offset at the leading edge can be formulated as

$$\widehat{\phi}_s = \frac{\phi_p + \phi_c}{2}. \quad (18)$$

At the same time, the sampling frequency offset at the trailing edge $\widehat{\phi}_e$ is determined as

$$\widehat{\phi}_e = 2\widehat{\phi}_c - \widehat{\phi}_s, \quad (19)$$

where $\widehat{\phi}_p$ and $\widehat{\phi}_c$ represent the average fine timing offset estimate from (17). It should be stressed that the estimation accuracy of these two parameters plays an important role in increasing the ability to compensate for the Doppler shift and its residual effects in the subsequent stages.

4.3. Tracking the Doppler Shift. If the relative velocity between the transmitter and receiver during the packet time is constant, that is, for zero acceleration, then the Doppler shift estimate computed can be used to compensate for the entire OFDM symbol. In time varying Doppler shift, however, a unique interpolation factor for the whole symbol does not hold due to the resulting nonnegligible sampling frequency errors which must be tracked. Therefore, the sampling frequency offset affects channel estimation, which is computed over pilot subcarriers, due to the different delays of the positions of these pilots. By searching for the delay in the 1st significant arrival of the estimated CIR [20], an approach to tracking the fractional sampling clock frequency offset due to a symbol timing error is possible. However, in the case of time-varying Doppler shift, it is necessary to estimate the sampling frequency offset frequently.

An alternative realistic Doppler shift estimator, which can be realized by adopting frequent estimation of this parameter during the symbol time in the time domain, is proposed here. In order to track the Doppler shift, it is necessary to derive a tracking step that corresponds to the sampling frequency offset change over $\widehat{\phi}_s < \widehat{\phi}_s + T_s < \widehat{\phi}_e$. In such a case, the tracking step is given as

$$\phi_{\text{step}} = \frac{\widehat{\phi}_e - \widehat{\phi}_s}{L_f}, \quad (20)$$

where L_f represents the up-sampled subcarriers. As shown in Figure 5, each OFDM symbol is identified by the two parameters of sampling frequency offset $\widehat{\phi}_s$ and $\widehat{\phi}_e$, based on the assumption that the speed changes linearly. Accordingly, the estimated timing offset at the leading edge is updated at each sample time k, based on the step in (20). At the same time, the integer Doppler shift can be computed as

$$\widehat{\Delta}(k) = \frac{L_f - \widehat{\Psi}(k)_{(\text{quant})}}{L_f}, \quad (21)$$

where Ψ is the sampling frequency offset initialized with $\widehat{\phi}_s$, and then updated at each sample as

$$\Psi(k) = \Psi(k-1) + \widehat{\phi}_{\text{step}}, \quad (22)$$

and $\widehat{\Psi}_{(\text{quant})} = \lceil \widehat{\Psi} \rfloor$ is rounded towards the nearest integer, respectively. This integer resampling factor is delivered to the sample-by-sample Lagrange quadratic interpolation unit, as shown in Figure 2(b), and the fractional part is dealt with as a carrier frequency offset. It should be stressed that the resolution of the interpolation factor in (21) is entirely dependent on the transmitted frame length.

4.4. Residual Doppler Shift Estimation. Efficient Doppler shift compensation relies on how accurately the resampling factor estimation reduces the residual Doppler. This residual Doppler has a direct impact on the performance of the receiver. Taking this effect into account involves finding the amount of the fractional part of the estimated samples that shifts the subcarrier spacing fractionally. This deviation can be modelled as $(\widehat{\Psi}(k) - \widehat{\Psi}_{(\text{quant})})$, and therefore

$$\widehat{\epsilon}(k) = \left[\widehat{\Psi}(k) - \widehat{\Psi}_{(\text{quant})}\right] \delta f \frac{f_c}{f_s}, \quad (23)$$

is the residual frequency estimate. The residual Doppler shift is not constant at each sample within the OFDM symbol, and thus it is dealt with by determining the standard deviation across the fractional part of the estimated Doppler shift. Once the Doppler shift and its residual have been estimated

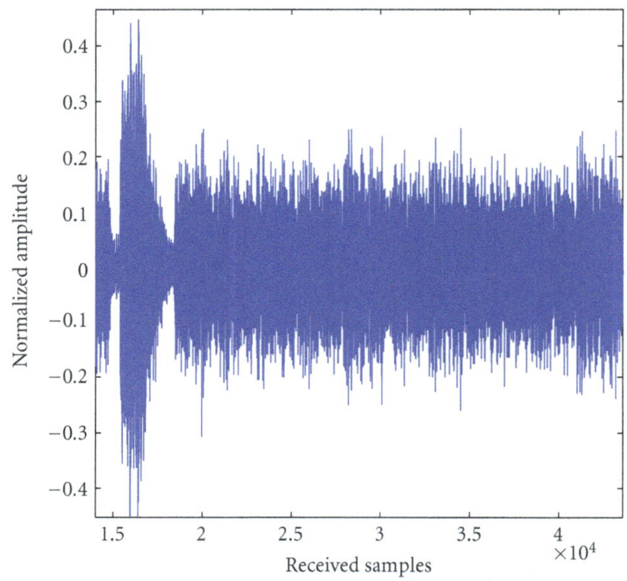

Figure 6: Received signal.

and compensated, the output signal $r(k)$ is delivered to the outer receiver in Figure 2(b). This signal is firstly down sampled and then its cyclic prefix is discarded. The PAPR phases $\mathbf{u_i}$ are removed prior to FFT demodulation. The zero forcing equalizer (ZFE) and least square (LS) method for channel estimation purposes are adopted by utilizing pilots which are embedded in a comb method. After removing the channel effect, the subsequent stage is BICM-ID.

4.5. Pilot-Based Channel Estimation. In the channel estimation of the OFDM symbol, a comb-type arrangement of the training sequence (pilot) is adopted. In this scheme, specific tone indices are allocated on all transmitted OFDM symbols and the rest for data transmission. Unlike a block-based training sequence, the comb type is quite convenient for fast fading channels. Additionally, with the comb type, all pilots and data are transmitted simultaneously on all symbols.

It is worth pointing out that in order to increase the accuracy of the channel estimation, the residual Doppler shift should be eliminated [21]. This is due to an induced ICI which destroys the orthogonality among subcarrier frequency components and ultimately the diagonal of the channel matrix. In OFDM systems, the advantage of increasing the symbol duration in reducing the ISI effect can conflict with increasing the ICI impact, as a consequence of subcarrier spacing reduction. Therefore, after resampling and CFO compensation, all subcarriers are orthogonal (i.e., ICI free).

The channel estimation was implemented using the following least square (LS) method [22]:

$$X(n) = X(mL + l) = \begin{cases} X_p(m), & l = 0 \\ X_d(m), & l = 1, \ldots, L - 1, \end{cases} \quad (24)$$

where $L = N_c/N_p$ and $X_p(n)$ is the nth pilot subcarrier value. Let $H_p(n)$ be the frequency response of the channel for $n = 0, \ldots, N_p - 1$ at pilot subcarriers. Let $Y_p(n)$ be the received pilot symbols after the FFT operation, the estimate of the channel at pilot subcarriers $\hat{H}_p(n)$ is given as

$$\hat{H}_p(n) = D\left[X_p(n)\right]^{-1} Y_p(n), \quad n = 0, \ldots, N_p - 1, \quad (25)$$

where $D[X_p(n)]$ is a diagonal matrix constructed using the known transmitted pilot symbols.

For subcarriers $n = 0, 1, 2, \ldots, N_c - 1$, LS channel estimation $\hat{\mathbf{H}}_{\mathrm{LS}}$ is written as

$$\hat{\mathbf{H}}_{\mathrm{LS}}[n] = \frac{\mathbf{Y[n]}}{\mathbf{X[n]}}. \quad (26)$$

The mean square error of the LS channel estimation is considered high when compared with the minimum mean-square error (MMSE) estimate [23]. However, LS is attractive in implementing real-time systems due to its simplicity. In order to increase the reliability of the channel estimation, an interpolation in frequency domain between each pilot and data subcarriers is adopted. It is well known that the LS is the first step of the channel frequency response estimation for the known pilots and should be followed by interpolation to obtain a nonpilot subcarriers frequency response.

5. Experimental Results

During the summer of 2009, an experiment was conducted in the North Sea to evaluate the system performance. In the trial setting, the transmitter and receiver were set at 10 and 5 m from the sea surface, respectively. The set-up ranges between the transmitter and the receiver were 200 m, 500 m, and 1000 m. The transmitter power was set to 180 dB re 1 μPa@1 m which is the equivalent of approximately 10 W. In the trial, transmission was organized in packets of equal duration, each containing one 50 ms LFM followed by a 12.5 ms silent period and then 10 CP-OFDM frames. A total of 8920 information bits were transmitted in each setting. A total of 20 packets of 2.795 s were sent. The carrier frequency was set to 12 kHz, whereas the sampling frequency was $4f_c$. 1024 subcarriers were employed and the system bandwidth was 4 kHz, which led to a subcarrier spacing of 3.90625 Hz. The guard interval was set at T_g =16 ms. Figures 6 and 7 show the channel measurements over a range of 1000 m. These figures show a received frame structure and the normalized CIR of a packet that exhibits maximum delay spread of the order of 6 ms, respectively. This multipath delay is equivalent to an ISI of 24 symbols for a system bandwidth of 4 kHz, and this delay spread is inversely proportional to the range. In addition to the silent period shown in Figure 6, the CP guard time also contributes towards reducing the ISI effect.

5.1. Proposed Receiver Performance. To evaluate the performance of the proposed system, the experimental results for both block-based and proposed techniques are depicted in Figure 8(a). The performances of both receivers are presented

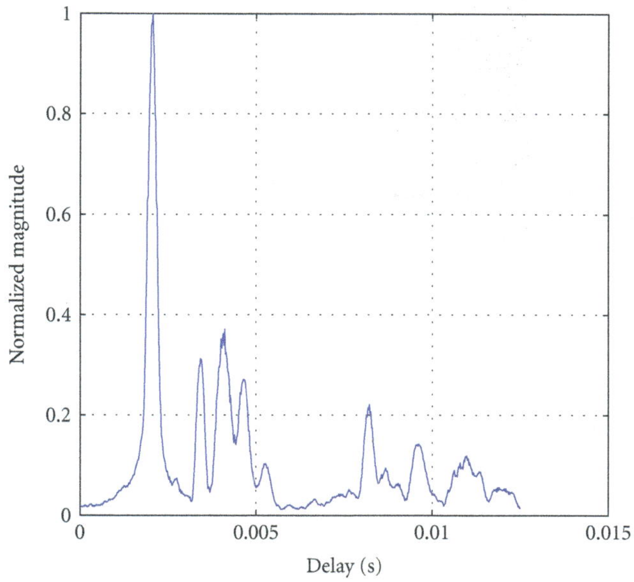

Figure 7: Sample of normalized channel impulse responses for 1000 m channel range.

Table 1: Average BER comparison of the experimental results at different settings of weighting coefficients between the proposed and block-based Doppler shift techniques for $N_c = 1024$.

Method	Error statistics	
	Errors	BER
Block	6503	0.0365
Proposed-set 1	772	0.004
Proposed-set 2	105	0.0006

Table 2: Performance of the experimental results between the improved and block-based Doppler shift techniques for $N_c = 1024$.

Packet index	1	2	3	4	5	6	7	8	9	10
Block	0	119	423	347	3	44	11	505	39	2443
Proposed	0	0	3	6	0	33	0	6	9	0
Improved	0	0	0	0	0	3	0	0	0	5
Packet index	11	12	13	14	15	16	17	18	19	20
Block	24	0	33	178	21	1702	21	0	471	119
Proposed	0	0	6	14	5	0	0	0	23	0
Improved	0	0	0	0	9	0	0	0	0	0

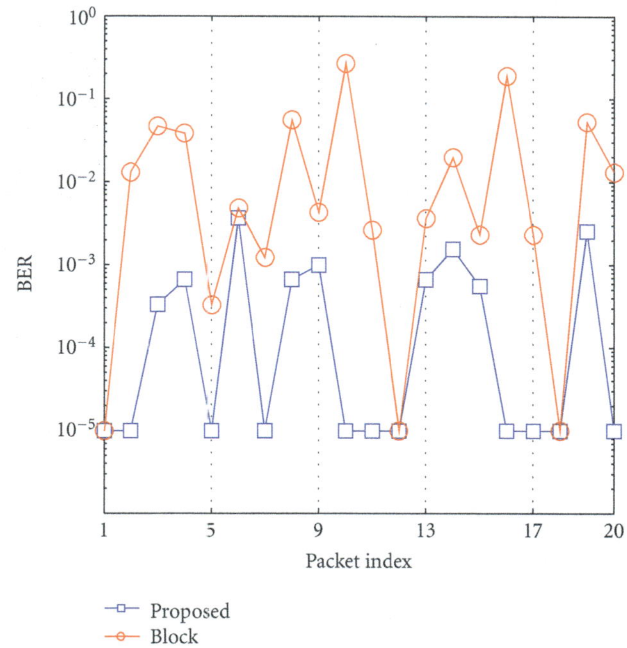

(a) BER for each packet over 1000 m range, where the label 10^{-5} represents zero error

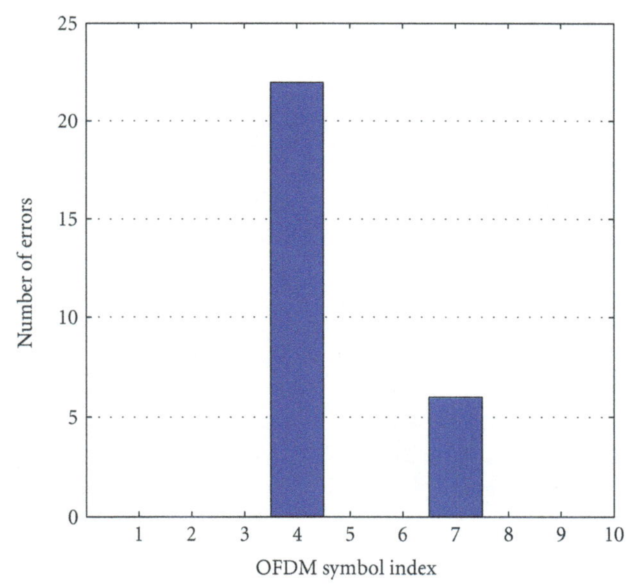

(b) Error statistics for packet 6; 2 out of 10 have decoding errors

Figure 8: Performance of the proposed system at 1024 subcarriers.

in terms of bit error rate (BER). In the block-based method, single resampling is performed for the received signal. It can be seen that for all packets the proposed technique outperforms the block-based method. This is further clarified in Table 2 which shows that the proposed technique achieves acceptable performance in reducing errors in all packets compared with the block technique. Error statistics for both schemes are presented in Table 1. It can be seen that compensating the time-varying Doppler scale and its residual leads to a reduction in the BER from 0.0365 to 0.0006, which is equivalent to 98.4%. However, Figure 8 shows high decoding error in packet 6.

In Figure 8, the bit errors are high only in two blocks within packet 6, as shown in Figure 8(b). This is due to the noise effect which affects the Doppler scale estimation when estimating the timing offset. Evidence for this is shown in Figures 9(b) and 9(d), where in packet 6 there is a mismatch in estimating the speed at the end of symbol 3 and at the start of symbol 4. Therefore, a decoding error results in symbol 4. Furthermore, it can be seen from Figure 9(a) that there is a relatively high deceleration of $\sim 0.9\,\text{m/s}^2$ during the symbol time, which adds an error in approximating the correlation-based Doppler scale estimation. This result shows that there

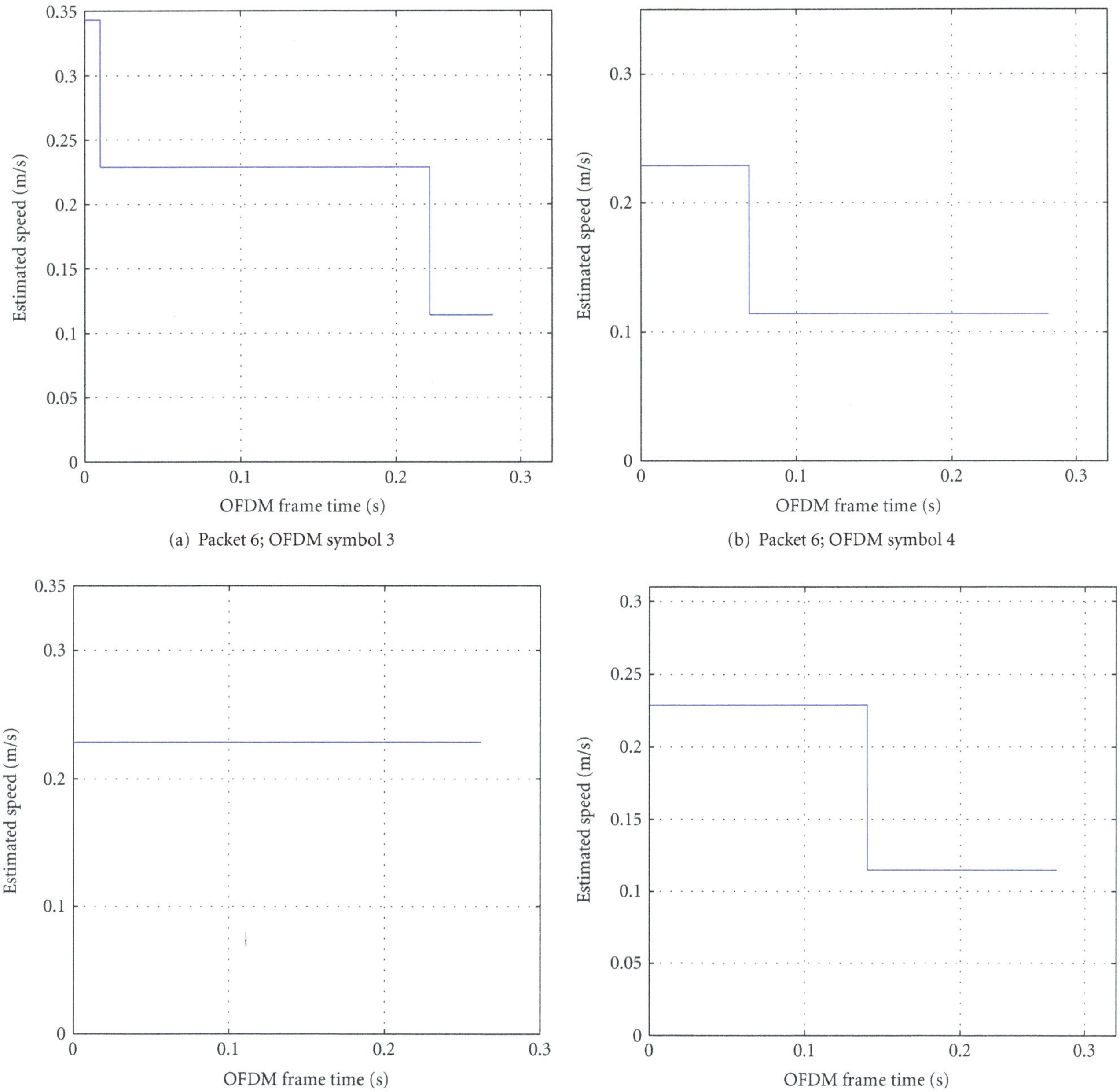

(a) Packet 6; OFDM symbol 3

(b) Packet 6; OFDM symbol 4

(c) Packet 6; OFDM symbol 6

(d) Packet 6; OFDM symbol 7

Figure 9: Estimated speed variation during OFDM symbol.

is a limitation on the acceleration that can be adopted in this algorithm.

Figure 9 demonstrates that the adopted system is capable of precisely tracking the speed variation in each symbol. Particularly, in Figure 9(a), the speed in symbol 3 of packet 6 has been changed three times during 0.256 s, whereas in Figure 9(c) the speed is constant. However, changing the direction of velocity within the packet period, along with higher acceleration, can produce higher intercarrier interference levels in the system. The source of this noise is the mismatch introduced by the transition from acceleration to deceleration, or vice versa. The proposed system detects this critical point through the CP correlation-based Doppler scale estimation, and the linear expectation has no effect on this scenario. However, linear expectation reduces the channel and/or noise effect on the CP correlation. Consequently, accurate Doppler scale estimation is obtained.

Figure 10 shows the performance of bit-interleaved coded modulation with iterative decoding (BICM-ID) and ZFE in the experiment. In terms of Figure 10(a), the figure shows that the ZFE delivers reliable information to the decoder. The reliability depends on how accurate the Doppler shift compensation is. It was mentioned earlier that the channel estimation is affected by the presence of residual Doppler shift which can cause ICI, and, as a result, the orthogonality is destroyed. Consequently, the iterative decoding stage can

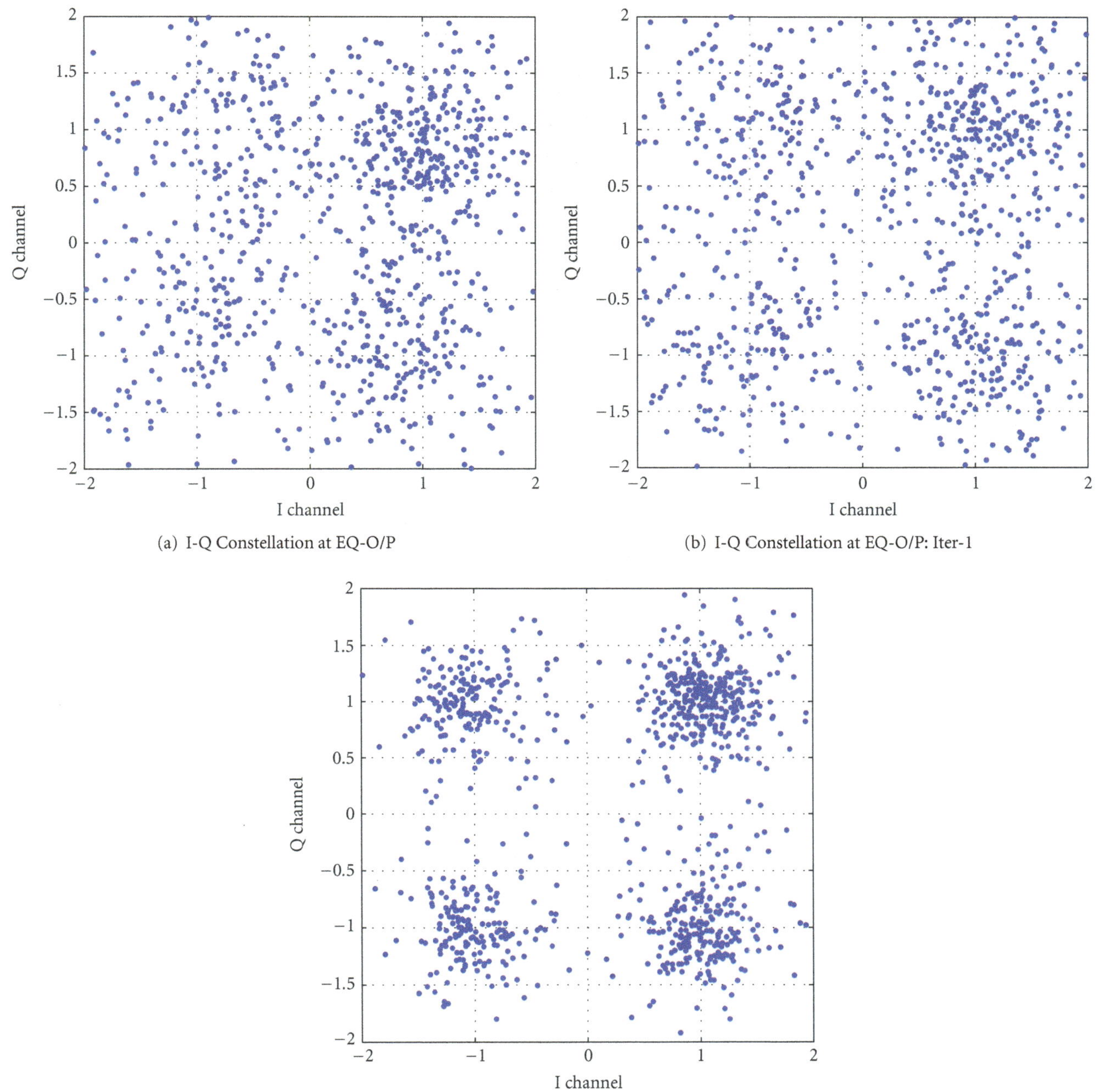

(a) I-Q Constellation at EQ-O/P

(b) I-Q Constellation at EQ-O/P: Iter-1

(c) I-Q Constellation at EQ-O/P: Iter-2

FIGURE 10: Constellation output from equalizer and iterative receiver.

generate unreliable LLRs [24]. Thus, it can be seen that there is an improvement in the second iteration in Figure 10(c) compared with the first iteration in Figure 10(b). At this stage, further iterations are pointless and no more gain is expected.

5.2. Effect of Weighting Coefficients. As mentioned in [13], the weighting coefficients play an important role in the accuracy of the Doppler scale estimation. For this reason, special settings of these parameters are required in order to achieve acceptable performance. It can be shown that there is a tradeoff between the value of the weighting coefficients and the receiver performance. To be more specific, by appointing the symbol timing offset, estimated by linear expectation, a lower weighting coefficient than correlation-based symbol timing offset estimation means there is a constant acceleration or deceleration between symbols, and vice versa. As shown in (17) and (20), the Doppler scale is approximated based on estimating the fine symbol timing offset and its tracking step is derived based on the sampling frequency offset at the start and end of the OFDM symbol. This means that the weighting coefficients have a direct effect on the estimation of the time varying Doppler scale $\Delta(t)$.

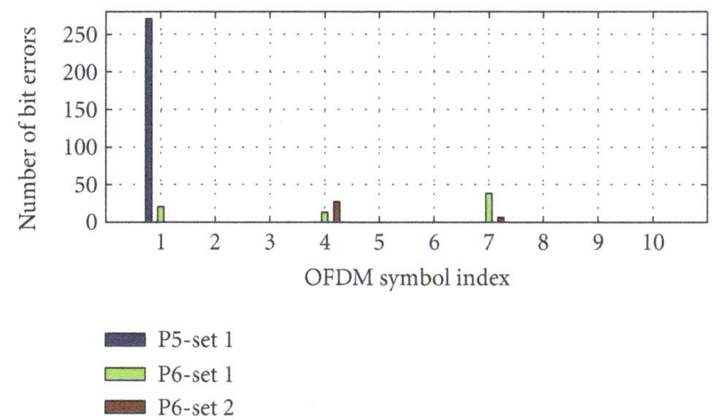

(a) I-Q Constellation at EQ-O/P

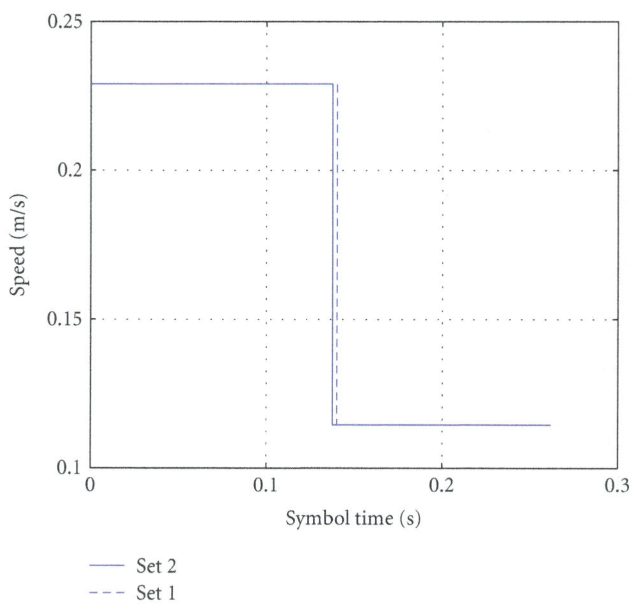

(b) I-Q Constellation at EQ-O/P: Iter-1

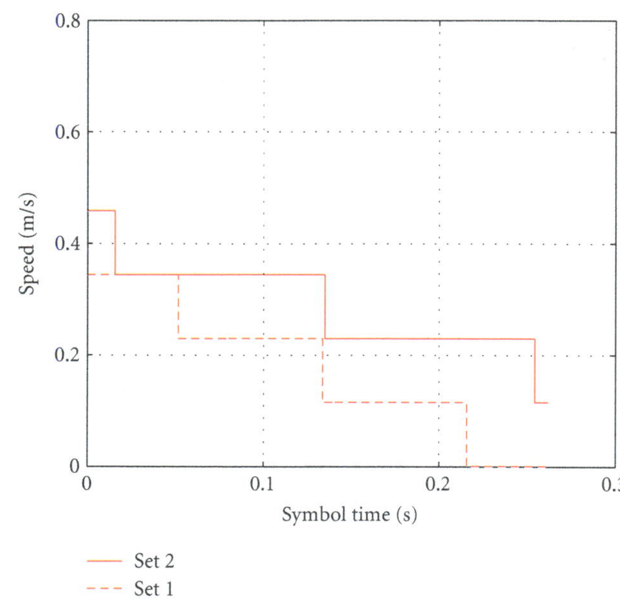

(c) I-Q Constellation at EQ-O/P: Iter-2

FIGURE 11: Effect of weighting coefficients on estimation.

Figure 11 shows two settings of these parameters and their effect on the performance of the receiver. In set 1, where $W_1 = 0.5$ and $W_2 = 0.5$, it can be seen that the receiver performance is poor. In Figure 11(a), it is obvious that packets 5 and 6 in set 1 exhibit a high BER of 271/8920 and 71/8920, receptively. The reason for this degradation is that increasing the weight of the linear expectation in a channel leads to significant acceleration that can cause maladjustment of the interpolation factor and make the tracking of the Doppler scale change coarsely. This is shown in Figure 11(c). Although both sets have the same slope, there is a mismatch between them at the start and end of speed estimation. In set 2, on the other hand, $W_1 = 0.85$ and $W_2 = 0.15$, there is a great improvement in the performance as shown in Figure 11(a), with 0 errors in packet 5 and 33 bits in packet 6. Table 1 shows the performance of the receiver for the subcarriers 1024 over a range of 1000 m using two different settings of the weighting coefficients. In set 2, it can be observed that the error decreases by about 86.4% compared with set 1.

5.3. Performance Evaluation with Improved Coarse Timing Estimation. As mentioned earlier, the impairments in the channel estimation due to synchronization failure will result in unreliable LLRs as a consequence of the Doppler effect. In contrast, estimating and compensating the Doppler scale precisely causes the received OFDM symbol to coincide with its transmitted period, thus improving the channel estimation and delivering reliable symbols to the decoder. Therefore, the target is to improve the Doppler scale estimation and ultimately reduce the burden on the channel estimation.

In order to extract the Doppler scale successfully, it is important to increase the reliability of estimating the symbol timing.

Considering the effect of acceleration on the chirp correlation is small, in the case of multiple OFDM symbols within a packet, the symbol timing error in each OFDM block is accumulated with acceleration during the packet time. Hence, adopting a single estimation of ζ for the whole packet is no longer accurate. Therefore, in order to mitigate the acceleration effect on the symbol timing error, ζ needs to be fine tuned. Performing the fine tuning necessitates updating the position of ζ after each symbol time. Let m, i denote the range of the timing offset around the leading and the trailing edge during the OFDM symbol, respectively. It follows that a two-dimensional timing function is written as

$$\Lambda(m,i) \triangleq \left| \sum_{n=0}^{N_g-1} r(\zeta+m+n)r(\zeta+n+N+i) \right| \qquad (27)$$
$$m \in \left\{-\frac{W}{2},\ldots,\frac{W}{2}\right\};\ i \in \left\{-\frac{\Upsilon'}{2},\ldots,\frac{\Upsilon'}{2}\right\}.$$

Then, $\widehat{\theta}_{m,i}$ can be estimated from obtaining the maximum peak of the multiplication, and it can be written as

$$\widehat{\theta}_{m,i} = \arg\max \max_{m,i} \Lambda(m,i) \qquad (28)$$
$$m \in \left\{-\frac{W}{2},\ldots,\frac{W}{2}\right\};\ i \in \left\{-\frac{\Upsilon'}{2},\ldots,\frac{\Upsilon'}{2}\right\},$$

and the fine tuned ζ' is obtained. The implementation of this fine tuning algorithm of the coarse packet synchronization can be summarized as follows:

(1) compute the coarse packet synchronization point ζ which represents the time position of the maximum peak of the chirp correlation,

(2) compute the timing function $\Lambda(m,i)$ for $m \in [-W/2, W/2], i \in [-\Upsilon'/2, \Upsilon'/2]$,

(3) choose the maximum of the maximum of $\Lambda(m,i)$ as the estimated packet timing offset,

(4) update ζ to be fine tuned which is given as

$$\zeta' = \zeta + \widehat{\theta}. \qquad (29)$$

It should be noted that a two-dimensional search (i.e., m and i) is included in the proposed timing function $\Lambda(m,i)$. This is the main difference from the single synchronization point estimation in [14], where only coarse estimation of the packet synchronization point is adopted. The first search parameter is m, corresponding to the first search region in the range around the coarse synchronization point ζ. Meanwhile, the second search parameter is i, corresponding to the range in the region around the tail of the OFDM symbol which yields the expected Doppler shift. Once the fine-tuned ζ' is obtained, the subsequent stage is the estimation of the first order moment $\widehat{\phi}$. In existing techniques, [6] and [14], due to the acceleration and the inherent ISI, there is a fluctuation in the maximum of the timing function, and the channel conditions have a direct effect on this maximum. Therefore, centroid-based localization is adopted to estimate $\widehat{\phi}$, because it reduces the position uncertainty caused by the fading channel, and the search range is built on the fine-tuned ζ', which can be written as

$$r_D \in \left[\zeta' + N_g + N - \left(\frac{\Upsilon}{2}\right) + i, \zeta' + N - \left(\frac{\Upsilon}{2}\right) + i\right], \qquad (30)$$

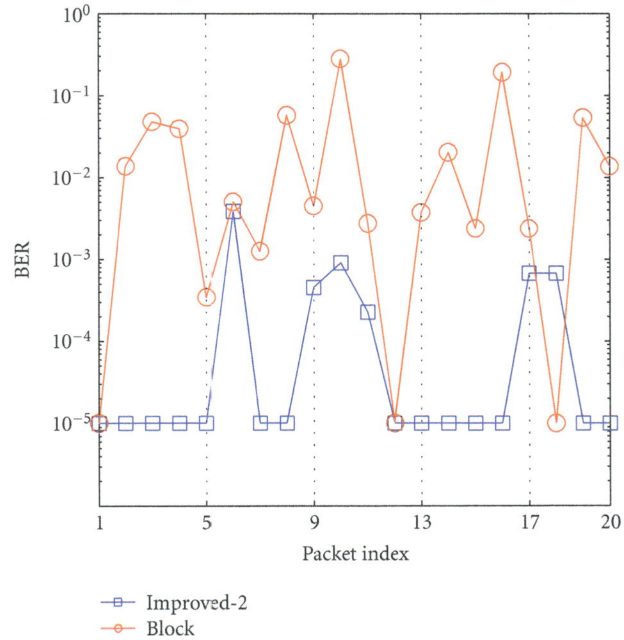

FIGURE 12: Performance of the proposed system with improved coarse timing estimation.

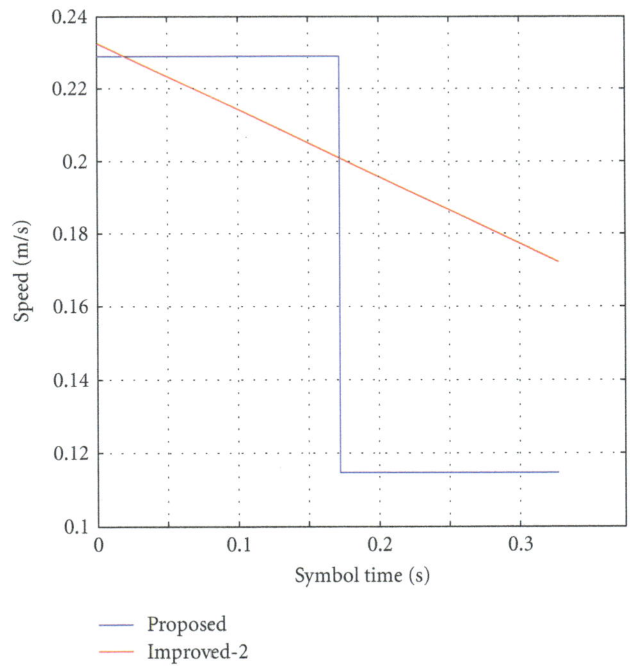

FIGURE 13: Improved time-varying speed estimation during OFDM symbol 7 of packet 6.

and the centroid-based first order moment $\widehat{\phi}_l$ is given as in [14]. Figure 12 shows that fine tuning this parameter results in reducing the BER. It can be inferred from this figure that adjusting the misalignments of the symbol timing due to the time-varying Doppler scale results in an improvement in the reliability of the resampling factor estimation, which in turn reduces the noise that accompanies accumulated errors from symbol to symbol within each packet and ultimately a reduction in BER is obtained.

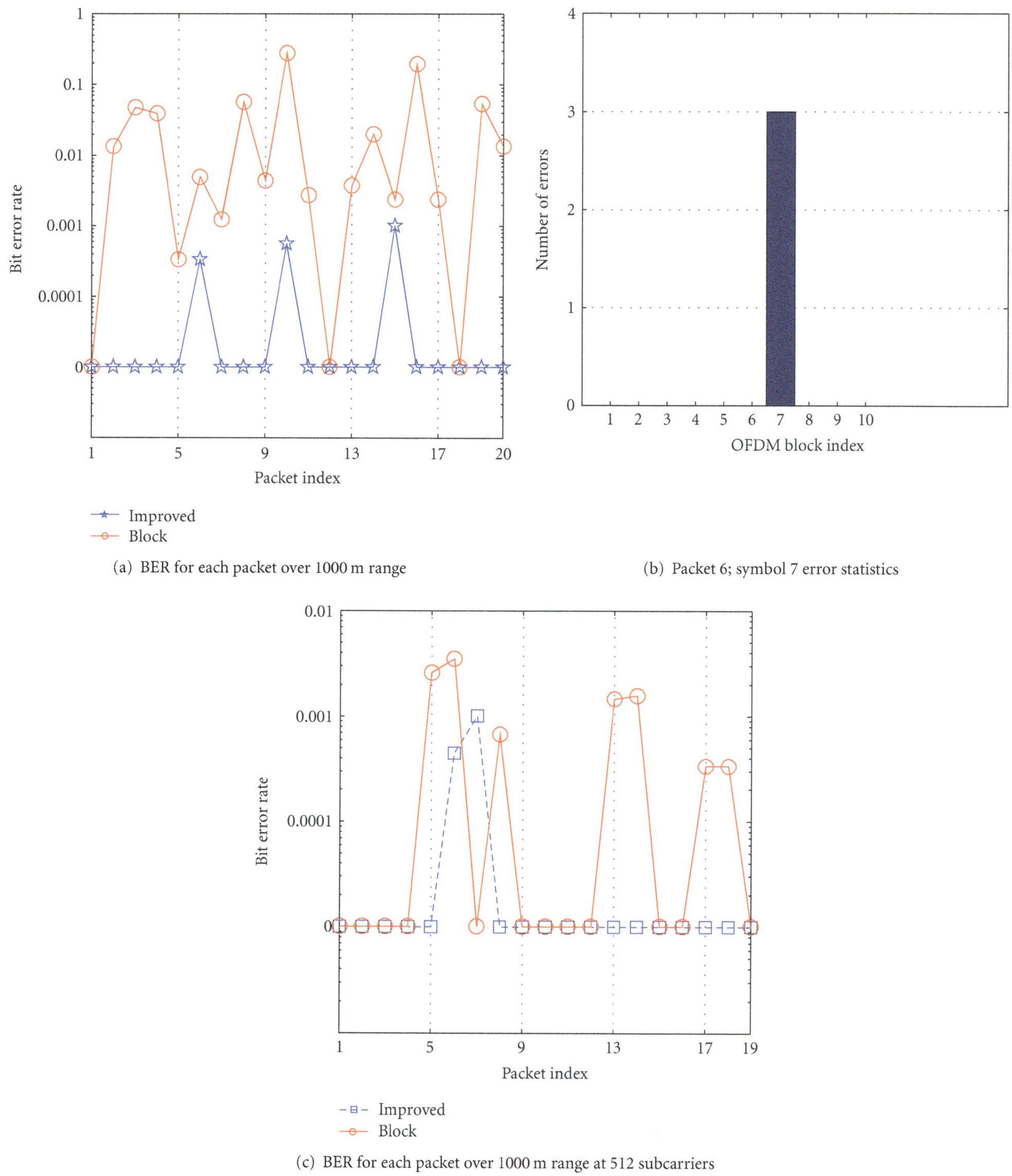

Figure 14: Performance of the improved proposed system.

5.4. Performance Evaluation Based on Two-Point Correlation. Fine tuning of the coarse symbol timing facilitates an alternative approach to estimating the first order moment of the correlation lag. The suggested approach here aims to increase the confidence of estimation by considering the first order moment that results from two correlation lags. The first correlation lag is estimated by means of centroid-based localization, in accordance with the anticipated window mentioned earlier. This type of correlation gives an accurate indication of the fractional part of the time-scale expansion/compression. However, the centroid-based localization is severely affected due to the velocity perturbation. This perturbation degrades the estimation performance of the timing function and ultimately $\hat{\phi}_l$. Therefore, an alternative

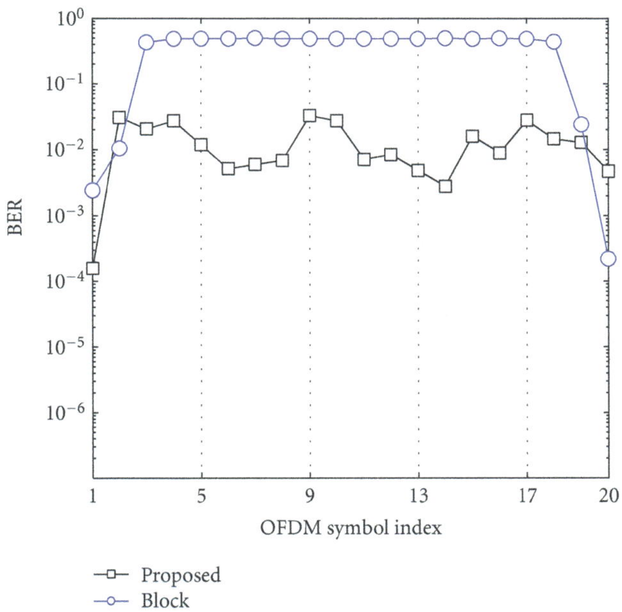

FIGURE 15: Performance comparison of block-based and proposed techniques.

approach has been adopted by involving another estimation point based upon full cross correlation of the CP with its replica. The addition of this correlation is based on the idea of increasing the certainty of the first order moment estimation. This correlation is based on the assumption that the OFDM timing is approximately aligned due to the fine tuning of the packet synchronization ζ'. By definition, the cross correlation between a pair of energy signals, $x[n]$ and $y[n]$, is given by [25];

$$r_{xy} = \sum_{n=-\infty}^{\infty} x[n]y\left[n-\chi\right], \quad \chi = 0, \pm 1, \pm 2, \ldots, \tag{31}$$

where the parameter χ is called lag and it indicates the time shift between the pair. Based on this theory, the time shift in samples for either expansion or compression can be measured with respect to a reference sequence length of the guard interval N_g. In the case of the existence of Doppler shift, the received samples are shifted to the right in expansion or left for compression with respect to the reference. To be more specific, once the start of the packet ζ is identified, it can be deemed that the symbol timing identification is reliable, and the correlation between the received CP and its replica is computed to measure the time shift in the samples as follows:

$$\Lambda_c \triangleq \left| \sum_{n=0}^{N_g-1} r(\zeta'+n)r\left(\zeta'+n+N-\chi\right) \right|, \qquad \chi = 0, \pm 1, \pm 2, \ldots. \tag{32}$$

Considering that the reference sequence of the transmitted CP is $N_{cp}N_s$, the first order moment of the Doppler shift ϕ_x can be approximated as

$$\widehat{\phi}_x = \arg\max \Lambda_c - N_{cp}N_s, \quad \chi = 0, \pm 1, \pm 2, \ldots. \tag{33}$$

Adopting such a scenario requires extraction of a fine tuned correlation lag. This necessitates involvement of two parameters of weighting coefficients to perform such a smoothing approach, as mentioned earlier. The coefficients W_1 and W_2 are empirically obtained from the experiment to accommodate the measured channel condition. Therefore, ϕ', which represents the fine tuned first order moment of the correlation lag, is given as

$$\phi' = \widehat{\phi}_x W_1 + \widehat{\phi}_l W_2. \tag{34}$$

This fine tuned parameter is then delivered to the Doppler extraction in Figure 2(b) in order to estimate the Doppler shift. Accordingly, the estimated Doppler shift, which comprises both an integer and fractional part, is considered and utilized for compensation. Therefore, the estimated resampling factor requires no extraction of the fractional part to estimate the residual Doppler shift, as shown in Figure 2(b); hence, the CFO is approximated as

$$\widehat{\epsilon} \simeq 0.5 f_c \delta f / f_s \simeq \frac{\delta f}{8}, \tag{35}$$

where $f_s = 4f_c$. For subcarrier spacing of 3.90625, as in the case of 1024 subcarriers, $\widehat{\epsilon}$ is 0.4883 Hz. These two-point estimations of $\widehat{\phi}_l$ and $\widehat{\phi}_x$, in conjunction with ϕ', contribute towards improving the Doppler shift estimation and thus eliminate the need to determine the CFO. Figure 13 demonstrates the implications of improving the Doppler shift estimation. It is obvious in this figure that there are two estimations that show the deceleration in velocities over the symbol time. With respect to the improved system, the gradient is estimated smoothly. This confirms that an accurate estimation of the drift in samples results in an accurate estimation and tracking of the time-varying Doppler shift. On the other hand, this figure illustrates that perturbations in estimating the variation of speed within the OFDM symbol can lead to inaccurate resampling factor estimation. In particular, it can be inferred from this figure that there is a time-varying Doppler shift during the symbol time which decelerates in the order of 0.25 m/s^2. This deceleration is estimated by smoothing ζ estimation. However, in the proposed system, the deceleration is approximated to 0.48 m/s^2 over the same symbol. For the sake of clarity, the "proposed system" refers to the system before the improvements and the "improved system" refers to the proposed system after improving ζ estimation. Table 2 illustrates the performance comparison between block-based Doppler compensation, time varying Doppler shift compensation, and its improvements. The achieved BER decreases significantly in the proposed system compared with the block-based approach. Likewise, there are additional improvements in the BERs of 83.8% compared with the proposed technique. This is shown in Figure 14(a), where the BER of packet 6 is reduced compared with Figure 8(a). Additionally, the error statistics of packet 6, shown in Figure 8(b) and Figure 14(b), confirm that estimating multilags contributes to an increase in the accuracy of the speed estimation.

As demonstrated in Figure 14(c) and Table 3, the experimental results show that the investigation was also

Table 3: Performance of the experimental results between the improved and block-based Doppler shift techniques for $N_c = 512$.

Packet index	1	2	3	4	5	6	7	8	9	10	11	12	13	14	15	16	17	18	19
Block	0	0	0	0	23	31	0	6	0	0	0	0	13	14	0	0	3	3	0
Improved	0	0	0	0	0	4	9	0	0	0	0	0	0	0	0	0	0	0	0

Table 4: OFDM symbol structure and the corresponding data rates.

N_c	N_d	N_p	N_b	Data rates (kb/s)
512	448	64	20	3.0833
1024	896	128	10	3.2794

Table 5: Complexity estimate.

Operator	Method	
	2D search + centroid	Correlation
+, −	$9\Upsilon' \times W + 8\Upsilon$	$N_g(\log_2 N_g + 1)$
×	$N_g \times \Upsilon 6' \times W + \Upsilon \times N_g$	$3N_g \log_2 N_g$

successful with 512 subcarriers, as it was able to improve performance by about 86%. This was an expected result, because reducing the symbol length entails increasing the subcarrier spacing and reducing the sensitivity to the Doppler shift. Additionally, reducing the symbol length enables more frequent tracking of the Doppler shift. However, severe consequences accompany this reduction in the symbol time, since it mitigates immunity against ISI, in addition to reducing the available bandwidth. This performance reveals that improving the synchronization and adopting smoothing produces low BER. Furthermore, compensating residual Doppler shift or CFO preserves the orthogonality of the subcarriers and ultimately contributes towards mitigating decoding errors. However, it is worthwhile mentioning that this approximation of the CFO cannot be extrapolated to all cases, as in the case of higher acceleration where a special signal processing method, such as an adaptive weighting coefficients selection and/or iterative-based estimation of the Doppler shift, should be adopted due to the effect of the time varying Doppler shift and the inherent ISI on the correlation peak. Another problem with this approach is that it fails to compensate for an abrupt change in the direction of velocity, as it needs at least two symbols to self-adapt to this sudden variation which causes a decoding error.

In terms of the achieved data rate, Table 4 presents two types of OFDM subcarrier allocation that account for the transmission overhead due to pilots, channel coding, and guard period.

Figure 15 shows, in terms of BERs, the performance comparison between the block-based approach and the proposed technique. For the block-based approach, two scenarios of the transmitted packet structure are investigated. The first structure includes 20 ms chirp, followed by a silent period then 10 CP-OFDM symbols. The second structure comprises only a single CP-OFDM frame. The former structure is investigated in the experiment; therefore the second structure is considered here for the purpose of the simulation. It can be seen that the performance of the block approach is poor in the case of multi-scale Doppler within the OFDM symbol. When the speed is low, as shown in the OFDM symbols indices 1 and 20, the block algorithm performance is approximately identical to that of the proposed scheme. However, as the speed increases, the BER also increases in the block-based approach, whereas the proposed algorithm demonstrates less performance error despite an escalation in speed. The degradation in the BERs in the proposed algorithm is due to the effect of the acceleration on the CP correlation.

Additionally, the complexity of the improved algorithm is summarized in Table 5. The complexity estimate includes a two-dimensional (2D) search algorithm mentioned in (32), centroid algorithm in [14], and cross correlation of the received CP samples. In this table, it is shown that the complexity is a linear function of the CP length and the parameters Υ, Υ', and W. It is worth to mention that the correlation complexity estimate is estimated via FFT-based correlation. It can be noticed that no division operation is required. Therefore, in terms of hardware implementation, this Doppler estimate is considered simple.

6. Conclusions

The performance of multi-scale Doppler shift compensation for an OFDM-based UWA communication system has been investigated. The algorithm accommodates for channels with linear acceleration during a packet of multiple OFDM frames. Unlike existing Doppler compensation methods, the proposed scheme is more pragmatic, as it considers the notion that the speed is changing linearly during the OFDM symbol time. Additionally, under the assumption of linear speed during the packet time, it has been shown that using the linear equation approach to predict the first order Doppler shift as a reinforcement parameter leads to acceptable performance over other techniques. Furthermore, it has been shown that employing weighted coefficients improves the performance as it fine tunes the estimated parameters. However, an approach to fine tuning these parameters adaptively and in accordance with the acceleration is required.

References

[1] A. F. Molisch, *Wireless Communications*, John Wiley, New York, NY, USA, 2005.

[2] B. S. Sharif, J. Neasham, O. R. Hinton, and A. E. Adams, "Computationally efficient doppler compensation system for underwater acoustic communications," *IEEE Journal of Oceanic Engineering*, vol. 25, no. 1, pp. 52–61, 2000.

[3] B. S. Sharif, J. Neasham, O. R. Hinton, A. E. Adams, and J. Davies, "Adaptive doppler compensation for coherent acoustic

communication," *IEE Proceedings*, vol. 147, no. 5, pp. 239–246, 2000.

[4] B. Li, S. Zhou, M. Stojanovic, L. L. Freitag, and P. Willett, "Multicarrier communication over underwater acoustic channels with nonuniform doppler shifts," *IEEE Journal of Oceanic Engineering*, vol. 33, no. 2, pp. 198–209, 2008.

[5] S. F. Mason, C. R. Berger, S. Zhou, and P. Willett, "Detection, synchronization, and doppler scale estimation with multicarrier waveforms in underwater acoustic communication," *IEEE Journal on Selected Areas in Communications*, vol. 26, no. 9, pp. 1638–1649, 2008.

[6] J. J. van de Beek, M. Sandell, and P. O. Börjesson, "ML estimation of time and frequency offset in OFDM systems," *IEEE Transactions on Signal Processing*, vol. 45, no. 7, pp. 1800–1805, 1997.

[7] B. C. Kim and I. T. Lu, "Parameter study of OFDM underwater communications system," in *Proceedings of the IEEE Oceans*, vol. 2, pp. 1251–1255, Providence, RI, USA, September 2000.

[8] K. Tu, T. M. Duman, J. G. Proakis, and M. Stojanovic, "Cooperative MIMO-OFDM communications: receiver design for doppler-distorted underwater acoustic channels," in *Proceedings of the 44th Asilomar Conference on Signals, Systems and Computers*, pp. 1335–1339, November 2010.

[9] J. Armstrong, "Analysis of new and existing methods of reducing intercarrier interference due to carrier frequency offset in OFDM," *IEEE Transactions on Communications*, vol. 47, no. 3, pp. 365–369, 1999.

[10] A. F. Molisch, M. Toeltsch, and S. Vermani, "Iterative methods for cancellation of intercarrier interference in OFDM systems," *IEEE Transactions on Vehicular Technology*, vol. 56, no. 4, pp. 2158–2167, 2007.

[11] K. Tu, D. Fertonani, T. M. Duman, M. Stojanovic, J. G. Proakis, and P. Hursky, "Mitigation of intercarrier interference for OFDM over time-varying underwater acoustic channels," *IEEE Journal of Oceanic Engineering*, vol. 36, no. 2, pp. 156–171, 2011.

[12] S. Yerramalli and U. Mitra, "Optimal resampling of OFDMm signals for multiscale-multilag underwater acoustic channels," *IEEE Journal of Oceanic Engineering*, vol. 36, no. 1, pp. 126–138, 2011.

[13] A. E. Abdelkareem, B. S. Sharif, C. C. Tsimenidis, and J. A. Neasham, "Adaptive doppler-shift compensation for OFDM underwater acoustic communications system," in *Proceedings of the Underwater Acoustic Measurements: Technologies and Results (UAM '11)*, Kos, Greece, June 2011.

[14] A. E. Abdelkareem, B. S. Sharif, C. C. Tsimenidis, J. A. Neasham, and O. Hinton, "Low-complexity doppler compensation for OFDM-based underwater acoustic communication systems," in *Proceedings of the IEEE Oceans*, pp. 1–6, Santander, Spain, June 2011.

[15] R. W. Bäuml, R. F. H. Fischer, and J. B. Huber, "Reducing the peak-to-average power ratio of multicarrier modulation by selected mapping," *Electronics Letters*, vol. 32, no. 22, pp. 2056–2057, 1996.

[16] M. Stojanovic, "Low-complexity OFDM detector for underwater channels," in *Proceedings of the MTS/IEEE Ocean Conference*, pp. 18–21, Boston, Mass, USA, September 2006.

[17] B. Li, S. Zhou, M. Stojanovic, L. Freitag, and P. Willett, "Non-uniform doppler compensation for zero-padded OFDM over fast-varying underwater acoustic channels," in *Proceedings of the IEEE Oceans*, pp. 1–6, Aberdeen, UK, June 2007.

[18] M. Speth, S. A. Fechtel, G. Fock, and H. Meyr, "Optimum receiver design for wireless broad-band systems using OFDM-part II," *IEEE Transactions on Communications*, vol. 49, no. 4, pp. 571–578, 2001.

[19] T. Pollet, P. Spruyt, and M. Moeneclaey, "The BER performance of OFDM systems using non-synchronized sampling," in *Proceedings of the Global Telecommunications(GLOBECOM '94)*, vol. 1, pp. 253–257, November 1994.

[20] B. Yang, K. B. Letaief, R. S. Cheng, and Z. Cao, "Timing recovery for OFDM transmission," *IEEE Journal on Selected Areas in Communications*, vol. 49, pp. 467–479, 2000.

[21] M. K. Ozdemir, Logus Broadband, wireless solutions Incorporation, and H. Arslan, "Channel estimation for wireless OFDM systems," *IEEE Communications Surveys and Tutorials*, vol. 9, no. 2, pp. 18–48, 2007.

[22] S. Coleri, M. Ergen, A. Puri, and A. Bahai, "Channel estimation techniques based on pilot arrangement in OFDM systems," *IEEE Transactions on Broadcasting*, vol. 48, no. 3, pp. 223–229, 2002.

[23] M. H. Hsieh and C. H. Wei, "Channel estimation for OFDM systems based on comb-type pilot arrangement in frequency selective fading channels," *IEEE Transactions on Consumer Electronics*, vol. 44, no. 1, pp. 217–225, 1998.

[24] C. P. Shah, C. C. Tsimenidis, B. S. Sharif, and J. A. Neasham, "Low-complexity iterative receiver structure for time-varying frequency-selective shallow underwater acoustic channels using BICM-ID: design and experimental results," *IEEE Journal of Oceanic Engineering*, vol. 36, no. 3, pp. 406–421, 2011.

[25] K. S. Mitra, *Digital Signal Processing A Computer Based Approach*, McGraw-Hill, New York, NY, USA, 2006.

6

Divide-by-Three Injection-Locked Frequency Dividers with Direct Forcing Signal

Antonio Buonomo and Alessandro Lo Schiavo

Seconda Università degli Studi di Napoli, Via Roma 29, 81031 Aversa, Italy

Correspondence should be addressed to Antonio Buonomo; antonio.buonomo@unina2.it

Academic Editor: Jose Silva-Martinez

Divide-by-three frequency dividers with direct forcing signal are analyzed, and the actual locking mechanism underlying their operation is highlighted. In particular, it is shown that the locking mechanism cannot be explained with the mixing between signals, as commonly made in the literature. An analytical procedure based on the averaging method is developed for solving the equation describing such dividers, and the first approximation to the oscillation in the locked states is predicted. The amplitude and phase of the output voltage in steady state as well as the locking range are derived in terms of the circuit parameters, obtaining useful design guidelines. The derived results are shown to be very close to SPICE simulations for a 0.13 um RF-CMOS process.

1. Introduction

Frequency dividers are key building blocks of RF integrated circuits, as they allow us to scale down the high frequency of the signal from the voltage controlled oscillator (VCO) to a relatively low frequency required for the phase locking in a Phase-Locked Loop (PLL). As an alternative to the widely used digital dividers, based on Current-Mode Logic (CML) flip flops [1], and to Miller dividers [2], some topologies of *LC*-CMOS analog dividers have been devised that allow higher operation frequencies with lower power consumption. These dividers are called Injection-Locked Frequency Dividers (ILFDs) as they rely on the well-known phenomenon of subharmonic synchronization (or injection-locking) that happens in a basic differential *LC* oscillator perturbed by an external periodic signal [3].

In applications, ILFDs with division ratios greater than two are often needed to reduce the overall power consumption and the chip area that the cascade of dividers with a low division ratio involves. Therefore, frequency dividers, able to divide by 3, 4, and more, were recently designed which, in addition, feature a wide locking range (*LR*) needed in practical PLLs. Moreover, the division by an odd number, and in particular for three, can be a particularly advantageous solution in some designs, as in multiband frequency synthesizers [4, 5]. In particular, the use of a divide-by-three ILFD enables a simple architecture for dual-band operation, when the ratio between the frequencies is three, as it happens for W and K bands [6]. For this reason, designing low-voltage divide-by-three ILFDs with a wide locking range has become a topic of relevant interest [7–9].

To perform the divide-by-three function, a variety of designs suitable for operation at high frequencies have been proposed, which include particular circuit arrangements aimed mainly at widening the lock range. They are basically made by a differential *LC* oscillator and an injection circuit providing a current signal that perturbs the oscillator. In divide-by-three ILFDs with direct forcing of the injection signal, the current signal is injected directly in the oscillator across the *LC*-tank and the locking takes place according to a mechanism different from the mixing between signals [10], usually invoked to explain the process of frequency division [11–13]. The purpose of this work is to complete the investigation in [10] with the aim to provide useful results for the design of these dividers. Using the behavioral model introduced in [10], which consists of a nonlinear differential equation of the second order, and the procedure of solution of that equation, based on the method of averaging, we were able

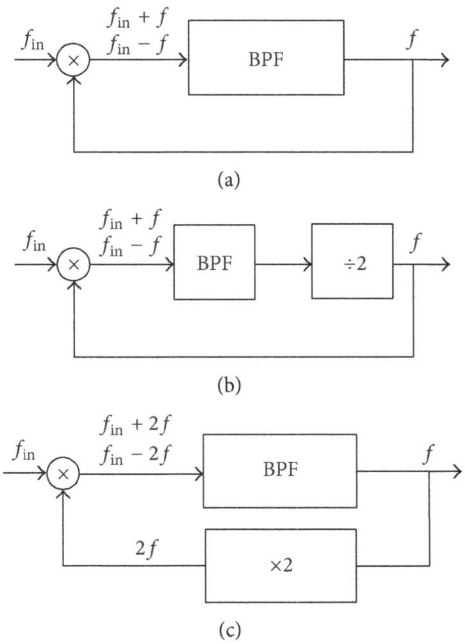

FIGURE 1: (a) Model of the Miller divide-by-two ILFD [2]. (b) Model of a divide-by-three ILFD based on a divide-by-two ILFD [7]. (c) Model of a divide-by-three ILFD based on a frequency multiplier [8].

to find the amplitude and phase of the output voltage in steady state, and the locking range in explicit form as a function of the relevant circuit parameters. Then, the dependence of the circuit performances on the circuit parameters is investigated. The application of the derived formulas leads to results that are in good agreement with SPICE simulations for a 0.13 um RF-CMOS process. A preliminary review of the preexisting results is also given in order to show how their limitations are overcome by presented results.

2. Principles of Operation of Divide-by-Three Circuits

The first analog frequency divider based on a mixer and a band-pass filter (BPF) was proposed by Miller and its block diagram is shown in Figure 1(a) [2]. It is easy to verify that it can operate as a divide-by-two frequency divider. Upon multiplication of the input and output signals, the mixer generates components at $f_{in} + f$ and $f_{in} - f$, where f is the frequency of the output signal. If the former component is suppressed by the low-pass filter but the latter is not, then $f_{in} - f = f$ and, hence, $f = f_{in}/2$.

The combined effect of a mixing nonlinearity and of a band-pass filter has become the common characteristic of all divide-by-two injection-locked frequency dividers used in practical applications [14, 15]. General models to analyze injection-locked frequency dividers were presented in papers [11, 12], which implicitly assume that the divider operation is based on a mixing effect.

Divide-by-three circuits have been developed starting from the approach used for divide-by-two frequency dividers,

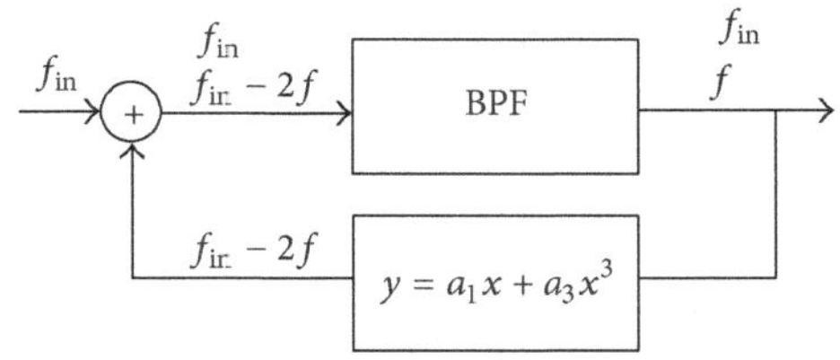

FIGURE 2: Model of a divide-by-three injection-locked frequency divider with a direct application of the forcing signal.

even if some modifications were needed in order to obtain a division by a number different from two. The approach in [7] is based on the model in Figure 1(b) and employs a divide-by-two frequency divider in the feedback loop. The mixer generates components at $f_{in} + f$ and $f_{in} - f$. The component $f_{in} + f$ is filtered out by the band-pass filter and the component $f_{in} - f$ is injected into the divide-by-two frequency divider. Thus, the output frequency can be derived as $(f_{in} - f)/2 = f$ and, hence, $f = f_{in}/3$. Consequently, the output frequency is locked at one-third of the input frequency.

However, the most widely used approach to realize a divide-by-three frequency divider exploits a circuit nonlinearity to produce a second harmonic of the output voltage, as shown in Figure 1(c) [8]. In this case, the input signal at frequency f_{in} is mixed with the feedback signal at frequency $2f$ to produce harmonic components at frequency $f_{in} \pm 2f$. However, only the desired component at frequency $f_{in} - 2f$ passes through the band-pass filter, while all other harmonic terms are suppressed. Thus, $f_{in} - 2f = f$ leads to $f = f_{in}/3$. Many improvements have been proposed in the literature, mainly aimed at widening the locking range through the use of additional inductors, even if in all the cases an input nonlinearity is present that performs a mixing between the input and the feedback signals.

Here, we show that a divide-by-three circuit can operate according to a different principle of operation, that does not employ a mixing between the input and the output signals, as it happens when a forcing signal is directly applied to the LC-tank. In the model in Figure 2, the input signal is added to the feedback signal and passes, even if attenuated, through the band-pass filter. Thus, the output signal has a frequency component at frequency f_{in} in addition to its main component at frequency f. If the feedback nonlinearity is a cubic polynomial, the feedback signal will have a component at frequency $f_{in} - 2f$, which will pass through the filter. Thus, imposing $f_{in} - 2f = f$, it results $f = f_{in}/3$.

Note that the model in Figure 2 can be seen as a particular case of the general model of ILFDs, presented in [11] and shown in Figure 3(a), when the nonlinearities are set equal to $i_{in}(v, v_{in}) = kv_{in}$ and $i_{nl}(v) = a_1v + a_3v^3$, and $H(s)$ represents the transfer function of a band-pass filter, usually made of an LC-tank. Taking into account that the model in Figure 3(a) is equivalent to the circuit in Figure 3(b), let us simulate this circuit with $i_{in} = kv_{in}$, and $i_{nl}(v) = a_1v + a_3v^3$, in order to verify that the proposed model is useful to implement a divide-by-three ILFD. The signal waveforms in Figure 4(a) show that

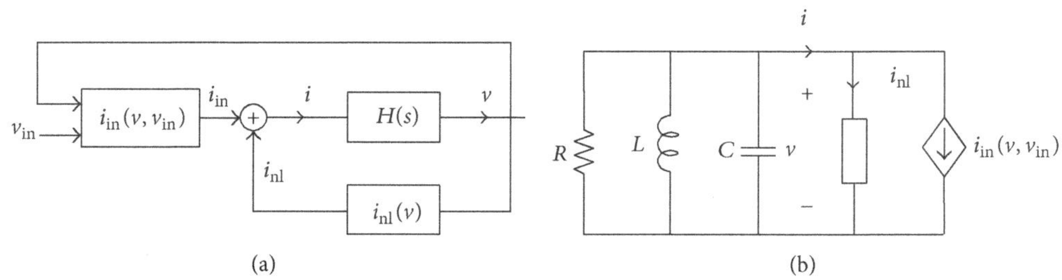

FIGURE 3: (a) General model of an injection locked frequency divider and (b) its representation as forced LC-tank [11].

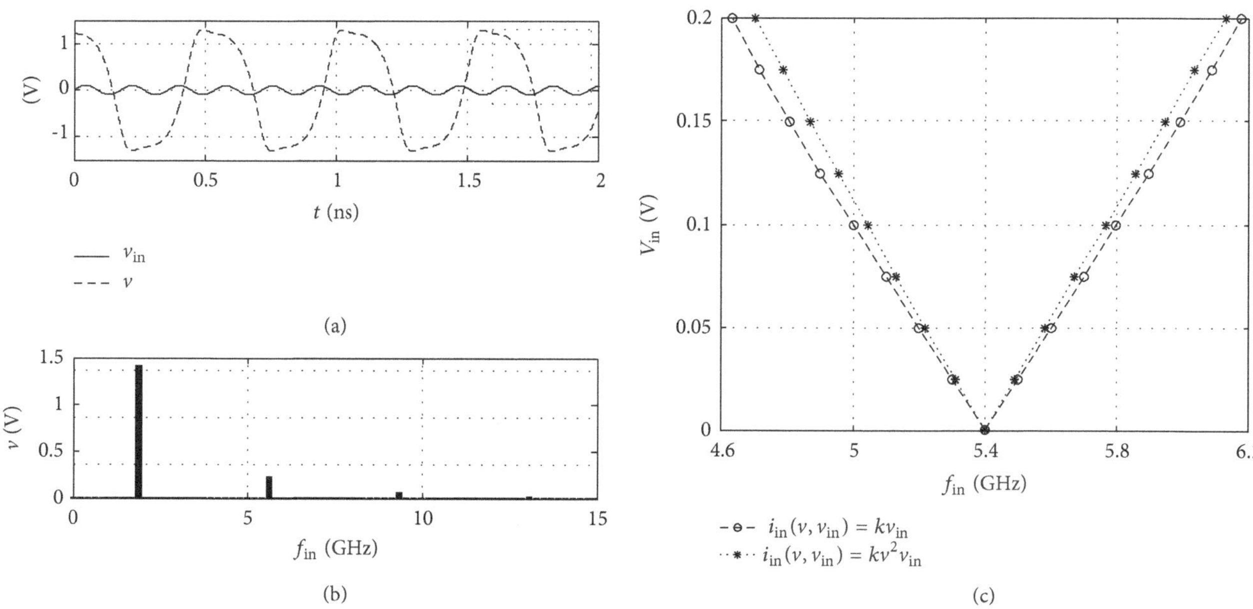

FIGURE 4: SPICE simulations of the circuit in Figure 3(b) with $i_{nl}(v) = a_1 v + a_3 v^3$, $i_{in}(v, v_{in}) = kv_{in}$, $a_1 = -12\,\text{mA/V}$, $a_3 = 8\,\text{mA/V}^2$, $k = -32\,\text{mA/V}^2$, $L = 10\,\text{nH}$, $R = 905\,\Omega$, $C = 635.45\,\text{fF}$, $V_{in} = 0.1\,\text{V}$, $f_{in} = 5.6\,\text{GHz}$. (a) Time waveforms of the input and output signals. (b) Frequency spectrum of the output signal. (c) Arnold tongues obtained by simulating the circuit in Figure 3(b) with $i_{in}(v, v_{in}) = kv_{in}$ and with $i_{in}(v, v_{in}) = kv^2 v_{in}$.

the output signal, v, has a fundamental component equal to one-third of the input frequency, and, thus, the circuit operates as a divide-by-three ILFD. Its frequency spectrum in Figure 4(b) confirms the presence of a third harmonic in the output signal, that is a component at frequency $f_{in} = 3f$ in addition to its main component at frequency f.

Model in Figure 3(b) gives us also the possibility of easily comparing the case when the injection nonlinearity includes a mixing nonlinearity and the case when a mixing is not present. To this end, we calculated the locking ranges for the divide-by-three mode of operation of the equivalent circuit in Figure 3(b) with $i_{in}(v, v_{in}) = kv_{in}$ (corresponding to the block diagram in Figure 2), and then with $i_{in}(v, v_{in}) = kv^2 v_{in}$ (corresponding to the block diagram in Figure 1(c)). The results reported in Figure 4(c) not only show that both circuits can operate as a divide-by-three ILFD but also using the same value of k, the locking range of the circuit without mixing, that is, $i_{in}(v, v_{in}) = kv_{in}$, is wider than the other one. Thus, we can conclude that dividers can be efficiently realized without performing a mixing between the input and the output signals.

3. Circuit Analysis and Design

The divide-by-three circuit in Figure 5 performs the frequency division without involving a mixing between the input and output signals, but is characterized by the direct application of the forcing signal on the tank of the LC oscillator. The circuit is formed by the parallel connection of an LC-tank, an active circuit consisting of two complementary pairs of cross-coupled devices (Figure 6(a)) and an injection circuit consisting of two complementary MOS switches (Figure 6(b)). Thus, the current i of the LC-tank consists of a component $i_{nl}(v)$ due to the active part and a component i_{in} due to the injection circuit. In general, the current i_{in} depends not only on the synchronization signal v_{in} but also on the voltage v and thus is written in the form $i_{in}(v, v_{in})$.

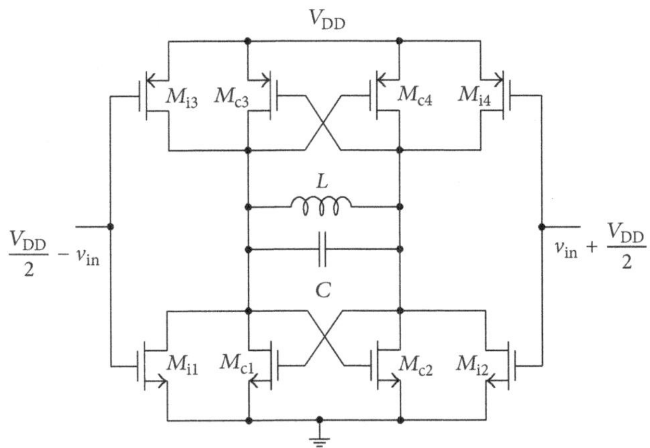

FIGURE 5: Circuit diagram of a divide-by-three LC-CMOS frequency divider with a direct forcing signal.

Analytical expressions of the functions $i_{nl}(v)$ and $i_{in}(v, v_{in})$ can be determined by assuming that the ILFD operates under a small-injection regime, so that the current $i(v, v_{in}) = i_{nl}(v) + i_{in}(v, v_{in})$ applied to the LC-tank can be written as $i(v, v_{in}) = i(v, 0) + g_1(v)v_{in}$ with $i(v, 0) = i_{nl}(v) + i_{in}(v, 0)$ and $g_1(v) = \partial i/\partial v_{in}|_{v_{in}=0}$. SPICE simulations of the circuits in Figure 6 for $v_{in} = 0$ show that the current $i(v, 0)$ can be well approximated by a cubic polynomial $i(v, 0) = a_1 v + a_3 v^3$ (Figure 7(a)) [16, 17]. On the other hand, the function $g_1(v)$, calculated through a finite difference approximation, can be modeled by a constant function $g_1(v) = k$ (Figure 7(b)). Thus, the current $i(v, v_{in})$ applied to the LC-tank can be written as

$$i(v, v_{in}) = a_1 v + a_3 v^3 + k v_{in} \tag{1}$$

and the circuit in Figure 5 is amenable to the block diagram in Figure 8, where $H(s)$ is the LC-tank transfer function, and to the equivalent circuit in Figure 3(b). As the block diagram in Figure 8 is reducible to the diagram in Figure 2, we conclude that the circuit in Figure 6 is able to operate as a divide-by-three ILFD.

The main design parameter of a frequency divider is the locking range, which can be analytically related to the circuit parameters as shown below. The circuit in Figure 3(b) is described by

$$\frac{d^2 v}{dt^2} + \frac{1}{RC}\frac{dv}{dt} + \omega_0^2 v = -\frac{1}{C}\frac{d}{dt} i(v, v_{in}), \tag{2}$$

where $\omega_0 = 1/\sqrt{LC}$ is the resonant frequency of the LC-tank. Assuming that the circuit operates as a divide-by-three ILFD under the action of the signal $v_{in} = V_{in}\cos(3\omega t)$, we seek a solution of (2) in the form $v = w + h$ where $w = W\cos(\omega t + \theta)$ is the first harmonic of v and h the sum of the remaining harmonics. Also the current $i(v, v_{in})$ in the LC-tank is decomposed into a component equal to its first harmonic and a component equal to the sum of the remaining harmonics, that is,

$$i(v, v_{in}) = i_1(w + h, v_{in}) + i_h(w + h, v_{in}). \tag{3}$$

Taking into account that the output harmonics of the nonlinearity $i(v, v_{in})$ depend slightly on the harmonics h of the voltage v, that is, $i_h(w + h, v_{in}) \cong i_h(w, v_{in})$, by virtue of (1) we obtain

$$i_h(w, v_{in}) = kV_{in}\cos(3\omega t) + \frac{1}{4}a_3 W^3\cos(3\omega t + 3\theta). \tag{4}$$

Note that i and v represent the input and the output quantities, respectively, of the band-pass filter in the model in Figure 2, which are decomposed in the fundamental component at frequency $\omega = 2\pi f = 2\pi(f_{in} - 2f)$ and in the harmonic components. Equation (4) shows that the harmonic component of the input quantity of the filter has frequency $3\omega = 2\pi f_{in}$, as shown in Figure 2.

By virtue of (3), (2) can be rewritten in the form

$$\left(\frac{d^2}{dt^2} + \frac{1}{RC}\frac{d}{dt} + \omega_0^2\right)(w + h) = -\frac{1}{C}\frac{d}{dt} i_1(w + h, v_{in}) - \frac{1}{C}\frac{d}{dt} i_h(w, v_{in}) \tag{5}$$

and can be decomposed into the following two equations for the fundamental and higher harmonics

$$\left(\frac{d^2}{dt^2} + \frac{1}{RC}\frac{d}{dt} + \omega_0^2\right) w = -\frac{1}{C}\frac{d}{dt} i_1(w + h, v_{in}), \tag{6}$$

$$\left(\frac{d^2}{dt^2} + \frac{1}{RC}\frac{d}{dt} + \omega_0^2\right) h = -\frac{1}{C}\frac{d}{dt} i_h(w, v_{in}). \tag{7}$$

We will first solve (7) for h and, then, exploiting the solution of (7), we will solve (6) for w. Thus, taking into account that the first term in (7) is greater than the second and third one, that is, $9\omega^2 \gg \omega_0^2 \approx \omega^2$, (7) can be reduced to

$$\frac{d}{dt}h = -\frac{1}{C} i_h(w, v_{in}) \tag{8}$$

and its solution provides the harmonics of the output voltage, that is,

$$h = -\frac{kV_{in}}{3C\omega}\sin(3\omega t) - \frac{a_3 W^3}{12C\omega}\sin(3\omega t + 3\theta). \tag{9}$$

Note that, taking into account that $v = w + h$ with $w = W\cos(\omega t + \theta)$, (9) implies that the output voltage contains a component at frequency $\omega = 2\pi f$ and a component at frequency $3\omega = 2\pi f_{in}$, as represented in the block diagram in Figure 2.

By using (9) in (1), the expression of the current applied to the LC-tank is obtained:

$$i(w + h, v_{in}) = a_1\left(W\cos(\omega t) - \frac{kV_{in}}{3C\omega}\sin(3\omega t) - \frac{a_3 W^3}{12C\omega}\sin(3\omega t + 3\theta)\right)$$

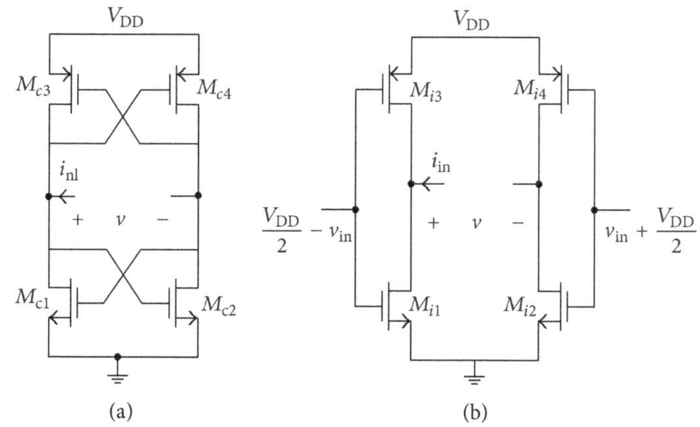

FIGURE 6: (a) Active part of the ILFD in Figure 5; (b) injection circuit of the ILFD in Figure 5.

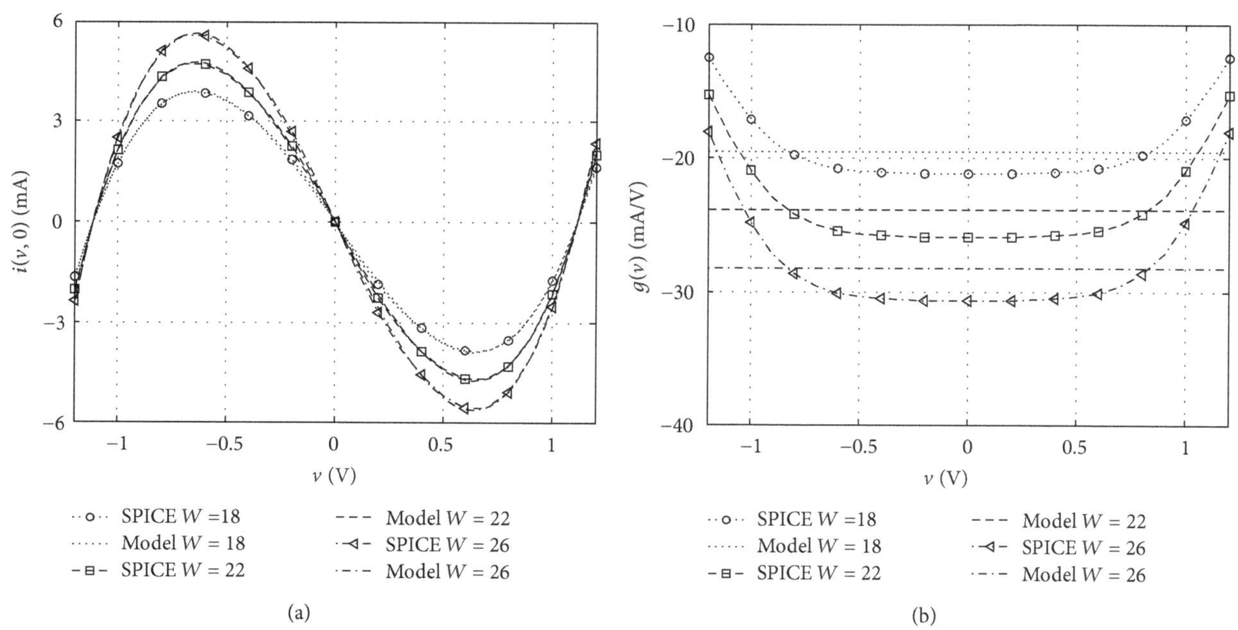

FIGURE 7: Current-voltage characteristics of the circuits in Figure 6. (a) Total current $i(v, 0) = i_{nl}(v) + i_{in}(v, 0)$ for $v_{in} = 0$. (b) Voltage-controlled transconductance $g_1(v)$. SPICE simulations are obtained by using BSIM3 transistor models (MOSIS-IBM 8RF 0.13 μ technology), with $V_{DD} = 1.2$ V. Analytical models are obtained by using polynomial approximations $i(v, 0) = a_1 v + a_3 v^3$ and $g_1(v) = k$.

$$+ a_3 \left(W \cos(\omega t) - \frac{kV_{in}}{3C\omega} \sin(3\omega t) - \frac{a_3 W^3}{12C\omega} \times \sin(3\omega t + 3\theta) \right)^3 + kV_{in} \cos(3\omega t). \tag{10}$$

Expression (10) allows us to calculate the fundamental harmonic of $i(w + h, v_{in})$, that is $i_1(w + h, v_{in})$, and in particular, the coefficients of cosine and sine Fourier components of the fundamental harmonic, that is,

$$I^c = \frac{1}{\pi} \int_0^{2\pi} i(w + h, v_{in}) \cos(\alpha)\, d\alpha,$$
$$I^s = \frac{1}{\pi} \int_0^{2\pi} i(w + h, v_{in}) \sin(\alpha)\, d\alpha \tag{11}$$

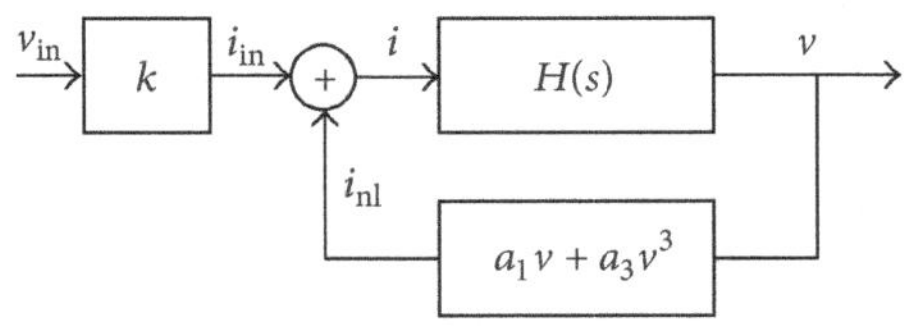

FIGURE 8: Block diagram of the injection locked frequency divider in Figure 5.

with $\alpha = \omega t + \theta$. It results in that

$$I^c = a_1 W \left(1 + \frac{a_3 k^2 V_{in}^2}{6 a_1 C^2 \omega^2}\right) + \frac{3}{4} a_3 W^3 \left(1 + 2\left(\frac{a_3 W^2}{12 C\omega}\right)^2\right) + \frac{a_3 W^2}{4C\omega} k V_{in} \left(\sin(3\theta) + 4\left(\frac{a_3 W^2}{12 C\omega}\right)\cos(3\theta)\right),$$

$$I^s = -\frac{a_3^2 W^5}{16 C\omega} - \frac{a_3 W^2}{4 C\omega} k V_{in} \cos(3\theta). \quad (12)$$

Taking into account that the third harmonic of the output voltage is quite smaller than the fundamental, that is $a_3 W^3/(12C\omega) \ll W$, it is possible to simplify (12), obtaining

$$I^c = a_1 W \left(1 + \frac{a_3 k^2 V_{in}^2}{6 a_1 C^2 \omega^2}\right) + \frac{3}{4} a_3 W^3 + \frac{a_3 W^2}{4C\omega} \times k V_{in} \sin(3\theta),$$

$$I^s = -\frac{a_3^2 W^5}{16 C\omega} - \frac{a_3 W^2}{4 C\omega} k V_{in} \cos(3\theta). \quad (13)$$

Once obtained the expression of $i_1(w + h, v_{in})$ in explicit form, or equivalently of I^c and I^s, (6) can be solved by using the averaging method [11, 18] and the following equations for the locked states are obtained:

$$W = -RI^c,$$
$$\omega = \omega_0 + \frac{1}{2CW} I^s. \quad (14)$$

Equation (13) reduces (14) to

$$W = -RI^c, \quad (15)$$

$$\omega = \omega_{fr} - \omega_0 \frac{a_3 W L}{8C} k V_{in} \cos(3\theta), \quad (16)$$

where

$$\omega_{fr} = \omega_0 - \omega_0 \frac{a_3^2 W^4 L}{32C}. \quad (17)$$

It is interesting to observe that if $V_{in} = 0$, that is, if the circuit operates as a free-running oscillator, (16) predicts that the output oscillation frequency ω is different from the tank resonant frequency ω_0 and it is equal to ω_{fr}. This is a consequence of the known nonlinear effect of harmonics

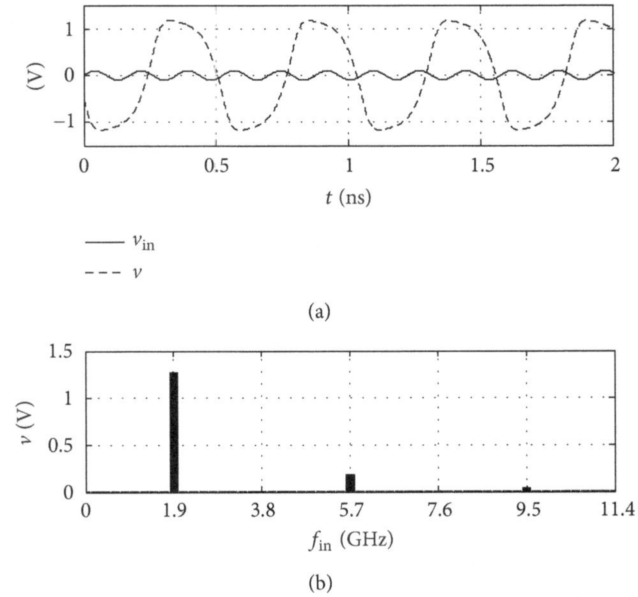

FIGURE 9: SPICE simulation of the circuit in Figure 5 obtained by using BSIM3 transistor models in MOSIS-IBM 8RF 0.13 μ technology and $W/L_c = W/L_i = 24/0.12$, $W/L_{PMOS} = 4.5W/L_{NMOS}$, $V_{DD} = 1.2$ V, $L = 10$ nH, $R = 905\,\Omega$. (a) Voltages at the input and output terminals. (b) Frequency spectrum of the output voltage.

on the oscillation frequency [19]. Moreover, (16) highlights an unknown aspect of the frequency entrainment process, namely, that the action of the external signal on the LC oscillator manifests itself by a frequency deviation from the actual frequency of the free-running oscillation. Note also that (17) can be used to estimate the equivalent capacitance C, which includes the device parasitics, starting from the free-running frequency.

Assuming that the frequency shift due to the harmonics does not significantly change in presence of the input signal, that is, assuming ω_{fr} as a constant in (16), we can solve (15) and (16) for W and θ in explicit form, obtaining

$$W = \sqrt{\frac{4(-aR-1)}{3Rc} - \frac{1}{6}\left(\frac{rV_{in}}{C\omega_0}\right)^2 - \frac{rV_{in}}{18C\omega_0}\sqrt{\frac{48(-aR-1)}{Rc} - q}}, \quad (18)$$

$$\theta = \frac{1}{3} \arccos\left(\frac{8C^2\omega_0}{cWrV_{in}}(\omega_{fr} - \omega)\right), \quad (19)$$

with

$$q = 7\left(\frac{rV_{in}}{C\omega_0}\right)^2 + \left(\frac{48C^2\omega_0}{crV_{in}}(\omega_{fr} - \omega)\right)^2. \quad (20)$$

The values of ω for which a solution (W, θ) does exist are the values of ω for which the locking occurs. The minimum

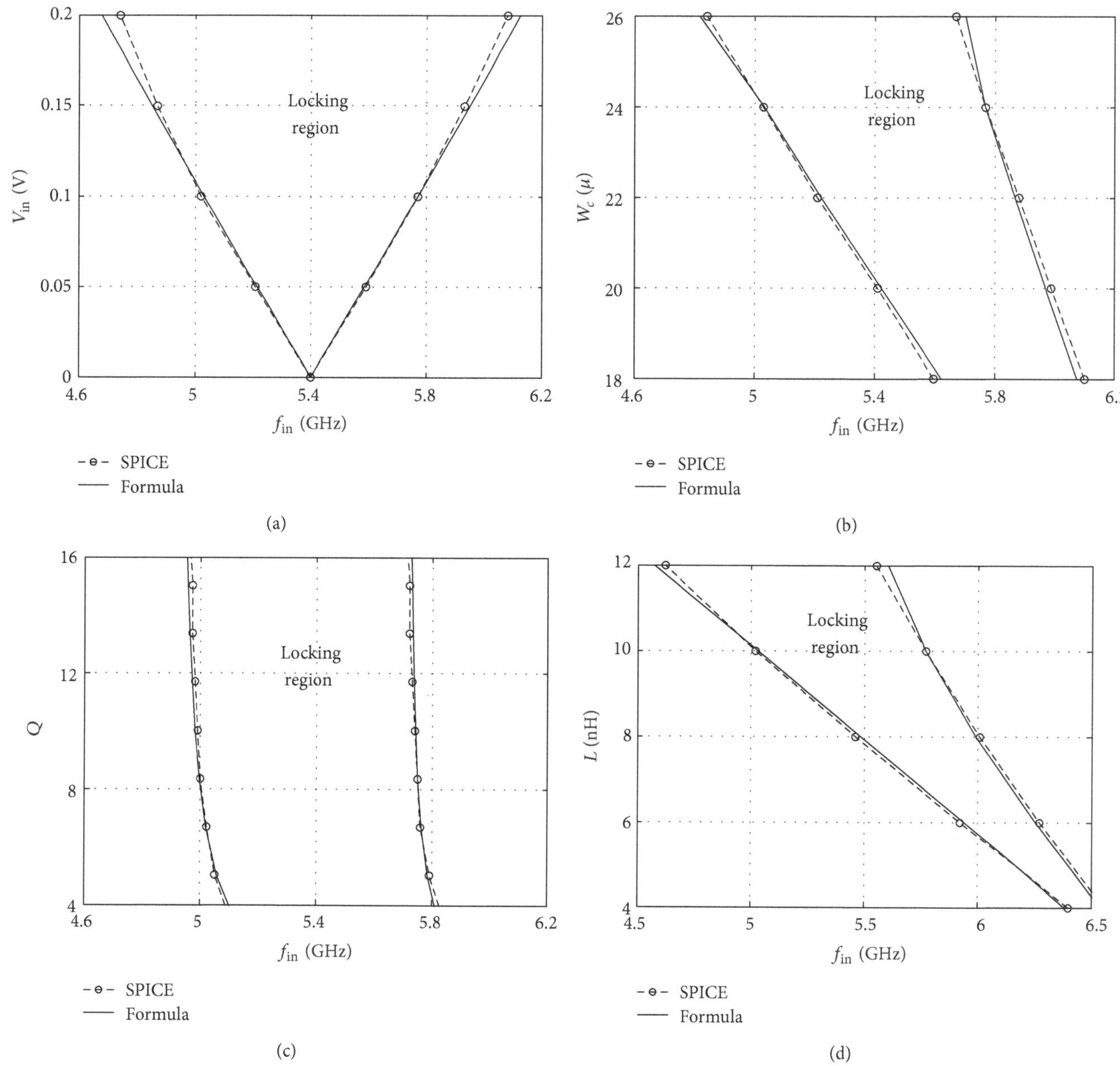

FIGURE 10: Locking regions for the divide-by-three ILFD in Figure 5, obtained by SPICE simulations (lines with circles) and by using analytical equations (lines without circles). For SPICE simulations, BSIM3 transistor models in MOSIS-IBM 8RF 0.13 μ technology with $V_{DD} = 1.2\,V$ have been used. (a) Locking range as a function of the input amplitude ($L = 10\,nH$, $R = 905\,\Omega$). (b) Locking range as a function of the transistor sizes ($W/L_c = W/L_i$, $L = 10\,nH$, $R = 905\,\Omega$). (c) Locking range as a function of the tank resistance ($W/L_c = W/L_i = 24/0.12$, $W/L_{PMOS} = 4.5W/L_{NMOS}$). (c) Locking range as a function of the L/C ratio keeping the LC product constant ($LC = 10\,nH\ 446.6\,fF$).

and the maximum values of ω define the output-referred locking range and are equal to

$$\omega_{min} = \omega_{fr} - \omega_0 \frac{kV_{in}Lr\sqrt{a_3\left|a_1 + R^{-1}\right|}}{4\sqrt{3}C},$$
$$\omega_{max} = \omega_{fr} + \omega_0 \frac{kV_{in}Lr\sqrt{a_3\left|a_1 + R^{-1}\right|}}{4\sqrt{3}C}, \quad (21)$$

where

$$r = \sqrt{1 - \frac{21La_3Rk^2V_{in}^2}{144C\left|a_1R + 1\right|}}. \quad (22)$$

The input-referred locking range is equal to the output-referred locking range multiplied by the division ratio $n = 3$. Taking into account that $r \approx 1$, (21) provides a very simple expression of the locking range as a function of the active and passive circuit parameters, which can be usefully employed for the design of ILFD with direct forcing signal. Note that all

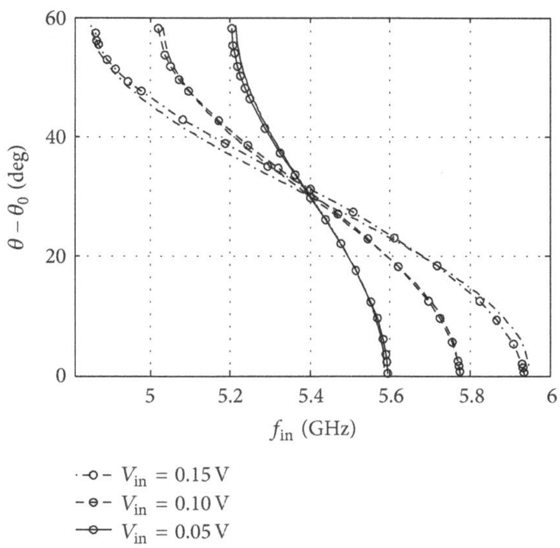

FIGURE 11: Phase difference between the input and output signals for the divide-by-three ILFD in Figure 5 with $L = 10$ nH and $R = 905\ \Omega$. SPICE simulations are obtained by using BSIM3 transistor models in MOSIS-IBM 8RF 0.13 μ technology with $V_{DD} = 1.2$ V, $W/L_c = W/L_i = 24/0.12$, $W/L_{PMOS} = 4.5W/L_{NMOS}$. Analytical results are obtained by using (19).

the formulas based on the mixing principle predicts that the locking range is zero in this case [11–13].

4. Circuit Simulation

SPICE simulations of the circuit in Figure 5 were performed by using BSIM3 models of active devices in 0.13 μm technology. Time waveforms of the input and output voltages, reported in Figure 9(a), demonstrated that the circuit under study can actually operate as a divide-by-three ILFD.

Moreover, SPICE simulations were performed to investigate the dependence of the locking range on the circuit parameters and to determine the accuracy of analytical expressions deduced in previous section. The locking region in the (f_{in}, V_{in}) plane is reported in Figure 10(a) and the modification of the locking region as a function of the transistor sizes of the tank quality factor and of the inductor values is reported in Figures 10(b), 10(c), and 10(d), respectively. The comparison of the SPICE results with the results obtained by analytical expression (21), also reported in Figure 10, shows a good accuracy of (21), which can be effectively used to design the ILFD in Figure 5.

It is interesting to observe that, according to SPICE simulations and to the prediction of (21), the width of the frequency range of locking increases with the transistor size, which is proportional to a_1 and a_3, but the center of the locking range moves toward lower frequencies due to the parasitic capacitances (Figure 10(b)). Moreover, the width of the locking range does not depend on the quality factor of the tank (Figure 10(c)) and, thus, it is not necessary to keep low the quality factor in order to widen the locking range as it happens in other ILFD topologies. Finally, we note that the increasing of the L/C ratio for a constant LC product (Figure 10(d)) widens the locking range but shifts it toward lower frequencies due to nonlinear effects, as predicted by (17).

Finally, the phase relationship between the input and output signals was investigated. SPICE simulations reported in Figure 11 show that a wide variation is present when the input frequency varies within the locking range. Analytical results obtained by applying (19) and reported in Figure 11 show that presented formula provides an acceptable approximation to the numerical results. However, formulas are not able to predict a constant phase offset $\theta_0 = 7^\circ$, that is, due to the reactive behavior of nonlinearities, which were here assumed to be memoryless. Thus, we conclude that, even if small discrepancies are present between numerical and analytical results, formulas are able to capture the main characteristics of the direct forcing injection, providing fairly accurate approximation to numerical results.

5. Conclusions

We have shown that in the injection-locked dividers wherein the external synchronization signal is applied directly to the LC-tank, the synchronization mechanism takes place according to a process that cannot be explained by the simple mixing between the input and output signals. This is the case of some divide-by-three circuits, of which we analyzed in detail a particular implementation. Then, a method for the approximate calculation of oscillation in the synchronous states was exposed that allows us to determine the amplitude and the phase of the oscillation and the interval of locking in explicit form as a function of the circuit parameters. These expressions, which are relatively simple and useful for design, provide sufficiently accurate results as shown by numerical simulations.

References

[1] J. Lee and B. Razavi, "A 40-GHz frequency divider in 0.18-μm CMOS technology," in *Proceedings of Symposium on VLSI Circuits*, pp. 259–262, June 2003.

[2] R. L. Miller, "Fractional-frequency generators utilizing regenerative modulation," *Proceedings of the IRE*, vol. 27, pp. 446–457, 1939.

[3] T. Shibasaki, H. Tamura, K. Kanda, H. Yamaguchi, J. Ogawa, and T. Kuroda, "20-GHz quadrature injection-locked LC dividers with enhanced locking range," *IEEE Journal of Solid-State Circuits*, vol. 43, no. 3, pp. 610–618, 2008.

[4] C.-F. Liang, S.-L. Liu, Y.-H. Chen, T.-Y. Yang, and G.-K. Ma, "A 14-band frequency synthesizer for MB-OFDM UWB application," in *Proceedings of IEEE International Solid-State Circuits Conference (ISSCC '06)*, pp. 113–126, February 2006.

[5] S. Saeedi, M. Atarodi, and M. S. Bakhtiar, "A divide-by-3 frequency divider for I/Q generation in a multi-band frequency synthesizer," in *Proceedings of IEEE Asia Pacific Conference on Circuits and Systems (APCCAS '08)*, pp. 1383–1386, December 2008.

[6] V. Jain, B. Javid, and P. Heydari, "A BiCMOS dual-band millimeter-wave frequency synthesizer for automotive radars,"

IEEE Journal of Solid-State Circuits, vol. 44, no. 8, pp. 2100–2113, 2009.

[7] T.-N. Luo, S.-Y. Bai, and Y.-J. E. Chen, "A 60-GHz 0.13-μm CMOS divide-by-three frequency divider," *IEEE Transactions on Microwave Theory and Techniques*, vol. 56, no. 11, pp. 2409–2415, 2008.

[8] Y.-T. Chen, M.-W. Li, H.-C. Kuo, T.-H. Huang, and H.-R. Chuang, "Low-voltage K -band divide-by-3 injection-locked frequency divider with floating-source differential injector," *IEEE Transactions on Microwave Theory and Techniques*, vol. 60, no. 1, pp. 60–67, 2012.

[9] I.-T. Lee, C.-H. Wang, and S.-I. Liu, "3.6mW D-band divide-by-3 injection-locked frequency dividers in 65nm CMOS," in *Proceedings of the 7th IEEE Asian Solid-State Circuits Conference (A-SSCC '11)*, pp. 93–96, November 2011.

[10] A. Buonomo and A. Lo Schiavo, "A deep investigation of the synchronization mechanisms in LC-CMOS frequency dividers," *IEEE Transactions on Circuits and Systems-I: Regular Papers*, 2013.

[11] A. Buonomo and A. Lo Schiavo, "Analytical approach to the study of injection-locked frequency dividers," *IEEE Transactions on Circuits and Systems-I: Regular Papers*, vol. 60, no. 1, pp. 51–62, 2013.

[12] S. Verma, H. R. Rategh, and T. H. Lee, "A unified model for injection-locked frequency dividers," *IEEE Journal of Solid-State Circuits*, vol. 38, no. 6, pp. 1015–1027, 2003.

[13] A. Buonomo and A. Lo Schiavo, "Nonlinear dynamics of divide-by-two injection-locked frequency dividers in locked operation mode," *International Journal of Circuit Theory and Applications*, 2013.

[14] A. Amann, M. P. Mortell, E. P. O'Reilly, M. Quinlan, and D. Rachinskii, "Mechanism of synchronization in frequency dividers," *IEEE Transactions on Circuits and Systems I: Regular Papers*, vol. 56, no. 1, pp. 190–199, 2009.

[15] S. Daneshgar, O. de Feo, and M. P. Kennedy, "Observations concerning the locking range in a complementary differential LC injection-locked frequency divider-part II: design methodology," *IEEE Transactions on Circuits and Systems I: Regular Papers*, vol. 58, no. 4, pp. 765–776, 2011.

[16] A. Buonomo and A. Lo Schiavo, "Finding the tuning curve of a CMOS—LC VCO," *IEEE Transactions on Circuits and Systems II: Express Briefs*, vol. 55, no. 9, pp. 887–891, 2008.

[17] A. Buonomo and A. Lo Schiavo, "On the theory of quadrature oscillations obtained through parallel LC VCOs," *IEEE Transactions on Circuits and Systems I: Regular Papers*, vol. 57, no. 9, pp. 2509–2519, 2010.

[18] A. H. Nayfeh and D. T. Mook, *Nonlinear Oscillations*, John Wiley & Sons, New York, NY, USA, 1979.

[19] A. Buonomo and A. Lo Schiavo, "A method for analysing the transient and the steady-state oscillations in third-order oscillators with shifting bias," *International Journal of Circuit Theory and Applications*, vol. 29, no. 5, pp. 469–486, 2001.

Noise Radar Technology as an Interference Prevention Method

Gaspare Galati and Gabriele Pavan

Department of Electronic Engineering, Tor Vergata University, Via del Politecnico, 1-00133 Rome, Italy

Correspondence should be addressed to Gaspare Galati; gaspare.galati@uniroma2.it

Academic Editor: Sandra Costanzo

In some applications, such as automotive and marine/navigation, hundreds of radars may operate in a small environment (e.g., a road complex or a strait) and in an allocated frequency band with limited width. Therefore, a compatibility problem between different radars arises that is not easily solved by time, frequency, space, or polarization diversity. The advent of fast digital signal processing and signal generation techniques makes it possible to use waveform diversity to solve this problem that will be exacerbated in the next future. Ideal waveforms for the diversity are supplied by Noise Radar Technology (NRT), whose application is promising in some military applications as well as in the civilian applications considered in this paper. In addition to being orthogonal as much as possible, the random signals to be transmitted have to satisfy requirements concerning side lobe level and crest factor, calling for novel, original design and generation processes.

1. Introduction

In spite of its military origin, dated back to the 1930s, radar technology is being more and more applied in the civilian domain for the safety of air, sea, and land traffic. In sea and road applications, this technology is entering the mass market with a significant yearly reduction of costs, mainly due to the larger and larger integration of the hardware and to the scale economy. For example in the automotive radar the factory cost has fallen by one order of magnitude (from hundreds to tens of dollars) in a few years. A similar situation is to be expected in the field of marine radar [1, 2], present in most vessels, as they (X-band radar) are mandatory for all passenger vessels and for all vessels above 500 tons GT (Gross Tonnage), with a double-band requirement (X-band radar + S-band radar) above 3000 tons GT. The allocated bands are 9.3 to 9.5 GHz (X) and 2.9 to 3.1 GHz. Most marine radars use the simple and cheap magnetron transmitter, a 1940s technology, generating short pulses (from 50–80 ns in the short-range mode to 800–1000 ns in the long-range mode, up to 96 NM) and peak power from a very few kW to 12.5–25 kW (and 50 kW in coastal applications for Vessel Traffic Systems).

The large number of marine radars in a given area, simultaneously present above the horizon, makes the mutual interference a very likely event. For a general overview of potential solution to this problem, see [3, 4]. In marine radars the most used solution is based on their relatively short pulses that can be suppressed (video blanking or thresholding) because they are received asynchronously with respect to the transmission from the victim (the own) radar. Such a suppression causes tolerable detection losses for small targets and negligible losses for large targets.

As in many other radar applications, for example, in Air Traffic Control (ATC), solid-state transmitter technology is attractive for marine radars. It permits (i) the implementation of highly stable, coherent Rx-Tx chains compatible with Doppler processing, (ii) the transmission of sophisticated waveforms [5], (iii) a better availability, and (iv) the absence of high voltages. Unlike ATC radars that have been using solid state transmitters for over two decades, solid-state marine radars are appearing on the market just now, most of them, in the S-band. While maintaining approximately the same average transmitted power as magnetron transmitters, these new (or experimental) solid-state marine radars have three orders of magnitude larger duty cycle, that is, in the range of 10%–12%. These long pulses (about 10 or 20 microseconds, in some cases up to 100 microseconds) are not compatible with the current interference suppression techniques. In the remaining part of this paper a new technique is proposed to solve this interference problem, whose general framework is synopsized in the appendix.

2. Signal Requirements and Advantages of Noise Radar Technology (NRT)

Mitigation techniques for interferences from similar radars can be based on diversity in (i) frequency, (ii) polarization, (iii) direction (antenna pattern adaptation), and (iv) waveform. In the marine radar context, as considered here, the limited allocated bandwidth (only 200 MHz in X-band and the same in the S-band) and the use of standard antennae with fixed polarization prevent the use of (i), (ii), and (iii). Therefore a possible solution has to be based on waveform diversity, that is, the transmission (by the various radars in potential interference conditions) of pairwise orthogonal waveforms, such that the matched filter output is mainly due to the "own" waveform, with a noise floor due to the "other radars" waveforms. The large number of radars in busy maritime environments such as straits and sea areas around harbors, makes it necessary to exploit a large number of orthogonal waveforms, which can only be obtained by polyphase codes [3] or, as proposed here, by suitable processing of samples of random processes, leading to Noise Radar Technology.

The main requirements for a pair of signals with complex envelope $s_i(t)$ and $s_j(t)$ with $i, j = 1, \dots, M$, pulsewidth T, and the same power are defined by the following:

(i) Peak Side Lobe Ratio (< −30 dB):

$$\text{PSLR} = \frac{\max_i(s_i)}{\max_k(m_k)}, \quad (1)$$

where s_i = side lobe samples and m_k = main lobe samples.

(ii) Crest Factor (C) or Peak-to-Average Ratio (PAR), that is, the peak amplitude of the waveform divided by the rms value of the waveform $s(t)$:

$$C = \frac{\max(|s(t)|)}{\sqrt{(1/T)\int_0^T |s(t)|^2 dt}}. \quad (2)$$

(iii) Mean Envelope-to-Peak Power Ratio:

$$\text{MEPPR} = \frac{(1/T)\int_0^T |s(t)|^2 dt}{\max\left(|s(t)|^2\right)}. \quad (3)$$

It results in EPPR $= 1/C^2$.

(iv) Normalized cross-correlation:

$$r_{ij}(t) = \frac{\left|R_{ij}(t)\right|}{\left|R_{i(j)}(0)\right|}, \quad (4)$$

where $R_{ij}(t) = \int s_i^*(\theta)s_j(t+\theta)d\theta$, $i \neq j$; it measures the orthogonality, and the desired value is $r_{ij}(t) < -30$ dB.

(v) Spectral band occupancy: sometimes this item is overlooked, especially when noise-like waveforms are concerned, but it is of paramount importance in most real-world radars; see [4] and the appendix.

Good candidate deterministic signals that satisfy the orthogonality requirements are the well-known "up" and "down" chirps (Linear-FM and Non-LFM) [3, 5], but in this case only pairs of signals (not the needed multiplicity M) can be obtained. To obtain M signals Costas codes, All top sequences can be considered, and polyphase codes represent a possible solution [3, 6, 7]. More recent research on orthogonal signals proposed the use of normal or interleaved OFDM techniques [8]. The main limitation of the OFDM approach is due to the nonconstant envelope of the signals, that is, MEPPR < 1, which means that the transmitter does not work at its maximum power.

Another class of waveforms, that is, the phase noise signals [9], has two main advantages as compared to the signals introduced before. The former is the possibility to generate a large number of orthogonal signals, which is of great importance in the present case. The latter, applicable to defense applications (e.g., to the navigation radars of military vessels), is about the detectability; in fact they are random signals so they limit the detection, the identification, and the eventual spoofing of the signal. Such a limitation is of great importance in many military applications which require low detectability of the active system. Finally the MEPPR (and the crest factor) can be very close to the unity.

3. Methods for Generating Noise Radar Signals

For a phase noise signal the complex envelope can be written as

$$s(t) = A \cdot \exp\{j\varphi(t)\} \cdot \text{rect}_T(t), \quad (5)$$

where A is the constant amplitude, $\text{rect}_T(t)$ is 0 outside the interval $[-T/2, +T/2]$ and 1 inside it (with T being the pulse length), and $\varphi(t)$ is the phase process modulating the noise signal $s(t)$. In the following we present three methods to generate the phase noise signals highlighting their strengths and weaknesses.

3.1. Random Phase Modulation. In [9] Axelsson supposed for $\varphi(t)$ a zero-mean Gaussian process with root mean square (rms) σ and a given power with density spectrum within the band b. He showed that the normalized autocorrelation function of the signal $s(t)$ can be written in a closed-form expression as

$$R(\tau) = \exp\left\{-\sigma^2\left[1 - \rho(t)\right]\right\}, \quad (6)$$

where $\rho(\tau)$ is the correlation coefficient of $\varphi(t)$. For example, $\rho(\tau) = \sin(\pi b\tau)/(\pi b\tau)$ for a constant spectrum within the band b. Of course, $R(\tau)$ depends on the bandwidth b, on the pulse length T, and on the rms phase fluctuation σ. The bandwidth b is related to the width of the main peak and defines the range resolution. An increase of T, and

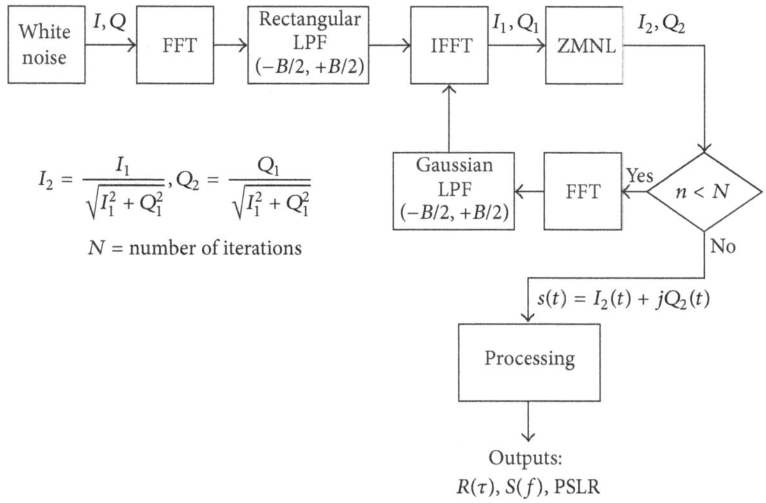

FIGURE 1: Block diagram of the iterative algorithm.

consequently of the compression ratio (the time-bandwidth product of the generated signal), causes a reduction in the range side lobe level, whereas the main lobe width remains fixed being independent of T. Finally the rms σ has two different effects. The former is on the side lobe level: an increase of σ causes a decrease of the side lobe level with improvement in the PSLR. The latter concerns the resolution. The rms value in fact establishes a connection between the bandwidth of the modulated signal and the bandwidth of the modulating signal. That is, when σ increases the final bandwidth increases too. In [9] a simple relation between the rms bandwidth of the phase modulated signal (B_{rms}) and the rms bandwidth of the phase modulating noise (b_{rms}) has been found:

$$B_{\text{rms}} = \sigma \cdot b_{\text{rms}}. \quad (7)$$

On the other hand, concerning the side lobe suppression, the expression of the autocorrelation function introduced in [9] would show a progressive improvement of the side lobe suppression as σ increases. However the periodic nature of $\varphi(t)$ with a folding in the $[-\pi, +\pi]$ interval has been neglected in [9], and in reality the model can be used only for values of σ significantly smaller than π. The Gaussian noise, used to modulate the signal phase, is to be compared with an *uniform* distribution in the range $[-\pi, +\pi]$ with a standard deviation of $(\pi/\sqrt{3}) \cong 1.8$ rad. Therefore, if σ is large ($\sigma > (\pi/\sqrt{3})$), the resultant phase is not Gaussian distributed and the mathematical formulation introduced in [9] does not apply. On the other hand, in connection with a potential real application, it would be better to generate the signal through a white Gaussian process with its *in phase* and *in quadrature* components (I, Q) that are band limited as desired. This is described in the following.

3.2. An Iterative Algorithm to Generate Phase Noise Signals. To control the spectral width and to reduce the Peak Side Lobe Ratio (PSLR) of the generated phase noise signals, we propose an iterative algorithm based on alternative projections in frequency and in time domain (Figure 1). The input to the algorithm is a zero-mean white complex Gaussian process $(I + jQ)$. The filtering is implemented in frequency domain. First the input is band limited using a rectangular Low Pass Filter of band B. The output from this filter is projected in time domain by IFFT, and the complex signal (I_1, Q_1) is amplitude limited using a Zero-Memory-Non-Linearity (ZMNL). The output in time (I_2, Q_2) is reprojected in frequency and filtered through a Gaussian LPF of band B. The output is again re-projected in time, and the procedure is iterated for a predefined value of N. Typically to achieve the optimum (minimum) PSLR, some tens of iterations are needed (it depends on the input white noise sequence).

Considering for ZMNL a hard limiter, that is,

$$I_2 = \frac{I_1}{\sqrt{I_1^2 + Q_1^2}}, \qquad Q_2 = \frac{Q_1}{\sqrt{I_1^2 + Q_1^2}}, \quad (8)$$

the signal (I_2, Q_2) shows a unitary amplitude and its randomness is only due to the random phase of the signal. Averaging different autocorrelations we have computed the mean normalized autocorrelation and its standard deviation (Figure 2 for BT = 10000).

Varying the compression ratio up to 30000, in Figure 3 the mean PSLR is shown.

To reach PSLR lower than −30 dB, compression ratios over 5000 are needed. However for BT = 30000 the PSLR is −36 dB, less, for example, than the one obtained with Non-LFM using Hamming weighting (−42.8 dB). Figure 4 shows the mean spectrum for BT = 10000 (B = 100 MHz). It shows a good control of the spectral width due to the rectangular LPF. Finally in Figure 5 the cross-correlation (it measures the orthogonality) for a pair of generated phase noise signals is shown with BT = 10000. It is lower than −30 dB (as compared with the same compression ratio, for a pair of up and down chirps the cross-correlation is −43 dB for LFM and −40 dB for Non-LFM).

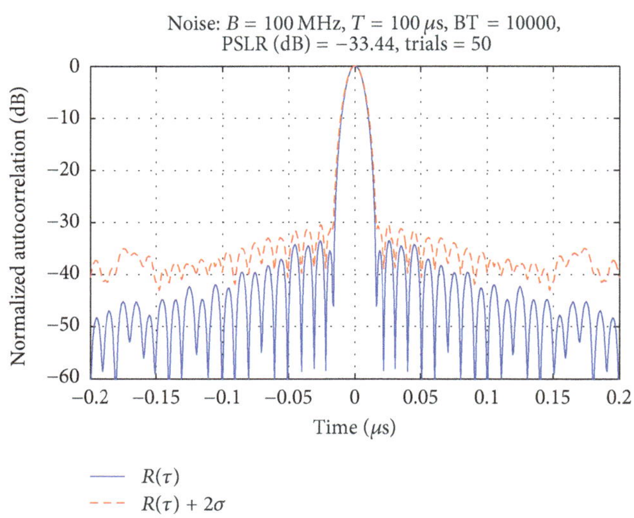

FIGURE 2: Mean normalized autocorrelation (obtained averaging 50 different autocorrelations). Dashed line: fluctuation at 2σ (σ denotes the standard deviation). Compression ratio BT = 10000, and the ZMNL is a hard limiter.

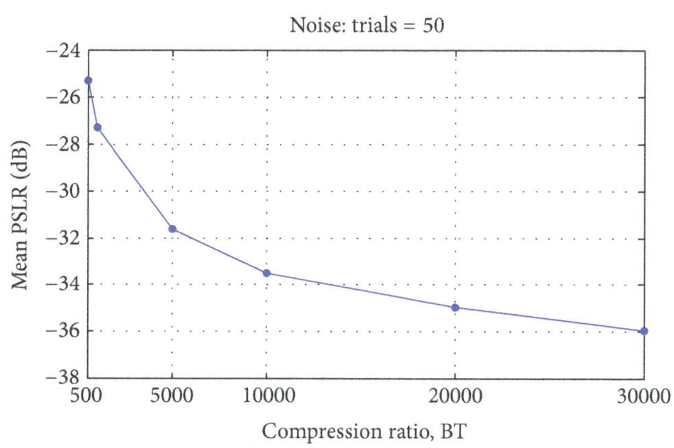

FIGURE 3: Mean PSLR (obtained averaging 50 different autocorrelations) versus the compression ratio BT, considering for ZMNL a hard limiter.

From the database of generated waveforms [10] a part is selected after analysis of the compliance with the required characteristics in terms of spectral occupancy and autocorrelation and cross-correlation functions. In order to further reduce the side lobe level (which is likely needed, especially for smaller values of the compression ratio BT), a mismatched filtering in reception may be needed [11, 12]. The filter coefficients must be computed for every selected waveform and associated with it for the ensuing usage. While permitting a significant suppression of the range side lobes, this technique calls for increased computation costs (the mismatched filter is typically three times longer than the matched filter) and larger memory requirements.

3.3. Closed Form by Inversion. A different, closed form generation approach is based on the following considerations.

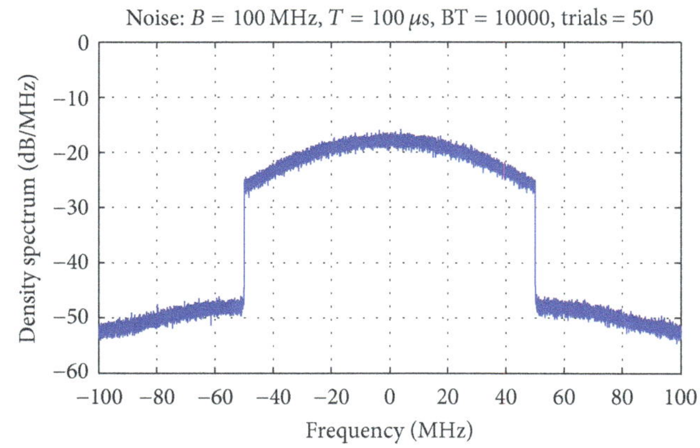

FIGURE 4: Mean density spectrum (obtained averaging 50 different spectra) for a pair of phase noise signals. Compression ratio 10000, considering for ZMNL a hard limiter.

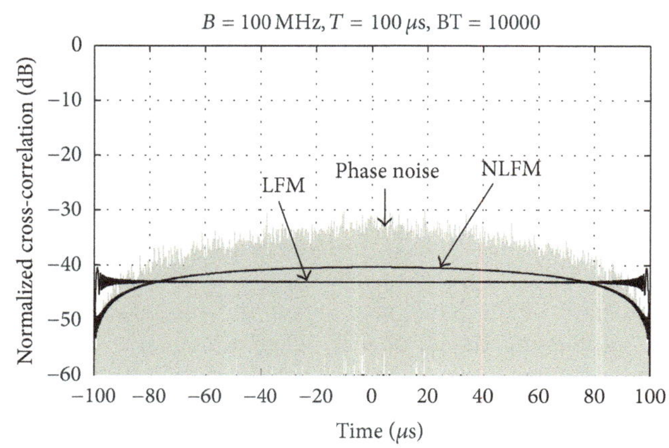

FIGURE 5: Normalized cross-correlation for a pair of generated noise phase signals by the iterative algorithm considering for ZMNL a hard limiter. Compression 10000.

For a real Gaussian process Van Vleck and Middleton [13] have shown that the autocorrelation coefficient (R_t with $t = t_2 - t_1$) of the output from a hard limiter is related with the input autocorrelation coefficient (here denoted r) by the well-known *arcsine law*:

$$R_t = \frac{2}{\pi} \cdot \arcsin(r). \tag{9}$$

Considering a complex Gaussian process, the correlation R_t after the hard limiter, that is, between

$$\frac{z_1^*}{|z_1|} = \frac{x_1 - jy_1}{\sqrt{x_1^2 + y_1^2}},$$
$$\frac{z_2}{|z_2|} = \frac{x_2 + jy_2}{\sqrt{x_2^2 + y_2^2}}, \tag{10}$$

is

$$R_t = E\left\{\frac{z_1^* z_2}{|z_1||z_2|}\right\}$$
$$= E\left\{\frac{x_1x_2 + y_1y_2 + j(x_1y_2 - x_2y_1)}{\sqrt{x_1^2 + y_1^2}\sqrt{x_2^2 + y_2^2}}\right\} = u + jv, \quad (11)$$

where $E\{\cdot\}$ is the statistical mean operator. Supposing a symmetrical power density spectrum with respect to the origin, the correlation is real and $v = 0$. Equation (11) has been evaluated, and it results in

$$R_t = b_0 r + \sum_{n=1}^{\infty} b_n \cdot r^{2n+1} \quad (12)$$

with

$$b_n = \frac{(2n-1)^2}{4n(n+1)} \cdot b_{n-1}, \quad b_0 = \frac{\pi}{4}, \quad n = 1, 2, 3, \ldots. \quad (13)$$

Then R_t can be expressed as a sum of odd powers of r, where the coefficients b_n are very similar to those evaluated for the *arcsine law*:

$$b_n = \frac{(2n-1)^2}{2n(2n+1)} \cdot b_{n-1}, \quad b_0 = \frac{2}{\pi}, \quad n = 1, 2, 3, \ldots. \quad (14)$$

Figure 6 shows R_t versus the input correlation r for real and complex Gaussian process. Inverting (12) it is possible to predistort the input autocorrelation to the hard limiter to obtain a desired R_t.

In such a way the requirements of Sections I and II can be met with no need for iterations. In fact, (a) the output autocorrelation is chosen in order to satisfy the PSLR requirement and the spectral band requirement; (b) the MEPPR requirement is satisfied by the hard limiter, and (c) the orthogonality is obtained by the randomness of the white Gaussian input sequence; and may be enhanced by proper choices of the generated output sequences.

4. An Architecture for Next Generation Marine Radar Based on NRT

The general architecture of a proposed, novel class of solid-state, coherent, NRT-based marine radar is shown in Figure 7, from [10]. Basically, this is a coherent radar transmitting preregistered pseudorandom signals and using a correlation receiver (or a mismatched, range-side lobe suppression receiver). In order to reduce the interference level between these marine radars exploiting the orthogonality of the transmitted waveforms, each radar must have a large number of them (order of hundreds of millions) stored in a mass memory; see Figure 7.

These waveforms are result of a generating procedure shown in Figure 1, as explained before. Thanks to this procedure, all of them have a crest factor close to the unity and their power spectral density (PSD) respects the spectral mask of the international regulations (see the appendix). For instance, let us consider a waveform, with a bandwidth of 80 MHz and a centre frequency of 9410 MHz, that is, a spectrum from 9.37 to 9.45 GHz with EIRP per MHz equal to 75 dBm. Outside the interval (the "Necessary Band") from 9.37 to 9.45 GHz its PSD will have to decay at 20 dB/dec, arriving at 35 dBm EIRP below 9.17 GHz and above 9.64 GHz. Only a part of the generated waveforms, that is, those with an acceptable PSR, is selected to be stored in the mass memory.

As in solid-state marine radar using deterministic waveforms and pulse compression, [14, 15], in the proposed marine

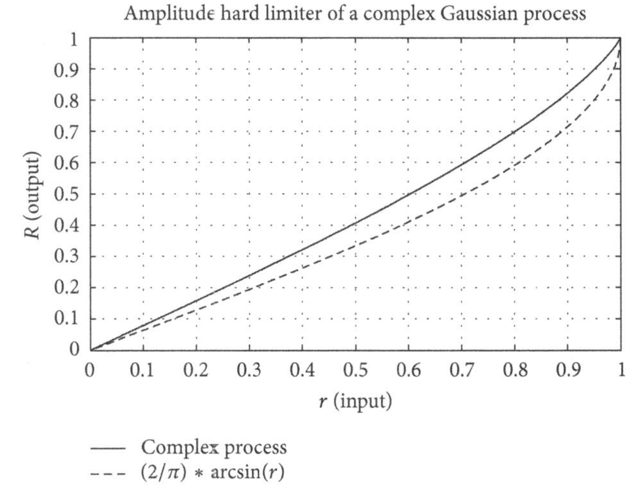

FIGURE 6: Output autocorrelation (R) from a hard limiter versus the input autocorrelation (r).

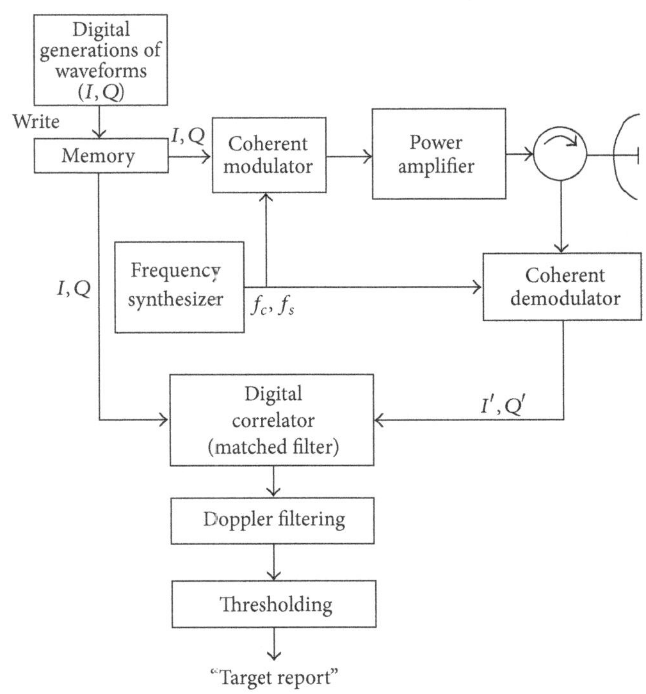

FIGURE 7: General architecture of solid-state, coherent, NRT-based marine radar.

noise radar three sets of length will be used: a short pulse for sort range, a medium-duration waveform for medium range, and a long waveform for the maximum range of 96 NM.

5. Plans for Future Work and Conclusion

It has been recognized (see for instance [14]) that the long pulses of solid-state marine radar will create formidable problems of electromagnetic comparability to (a) traditional magnetron radar as well as to (b) other solid-state marine radars. Usage of pseudorandom, orthogonal signals will alleviate case (b) by a figure close to the value of the compression factor (the BT product). Further analysis is needed for case (a). For example, in [14] it is stated that "*characteristic pulsed interference was observed, caused by the long pulses disabling pulse cancellation logic in the magnetron radar. Such interference on established radar systems raised concerns to the system's further development* ...," and "*A novel 3 pulse waveform has been designed and tested for solid-state marine radar. The waveform's spectral and range performance are excellent,* ... *However the interference effects that such a radar might cause on existing marine radars may be catastrophic. Some of these effects could be reduced by techniques that were investigated, however no techniques were found that gave confidence to proceed with a solid state marine radar product based on this waveform.*" The damaging signals are of two types: direct (antenna to antenna) and indirect or bistatic clutter (antenna to reflecting surfaces to antenna). Summing up, further analysis, simulation, and trials are needed to establish the cost-effectiveness of solid-state S- and X-bands marine radars to be used in crowded environments as well as the electromagnetic compatibility between them and, presently more important, with legacy marine radar. The Noise Radar Technology is a promising tool to solve these problems.

Appendix

Interference in Marine Radar

Recommendations and rules for avoiding and/or mitigating the interferences related to radar systems have been issued by the International Electro Technical Committee [16] and by the International Telecommunication Union [17, 18]. Generally, distinction is made between (i) the "Necessary Band," related to the shape and modulation of the radar signal as designed to satisfy the radar performance, (ii) the "Out of Band" (OoB) domain that contains the spectral residuals of the modulation process for the radar signal, and (iii) the "spurious" domain that contains the harmonic, parasitic, intermodulation and frequency conversion effects. Methods are suggested to compute the Necessary Band from the pulse length and its rise and fall time or from the frequency span in the cases of linear or nonlinear frequency modulation (chirp signals). Roughly, the Necessary Band for a simple rectangular pulse of duration T is about $1.8/T$, and for a linear chirp of frequency deviation Δf is about $2\Delta f$. A −40 dB band is defined in the radar emission mask; in some significant cases, it results in $B_{-40} = 2.5B_{\text{Necessary}}$ and the transition between them has a 20 dB/decade roll-off.

References

[1] A. Bole, B. Dineley, and A. Wall, *Radar and ARPA Manual*, Elsevier, New York, NY, USA, 2nd edition, 2005.

[2] J. N. Briggs, *Target Detection by Marine Radar*, vol. 16 of *IEE Radar, Sonar and Navigation Series*, The Institute of Electrical Engineers, London, UK, 2004.

[3] S. Harman, "Spectral sharing with radar," in *Principles of Waveform Diversity and Design*, M. Wicks, M. Wicks, E. Mokole, S. Blunt, R. Schneible, and V. Amuso, Eds., Part 6: Spectrum Management, Chapter 35, SciTech, Edison, NJ, USA, 2010.

[4] J. T. Ascroft, "A study into techniques for improving radar spectrum utilization," Project AY, 4490, 2004, http://citeseerx.ist.psu.edu/viewdoc/download?doi=10.1.1.127.6876&rep=rep1&type=pdf.

[5] N. Levanon and E. Mozeson, *Radar Signals*, IEEE Press, Wiley-Interscience, Jhon Wiley & Sons, New York, NY, USA, 2004.

[6] J. P. Costas, "A study of detection waveforms having nearly ideal range-doppler ambiguity properties," *Proceedings of the IEEE*, vol. 72, no. 8, pp. 996–1009, 1984.

[7] W. O. Alltop, "Complex sequences with low periodic correlations," *IEEE Transactions on Information Theory*, vol. 26, no. 3, pp. 350–354, 1980.

[8] P. van Genderen, O. A. Krasnov, Z. Wanga, and R. F. Tigrek, "Some OFDM waveforms for a fully polarimetric weather radar," *International Journal of Microwave and Wireless Technologies*, vol. 4, no. 02, pp. 163–169, 2012.

[9] S. R. J. Axelsson, "Noise radar using random phase and frequency modulation," *IEEE Transactions on Geoscience and Remote Sensing*, vol. 42, no. 11, pp. 2370–2384, 2004.

[10] G. Galati, "Radar coerente," Patent Application N. RM, 2013 A000290, Roma, Italy, 2013.

[11] A. S. Mudukutore, V. Chandrasekar, and R. Jeffrey Keeler, "Pulse compression for weather radars," *IEEE Transactions on Geoscience and Remote Sensing*, vol. 36, no. 1, pp. 125–142, 1998.

[12] H. Urkowitz and N. J. Bucci, "Doppler-Tolerant range sidelobe suppression for meteorological radar," *IEEE Transactions on Geoscience Electronics*, vol. 4, pp. 206–208.

[13] J. H. Van Vleck and D. Middleton, "The cpectrum of clipped noise," *Proceeding of the IEEE*, vol. 54, pp. 2–19, 1966.

[14] S. Harman, "The performance of a novel Three-Pulse radar waveform for marine radar systems," in *Proceedings of the 5th European Radar Conference (EuRAD '08)*, pp. 160–163, Amsterdam, The Netherlands, October 2008.

[15] A. Nelander and Z. Tóth-Pál, "Modular system design for a new S-band marine radar," in *Proceedings of the International Radar Conference Surveillance for a Safer World (RADAR '09)*, Bordeaux, France, December 2009.

[16] International Electro Technical Commission, "Shipborne radar," IEC, 62388 Ed. 1. 0, 2007.

[17] ITU International Telecommunication Union, "Unwanted emission in the out of band domain," Recommendation ITU-R SM, 1541, 2001.

[18] ITU International Telecommunication Union, "Efficient use of the radio spectrum by radar stations," Recommendation ITU-R M, 1327-1, 1998–2003.

8

Hyperspectral Anomaly Detection: Comparative Evaluation in Scenes with Diverse Complexity

Dirk Borghys,[1] Ingebjørg Kåsen,[2] Véronique Achard,[3] and Christiaan Perneel[4]

[1] *Department CISS, Royal Military Academy, 2007 Brussels, Belgium*
[2] *Land and Air Systems Division, Norwegian Defence Research Establishment (FFI), 2007 Kjeller, Norway*
[3] *Theoretical and Applied Optics Department, French Aerospace Laboratory (ONERA), FR-31055 Toulouse Cedex 4, France*
[4] *Department of Mathematics, Royal Military Academy, Brussels, Belgium*

Correspondence should be addressed to Dirk Borghys, dirk.borghys@gmail.com

Academic Editor: Xiaofei Hu

Anomaly detection (AD) in hyperspectral data has received a lot of attention for various applications. The aim of anomaly detection is to detect pixels in the hyperspectral data cube whose spectra differ significantly from the background spectra. Many anomaly detectors have been proposed in the literature. They differ in the way the background is characterized and in the method used for determining the difference between the current pixel and the background. The most well-known anomaly detector is the RX detector that calculates the Mahalanobis distance between the pixel under test (PUT) and the background. Global RX characterizes the background of the complete scene by a single multivariate normal probability density function. In many cases, this model is not appropriate for describing the background. For that reason a variety of other anomaly detection methods have been developed. This paper examines three classes of anomaly detectors: subspace methods, local methods, and segmentation-based methods. Representative examples of each class are chosen and applied on a set of hyperspectral data with diverse complexity. The results are evaluated and compared.

1. Introduction

Many types of anomaly detectors have been proposed in literature [1, 2]. The most frequently used anomaly detector is the (spectral only version of the) Reed-Xiaoli (RX) detector [3] that is often used as a benchmark to which other methods are compared. The RX detector characterizes the background by its spectral mean vector μ_B and covariance matrix Σ_B. The actual detector calculates the Mahalanobis distance between the pixel under test r and the background as follows:

$$D_{\mathrm{RX}} = (r - \mu_B)^T \Sigma_B^{-1} (r - \mu_B). \quad (1)$$

The global RX detector characterizes the background of the complete scene by a single multivariate normal probability density function (pdf). In many scenes, this model is not adequate. For that reason, several variations of the global RX detector have been proposed in literature [1, 2, 4–12]. They can be sub-divided into three classes: subspace methods, local methods, and segmentation-based methods. In complex scenes the latter category was shown to be very effective and several segmentation-based anomaly detectors (SBAD), not necessarily based on RX, have recently been proposed [13–20]. The aim of the current paper is to compare the results obtained by different types of anomaly detectors in scenes characterized by different types of background. In particular, two rural scenes with subpixel anomalies, a rural scene with some of the targets in shadow, and an urban scene were considered. Representative examples of each of the three previously mentioned classes of anomaly detectors were included in the comparison. In previous work [21], we noted the importance of data reduction and preprocessing on anomaly detection results. The current paper therefore also presents a comparison of results obtained by the different detectors after applying different preprocessing methods.

The evaluation of the detection results is mainly based on receiver-operating characteristic (ROC) curves. For spatially fully resolved targets, the false alarm rate at first detection was also considered. For the two scenes with extended targets, besides an objective evaluation, a more subjective evaluation is also presented. The rest of the paper is organized as follows. Section 2 presents the used datasets; in Section 3 the examined anomaly detection methods are briefly presented; Section 4 presents the different preprocessing methods that have been applied to the data. The last two sections of the paper present the results and the conclusions. The appendix presents a brief exploratory data analysis that mainly aims at verifying to what extent the different used datacubes comply with the assumption of global or local unimodal multivariate normality.

2. Overview of the Dataset

The analysis was performed on a set of hypercubes of scenes with various backgrounds and representative of three scenarios as follows:

(i) a rural environment with subpixel targets (CAM and OSLO1),

(ii) a rural environment with some of the targets in shadow (BJO),

(iii) an urban environment (OSLO2).

Table 1 presents an overview of the used dataset. The first two datacubes are real hyperspectral images in which a matrix of anomalies was inserted artificially. Figure 1 shows RGB composites of these images on which the targets have been superimposed.

The results shown in this paper were obtained with 10% mixing ratio subpixel anomalies for the CAM scene. For OSLO1, the mixing ratio was varied from 100% to 10%. The inserted anomalies are spectra of a green paint (CAM) and a green fabric (OSLO1).

The BJO image (Figure 2(a)) was acquired over a natural scene with an agricultural region and a small forest near the village of Bjoerkelangen in Norway. The figure shows the target locations with light blue colored rectangles representing the target sizes superimposed on the RGB composite of the scene. Fourteen targets composed of different types of material and with different colors were laid in the scene during the image acquisition. Targets T3–T7 were in shadow. T3 was in deep shadow between the trees, and the four others were in the shadow at the edge of the forest. Table 2 presents the dimensions and material types of the different targets.

The OSLO2 scene (Figure 2(b)) is part of the center of Oslo. In this scene, four targets (T1–T4) were laid out. Their respective dimensions in the image are T1: 5×10, T2: 5×9, T3: 2×6, and T4: 6×7 pixels. Targets T2 and T3 are pieces of green fabric and the other two of a blue plastic. T1–T3 were laid out on the grass in a park, and T4 was put on an asphalt background in the shadow from a building.

The CAM image was rectified and atmospherically corrected. The images BJO, OSLO1, and OSLO2 were not rectified before processing and all processing on these scenes was applied to radiance data, that is, without applying any atmospheric correction.

3. Anomaly Detection Methods

Besides global RX, representative examples of three categories of anomaly detectors are examined in the paper. Figure 3 presents an overview of the selected detectors in the three classes. As can be seen from the figure, many of the investigated methods are RX-based, but for the subspace detection methods and in particular for the SBAD methods, some anomaly detectors that are not related to RX have also been included. The different detectors are briefly described below.

3.1. Subspace Methods. The subspace methods are global and have in common that they apply principal component analysis (PCA) or singular value decomposition (SVD) to the datacube. The first PCA/SVD bands are supposed to represent the background and they are eliminated in different ways by the various subspace methods. Subspace anomaly detectors are thus global anomaly detectors applied on a spectral subset (subspace). For all of the subspace methods, the only parameter is the number of PCA or SVD bands (n_b) that is considered to represent the background. If this number is set too high, targets will disappear in the background, if it is too low, too many false alarms will remain. Automatically determining an optimal value for the dimension of the background subspace remains a current research topic.

3.1.1. Subspace RX (SSRX). In SSRX, the global RX is applied on a limited number of PCA bands. The first PCs are discarded in SSRX.

3.1.2. RX after Orthogonal Subspace Projection (OSPRX). In OSPRX, the first PCA/SVD components define the background subspace and the data are projected onto the orthogonal subspace before applying the RX detector [2, 22].

In the current paper, the SVD of the global spectral covariance matrix Σ is used. Because Σ is positive definite, the SVD is equivalent to the following eigenvector/eigenvalue decomposition:

$$\Sigma = U \Lambda U^T, \quad (2)$$

where U is the matrix of eigenvectors of the decomposition and Λ the diagonal matrix with decreasing eigenvalues. The projection operator P_{SVD} is defined as a function of the first n_b eigenvectors (columns of U), corresponding to the highest eigenvalues, $W = U(1 \ldots n, 1 \ldots n_b)$ as follows:

$$D_{\mathrm{OSPRX}}(r) = r^T (I - P_{\mathrm{SVD}}) r = r^T \left(I - W W^T \right) r, \quad (3)$$

with I the $n \times n$ identity matrix. n is the number of channels in the datacube, and n_b the number of channels used to model the background subspace ($1 \leq n_b < n$).

3.1.3. RX after "Partialling Out" the Clutter Subspace (PORX). In this method, the effect of the clutter in a pixel is removed (partialled out) component-wise by predicting each of its

Table 1: Overview of the used dataset.

Name	Site	Sensor name	Number of bands n	Spectral range (in μm)	Spatial resolution (in m)	Image size (in pixels)	Number of targets	Total target size (in pixels)
CAM	Camargue (Fr)	Hymap	126	0.44–2.45	4	150 × 100	45	45
OSLO1	Oslo (No)	HySpex	80	0.410–0.984	0.25	286 × 287	81	81
BJO	Bjoerkelangen (No)	HySpex	80	0.410–0.984	0.25	700 × 1600	14	574
OSLO2	Oslo (No)	HySpex	80	0.410–0.984	0.25	700 × 1600	4	45

Table 2: Material types and sizes (in pixels) of the different targets in the BJO image.

T1	T2	T3	T4	T5	T6	T7	T8	T9	T10	T11	T12	T13	T14
Green car	Red car	Cloths	Cloths	Painted boards			Cloths	Cloths	Painted boards				People
				Paint 1	Paint 2	Paint 3			Paint 1	Paint 2	Paint 3	Paint 4	
8 × 22	5 × 11	4 × 3	2 × 3	5 × 7	4 × 4	3 × 3	2 × 7	3 × 4	8 × 10	5 × 7	4 × 7	5 × 8	7 × 16

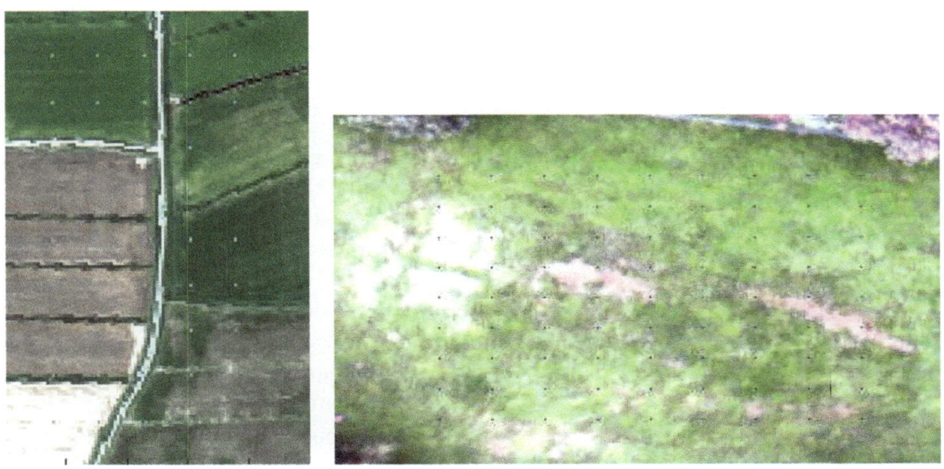

(a) (b)

Figure 1: RGB color composite of the CAM (a) and OSLO1 (b) datacubes with targets superimposed, respectively, in cyan and black.

(a) (b)

Figure 2: RGB composite of the BJO (a) and OSLO2 (b) datacubes with target locations indicated.

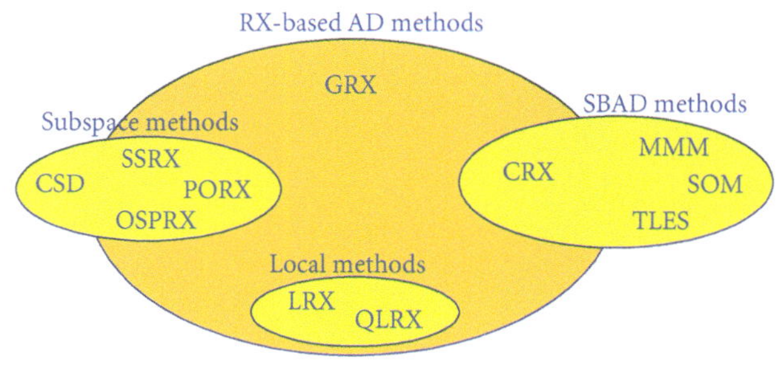

Figure 3: Schematic overview of the examined anomaly detection methods.

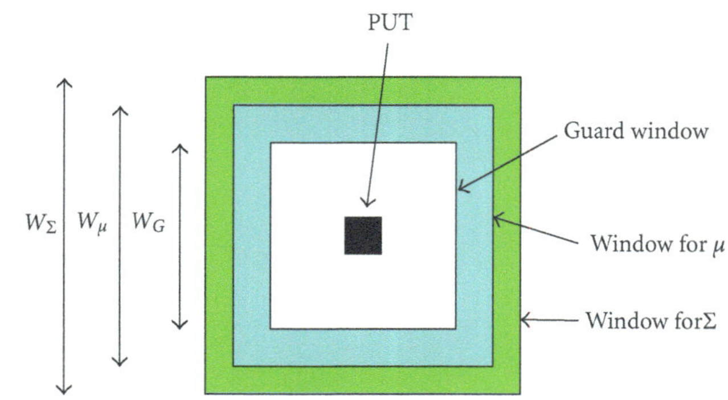

Figure 4: Sliding triple window used in the local AD methods.

spectral components as a linear combination of its high-variance principal components. The detector applies the RX detector on the residual. Details of the method can be found in [10].

3.1.4. Complimentary Subspace Detector (CSD). The CSD is not an RX-based method. In the CSD, the highest variance principal components are again used to define the background subspace and the other PCs, to define the target subspace (the complimentary subspace) [7]. The PUT is then projected on the two subspaces and the anomaly detector is the difference of the projection onto the target subspace and the background subspace as follows:

$$D_{\mathrm{CSD}}(r) = r^T P_t r - r^T P_b r, \tag{4}$$

where

$$\begin{aligned} P_b &= U(1\dots n, 1\dots n_b)U^T(1\dots n, 1\dots n_b), \\ P_t &= U(1\dots n, (n_b+1)\dots n)U^T(1\dots n, (n_b+1)\dots n). \end{aligned} \tag{5}$$

3.2. Local Methods. In the local anomaly detection methods, the statistics of the background are estimated locally in a window around the PUT. A double sliding window is used: a guard window and an outer window are defined, and the background statistics are determined using the pixels between the two (see Figure 4). Sometimes a triple window is used where the covariance matrix of the background is estimated in a larger window than the average local background spectrum.

3.2.1. Local RX (LRX). In LRX, the covariance matrix Σ_B and mean spectrum μ_B of the background are estimated locally in a triple window around the PUT. In the used implementation, the size of the guard window is a parameter from which the size of the two other windows is determined as a function of the number of bands in the image as follow:

$$\begin{aligned} W_\mu &= \min_{k\,\mathrm{odd}}\left(k^2 - W_G^2 \geq \sqrt{10n}\right), \\ W_\Sigma &= \min_{k\,\mathrm{odd}}(k^2 - W_G^2 \geq 10n). \end{aligned} \tag{6}$$

3.2.2. Quasi-local RX (QLRX). Quasi-local RX (QLRX) [9] offers a compromise between the global and local RX approach. In QLRX, the global covariance matrix Σ is decomposed using eigenvector/eigenvalue decomposition (2). The eigenvectors are kept in the RX, but the eigenvalues are replaced by the maximum of the local variance and the global eigenvalue as

$$\lambda^i_{\mathrm{QL}} = \left[\max\left(\lambda^i_{\mathrm{loc}}, \lambda^i_{\mathrm{glob}}\right)\right], \tag{7}$$

where i is an index denoting the pixel in the image. This means that the score of the detector will be lower at locations of the image with high variance (e.g., edges) than in more homogeneous areas. Spectral statistical standardization (see Section 4.4) is applied as a preprocessing step. The local variance is determined in a double sliding window.

3.3. Segmentation-Based Methods. In complex scenes, the hypothesis of a single multivariate normal distribution of background spectra is usually not verified, not even locally. For that reason several segmentation-based anomaly detectors have been proposed in literature. In this paper, four of these methods have been included in the analysis.

3.3.1. Class-Conditional RX (CRX). In CRX, the image is first segmented, the covariance matrix and mean within each class i (i.e., Σ_i and μ_i) are determined. The Mahalanobis distance between the PUT and each of the classes is calculated. The final result is the minimum of these distances:

$$D_{\mathrm{CRX}} = \min_i\left[(r-\mu_i)^T \Sigma_i^{-1}(r-\mu_i)\right]. \tag{8}$$

In the current paper, K-means clustering is used and the parameters of the method are the minimum number of pixels allowed in each class and the maximum number of classes used in the clustering. The number of classes follows from these parameters.

3.3.2. Method Based on Multivariate Normal Mixture Models (MMM) [13]. A Stochastic Expectation Maximization (SEM) algorithm [23] is used for fitting a multivariate normal mixture model to the image for describing the background. The anomaly detector detects pixels that have a low probability according to the fitted model.

The parameters of the method are the maximum number of mixture components and the termination threshold for the iterative parameters estimation method. The idea behind the mixture model is to use the mixture components (the multivariate normal pdfs) as base pdfs in a row expansion of the true pdf, which can then, in principle, have any shape.

3.3.3. Two-Level Endmember Selection Method (TLES). The principle of the TLES method [19] is as follows: a small scanning window (50 × 50 pixels) runs over the image and at each position of the window the principal background spectra are determined using a segmentation method based on endmember selection. Endmembers that correspond to at least a given percentage (MP) of the image tile are stored. At the end of the process, an endmember selection is again applied on the stored endmembers and linear unmixing is applied on the image. Anomalies correspond to pixels with a large residue after unmixing. In [19], N-FINDR was used as the endmember selection method. In the current paper, the minimum volume simplex analysis algorithm (MVSA) [24] was used because it was found to give better results. Parameters of the method are the threshold MP and the number of endmembers kept in the two stages of the algorithm.

3.3.4. Method Based on Self-Organizing Maps (SOM). A trained SOM is considered as a representation of the background classes in the scene. Anomalies are determined by computing the spectral distances of the pixels from the SOM units [16, 17]. The SOM was applied on the first PCA components and run using a square map consisting of NsxNs hexagonal cells. The SOM was optimized sequentially. The parameters of the method are Ns and the number of PCA bands used.

4. Preprocessing Methods

Before applying the actual anomaly detectors, some preprocessing methods were applied to the data. Three different types of preprocessing were applied.

The first type is data dimension reduction, which has two objectives. The first objective is to describe the background better and to obtain more reliable statistical estimation, especially when applying local methods where the number of samples to compute statistics from is low. Moreover, reducing data dimension allows to reduce the size of the windows for the local methods. This reinforces the local aspect of the method and reduces the risk that nearby targets overlap with the window used to compute background statistics. The second objective is to project the data on axes where the anomalies are enhanced, that is, the most separated from the background pixels. In this paper, we focus on two different methods, spectral binning which fulfills the first objective, and kurtosis-based dimension reduction, that attempts to fulfill both.

The second type of preprocessing aims to account for the effects of shadow. In this paper, a simplified approach consisting in square root transforming the data is used (Section 4.5).

Finally, some AD methods need some specific preprocessing that is described in Sections 4.3 and 4.4.

4.1. Dimension Reduction by Spectral Binning (SB). As noted in [25, 26], dimension reduction can improve hyperspectral anomaly detection performance substantially. We have applied a dimension reduction method based on spectral binning, similar to the method applied in [26]. The binning consists in averaging over groups of neighboring bands, down to a spectral resolution of about 30 nm. The binning tends to improve the signal-to-noise ratio by reducing the relative contribution of photon noise. When this can be done while preserving the relevant spectral features, the result is improved detection performance.

4.2. Kurtosis-Based Dimension Reduction (KDR). As anomaly detection aims to search for outliers, a projection that enhances outliers applied as a preprocessing can improve detection performance. It has been shown that kurtosis is very sensitive to outliers. In [27] data are projected on the (first) eigenvectors of the kurtosis matrix K:

$$K = \Sigma^{-1}\frac{1}{N}\sum_{i=1}^{N}(X_i - \mu)^t\Sigma^{-1}(X_i - \mu)(X_i - \mu)(X_i - \mu)^t, \quad (9)$$

where X_i is the ith element of X, the matrix of observations (the spectrum of the ith pixel), and μ and Σ are the spectral mean and covariance matrix of the datacube. N is the total number of pixels in the image. This method is mainly useful if the data are unimodally distributed, that is, in scenes characterized by a relatively homogeneous background (in this work, the CAM and OSLO1 images). Usually, only the first 3 to 5 kurtosis components are kept for further processing. This is the case for GRX, LRX and QLRX, MMM, and SOM. For the subspace methods and TLES, all kurtosis components were used.

4.3. Spectral Whitening. If the eigenvalues and eigenvectors of the covariance matrix of the complete image are, respectively, Λ and U, and μ is the average spectral vector of the image, then the spectral whitening of the pixel r is given by [6, 7]

$$r_W = U\Lambda^{-1}U^T(r - \mu). \quad (10)$$

After spectral whitening, a Gaussian distributed variable becomes spherically symmetric and this is sometimes beneficial for detection [6]. Whether whitening is beneficial for the anomaly detector depends on the AD method and the datacube. For CSD, spectral whitening is always applied. The other subspace methods were applied with and without whitening and the best result obtained is reported in this paper.

4.4. Spectral Statistical Standardization. The spectral statistical standardization converts each spectral band to have a zero mean and a standard deviation of one. This is necessary for the QLRX in order to make sure that the global eigenvalues and the local variances can be interchanged in the algorithm.

4.5. Square Root Transform. Detection performance is generally degraded in shadow, due, among other things, to low signal-to-noise ratio and distortions of the spectral signatures caused by secondary illumination, the adjacency effect, and path-scattered skylight. In addition, the large dynamic range of the data from scenes containing both sunlit and shadowed areas makes the data modeling task more difficult. In order to improve detection performance in shadow, different strategies can be applied: de-shadowing and illumination suppression for estimation of sunlit-like radiance in shadowed areas [28–31], sun/shadow segmentation and application of adapted modeling in the respective areas [32], or transformation of the data to account for the effects of shadow. We have square root transformed the data $r_{\text{sqrt}} = \sqrt{r}$. This reduces the dynamic range of the data, and, perhaps more importantly, makes the noise signal level independent [33], with benefits for data modeling through a suppression of the influence of noisy low-level signals. At low signal levels for homogeneous backgrounds, the dominating source of variation in the signal is Poisson distributed counting (photon) noise, and square root transforming the data yields approximate normality [34]. Square root transforming the data of course also affects the distribution in the, vast majority of, cases where scene clutter is the main source of signal variation, but in unpredictable ways for complex backgrounds.

5. Evaluation Method

Experimental ROC (receiver operating characteristic) curves, showing the detection rate (DR) versus the false alarm rate (FAR), are used to evaluate the results obtained with the various detectors. For the images with resolved targets, a pixel-based ROC curve is calculated for each target, whereas for the images with subpixel targets, an ROC curve is calculated based on all the targets in the image. DR is plotted versus the logarithm of the FAR (the resulting curve is referred to as a logROC), and the area under the logROC curve (the logAUC) is calculated and used as the measure of performance. The reason for using a logarithmic FAR scale is that it ensures equal weight across the range of FAR values.

For extended targets (in BJO and OSLO2), ROC curves give the detection performance for each pixel of the target with respect to the false alarm rate. For defense and security applications, it is also of interest to assess the performance at the first detection of a target. In this paper, for extended targets, we therefore also determined the false alarm rate at the first detection for each of the targets.

Besides these objective evaluation metrics, it is also interesting to look at the type of false alarms that the various detectors produce in the different scenes. Therefore for the "best" detectors, a detection image is shown corresponding to the threshold for which at least one pixel of the most difficult target is detected. This subjective result is shown for scenes with extended targets (BJO and OSLO2).

6. Results and Discussion

6.1. Implementation Issue: Parameter Selection. The different examined AD methods depend on different parameters. For some of the applied methods, the parameters were set according to experience, whereas for others, where we lack this experience and where no consensus exists in literature for setting the parameters, we chose the optimal parameter setting through an optimization process in order to make the comparison between the detectors fair.

For the local methods, the parameters are the dimensions of the guard window and the outer window(s). The guard window should be set to be larger than the largest target of interest expected to be present in the scene, and size of the outer window(s) is derived from it as explained in Section 3.2.1. For the two datasets with subpixel anomalies, the guard window was set to 1, and for the two other datasets, it was set to 15.

GRX has no parameters.

For subspace methods, the only parameter is the dimensionality attributed to the background subspace. Several methods have been proposed for estimating this "signal subspace," mainly for unmixing purposes [35–39]. The two latter focus on finding the signal subspace dimension in the presence of "rare signals." They are thus likely to add the signal components containing the target to the signal subspace and are therefore less relevant to the choice of the signal subspace in subspace anomaly detection. The remaining methods [35–37] give different results and none of the signal subspace dimensions estimated by these methods correlate in a consistent way with the optimal number of bands in the subspace detectors for the different scenes. A consistent way for identifying the proper dimensionality to use in modeling the background clutter for subspace anomaly detection has yet to be found, as mentioned in the conclusion of [10]. Because the aim of the paper is to compare the different algorithms in the different scenes, the dimensionality parameter is optimized for each detector/scene combination. The complete range of possible background dimensionality (1 to $n-1$) was explored and the results shown are the best results obtained by that algorithm.

Each of the SBAD methods has its own set of parameters. For CRX, two parameters are set: a maximum number of classes and a minimum number of pixels per class. This last parameter can be used to reduce the risk that anomalies form their own classes. Then the maximum number of classes can be set higher than the actual number of background classes. The minimum number of pixels in each of the classes is set to a low percentage (for instance 0.5%) of the total number of pixels. For MMM, the parameters are similar to those of CRX. For SOM and TLES, the parameters, described in Section 3.3, were varied in a reasonable range and the results shown are the best obtained for the examined range of parameters.

6.2. Results for Subpixel Detection in a Rural Environment

6.2.1. Results for CAM. Table 3 shows the logAUC results for the different detectors obtained in the CAM dataset with a 10% mixing ratio. Results are shown without prior data reduction and with two types of data reduction: spectral binning and kurtosis-based data reduction. We see from the

Table 3: LogAUC results for CAM scene with a mixing ratio of 10% for different types of data reduction.

AD method	Data reduction method No data reduction	Spectral binning	Kurtosis data reduction
GRX	0.595	0.637	*0.732*
SSRX	0.742	0.778	**1.000**
OSPRX	*0.931*	0.754	0.893
PORX	0.568	0.205	*0.957*
CSD	*0.868*	0.826	0.763
LRX	0.616	0.950	**1.000**
QLRX	0.720	0.828	**1.000**
CRX	0.683	0.830	**1.000**
TLES	0.128	0.268	*0.698*
SOM	0.116	*0.529*	0.120
MMM		**1.000**	**1.000**

Table 4: LogAUC results for OSLO1 scene with a mixing ratio of 33% for different types of data reduction.

AD	No data reduction	Spectral binning	Kurtosis data reduction
GRX	0.163	0.357	*0.390*
SSRX	0.292	*0.486*	0.394
OSPRX	0.383	*0.486*	0.391
PORX	0.322	*0.493*	0.391
CSD	0.246	0.391	*0.396*
LRX	0.821	0.983	**0.995**
QLRX	0.294	0.395	*0.473*
CRX	0.163	0.357	*0.449*
TLES	0.108	*0.466*	0.356
SOM	0.232	0.385	*0.463*
MMM	0.128	0.544	*0.597*

table that all SBAD methods, local methods, and GRX benefit from dimension reduction, more from kurtosis dimension reduction (KDR) than from spectral binning (SB) (with exception for SOM, which performs much better after SB than after KDR). For the subspace methods, the results are more diverse: OSPRX and CSD perform best on data with no dimension reduction, whereas for SSRX and PORX a large improvement in performance is observed after KDR.

Figure 5 shows a scatter plot of the CAM data with 10% mixing ratio subpixel targets in the 2D space defined by the first two kurtosis components. The figure shows that the targets are very well separated from the background after KDR transform in this scene. This clear separation is not observed on any of the PCA components or on spectral bands before or after spectral binning.

The best results for this scene are indeed obtained with SSRX, LRX, QLRX, CRX, and MMM after KDR. These detectors all achieve a logAUC of 1.0, which means that all targets have been detected with a false alarm rate that is smaller than 1/image size, that is, FAR $< 6.6 * 10^{-5}$. Since the

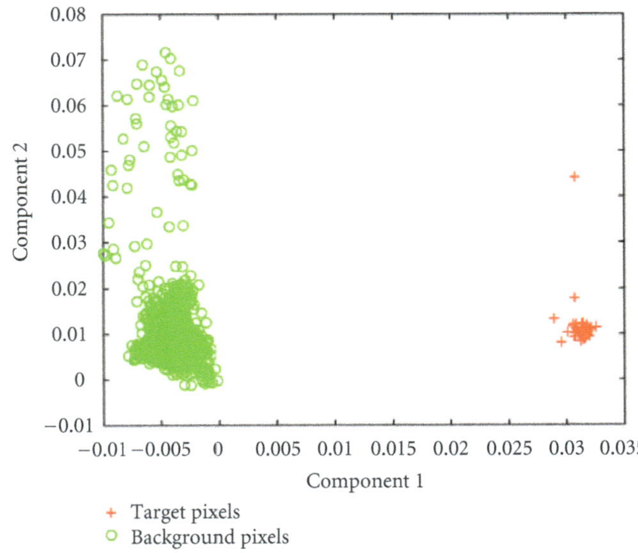

Figure 5: Projection of background (green) and target (red) pixels on the first two kurtosis component axes for CAM10.

results are saturated, we cannot properly distinguish between the methods.

6.2.2. Results for OSLO1. Table 4 shows the logAUC results for the different detectors obtained in the OSLO1 dataset with a 33% mixing ratio. Results are shown without prior data reduction and with spectral binning and kurtosis-based data reduction.

LRX clearly outperforms all other detectors. The assumption of local normality is very well met in this image (cf. Table 6 in the Appendix). LRX is also, in contrast to all other methods, able to model the background without influence of targets. For subpixel targets this is particularly true, since the guard window will always contain the whole target. As regards data reduction, for SBAD and local methods and GRX, the results are the same here as for CAM10: the methods all benefit from dimension reduction, and more from KDR than from SB (with exception for TLES which benefits more from SB than from KDR). In this dataset, we also observe an improvement in performance with dimension reduction for the subspace methods, but for these methods SB is generally more beneficial than KDR.

For OSLO1, the behavior of each detector as a function of the mixing ratio was also investigated. Figure 6 shows the logAUC results of the different detectors versus the mixing ratio for the OSLO1 datacube. In the experiment, the mixing ratio was varied from 100% (full pixel anomaly) to 10%. For creating the figure, the data reduction method that gives the best results was selected for each of the detectors. Results of global RX-based methods and CSD are shown as solid lines, LRX and QLRX results as dashed lines, and results of segmentation-based methods as dot-dashed lines. LRX clearly gives the overall best results, followed by SSRX for larger target portions and MMM for smaller. The result of

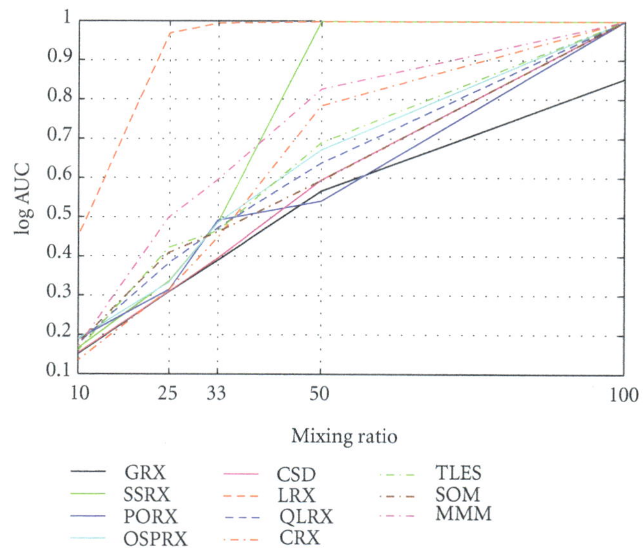

FIGURE 6: LogAUC versus mixing ratio for the different detectors in OSLO1.

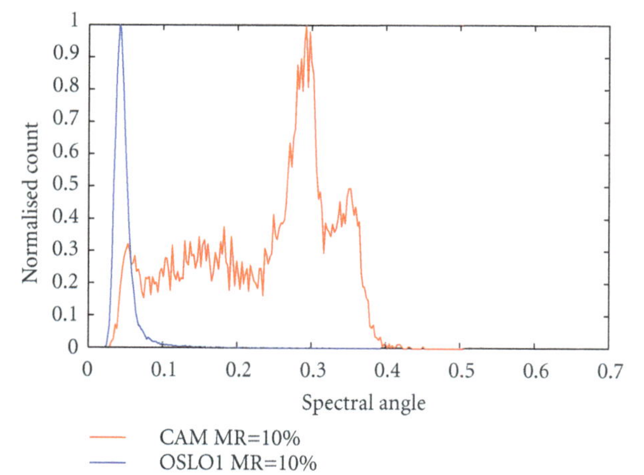

FIGURE 7: Normalized histograms of the spectral angle of the background pixels with respect to the average target spectrum in the CAM and OSLO1 scenes, at 10% mixing ratio.

SSRX for the 50% mixing ratio is very deviant, suggesting perhaps that subspace fit is somewhat random.

Contrary to the CAM scene, in OSLO1 the performance of the detectors at 10% mixing ratio is very low. The targets are more difficult to detect in the OSLO1 scene than in the CAM scene although the OSLO1 scene has a more homogeneous background and conforms well to the multivariate Gaussian assumption (see discussion of Table 5 in the Appendix). The OSLO1 scene is more difficult than the CAM scene because both the spectral angle and the Euclidean distance between the targets and the different background spectra are much smaller in the former. Figure 7 illustrates this point by means of the normalized histogram of the spectral angle of all background pixels with respect to the average target spectrum for both scenes at a mixing ratio of 10%. The figure shows that the spectral angle is indeed larger in the CAM scene than in the OSLO1 scene. The spread of the spectral angle in the CAM scene also illustrates the heterogeneity of the background.

6.3. Results for a Rural Environment with Some of the Targets in Shadow (BJO). Figures 8 to 10 show a graphical representation of the logAUC (a) and the logarithm of the false alarm rate at first detection (logFARAt1stDet) (b) for each of the detectors and for each target for the BJO scene. The colors represent the value of the respective performance metric. The color map is such that red corresponds to the best performance. The three different figures represent results after different types of preprocessing: Figure 8 shows results obtained without any preprocessing, Figure 9 results after spectral binning, and Figure 10 results after spectral binning and square root transform.

From the figures it is immediately clear that the targets in shadow (T3–T7) are more difficult to detect than the others. T3, hidden in the forest is the most difficult to detect. T2 (a red car) is the most easily detectable target.

We observe a slight improvement of performance from spectral binning for all detectors except TLES and SOM, which have their detection performance for targets in the sun substantially degraded by spectral binning. The reason why we—as opposed to what we saw in the previous datasets—only observe a slight improvement in performance from spectral binning, could be that the targets in sun are so easy to detect that we detect them anyhow, whereas the targets in shadow need some form of compensation for shadow in order to be detectable. The levels of the performance values indicate that this could be the case. For LRX, the size of the windows used to calculate the background statistics depends on the number of bands, and the results hence are not comparable across different numbers of bands.

Square root transforming the data generally improves the detection performance for targets in shadow substantially. It also improves the performance for targets in the sun for some detectors, but for others, mainly the subspace detectors and GRX, it reduces the performance for some sunlit targets—notably targets that are intensity anomalies (T8, T10, and T11), and that hence have their degree of anomality reduced when the data are square root transformed. The best results on this dataset are obtained with MMM, LRX, and CRX. MMM gives the overall best results, whereas LRX gives the best results for the targets in shadow: for targets T5 and T6, LRX gives significantly better results than MMM. LRX gives also slightly better results than MMM for T10 and T11, which are painted with the same paints as, respectively, T5 and T6. The logAUC and logFARAt1stDet results are globally inter-consistent, but they do show supplementary information. The results are consistent with the complexity of the scene and with the compliance with the multinormal distribution assumption locally shown in Table 6 of the Appendix.

In order to give an idea of the type of false alarms produced by the three best detectors, in Figure 11 the results of the three best detectors (MMM, LRX, and CRX after spectral binning and sqrt transform) are superimposed on a grayscale image of the BJO scene. The shown results are

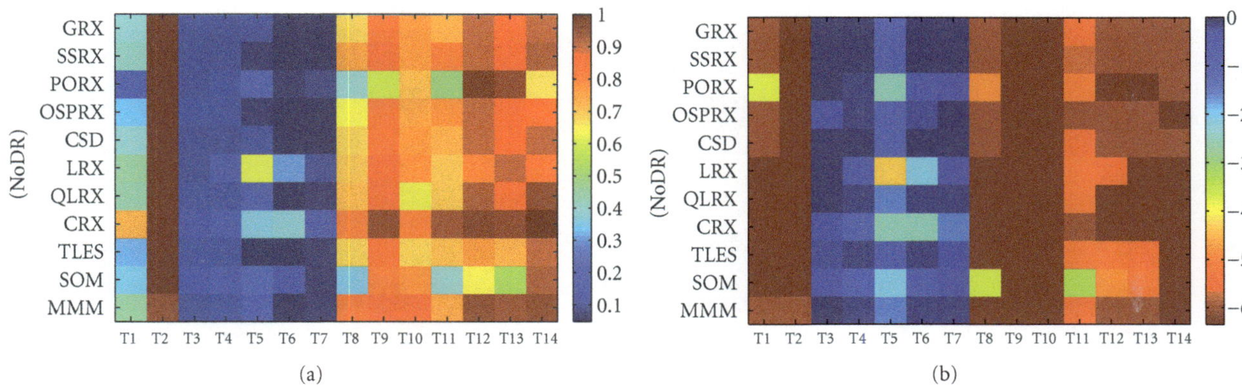

FIGURE 8: logAUC and logFARAt1stDet results per target without any data reduction for BJO.

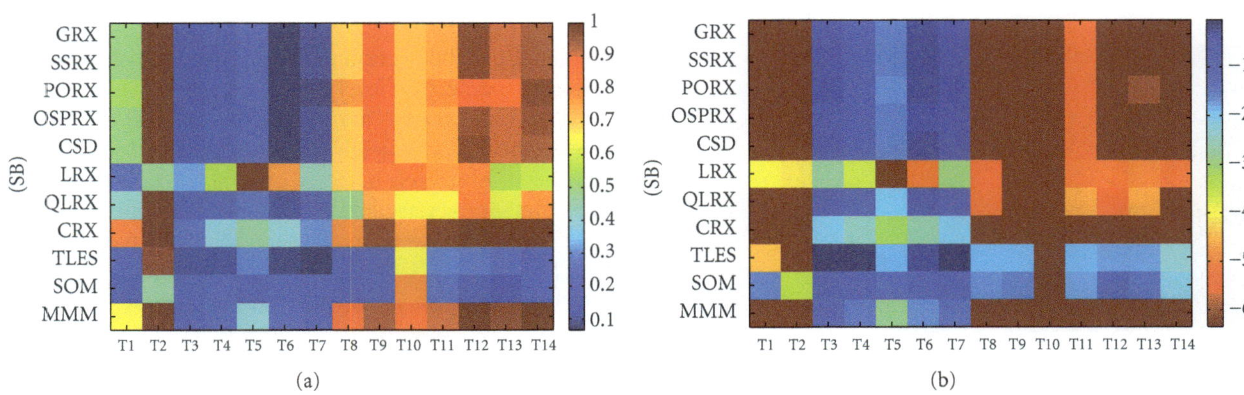

FIGURE 9: logAUC and logFARAt1stDet results per target after spectral binning for BJO.

thresholded detection results with the threshold set to the lowest first detection level for the true targets (i.e., the threshold for which at least one pixel of the most difficult target is detected). The figure shows that the false alarms produced by MMM mainly consist of isolated pixels in the forest and also some more extended false alarms at the top right of the image. LRX produces some small false alarms in the forest while CRX detects part of the stream as well as some detections in the forest. Most of the false alarms are detected by only one detector. On the other hand, for each target, except T7, an overlap in the detected zone for the three detectors is seen. The detectors are thus complimentary and fusing their results may be of interest. Likely causes of the complementarity of the results are that LRX are able to account more correctly for local illumination, whereas MMM/CRX are able to model locally heterogeneous background (forest) more precisely. The results for CRX indicate that too few classes are used: a large background structure like the stream is poorly modelled.

6.4. Results for the Urban Scene (OSLO2). Figure 12(a) shows the logAUC results for the OSLO2 scene. Figure 12(b) shows the logFARAt1stDet results. For creating the figure, the data reduction method that gives the best results was selected for each of the detectors. The best data reduction method was spectral binning for most detectors, but for OSPRX and QLRX the best results were obtained without data reduction, and for LRX, KDR gave the best results. The superiority of spectral binning over KDR for most detectors is to be expected because of the complexity of the scene (cf. Table 5 in the Appendix), and similarly the good performance of KDR for LRX—it can be attributed to local normality, see Table 6 in the Appendix. The mixed results for subspace detectors are consistent with the results for CAM and OSLO1. The result for QLRX, on the other hand, is not, but we should probably not read too much into this, since the detector more or less fails to detect the targets.

It can be seen that the values of logAUC are much lower than for the other datacubes. The maximum value obtained here is around 0.5. This is due to the complexity of the scene: the targets inserted into the scene are not the only anomalies. In an urban environment many objects can present an anomalous spectrum, for example, cars, special roof materials, and so forth. The comparison therefore only shows how well the different anomaly detection methods cope with urban "clutter."

From the two figures it can be seen that MMM gives the overall best results, followed by TLES and CRX. Of the two latter CRX gives the best results according to the logAUC metric and TLES according to the logFARAt1stDet metric. As could be expected the SBAD methods thus obtain the best results in the urban scene.

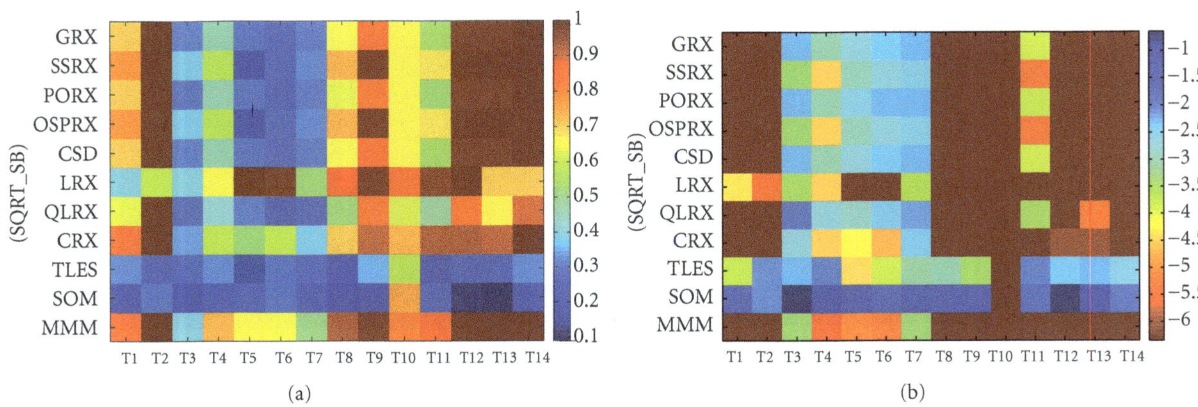

FIGURE 10: logAUC and logFARAt1stDet results per target after SQRT transform and spectral binning for BJO.

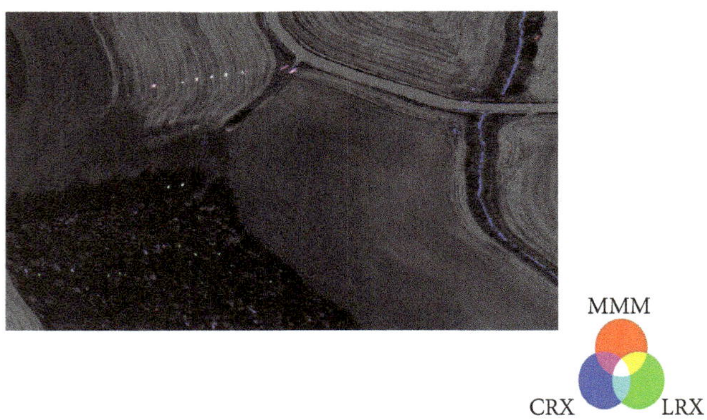

FIGURE 11: Color composite of the detection results for MMM, LRX, and CRX at lowest 1st detection threshold, superimposed on B/W image of BJO (R: MMM results, G: LRX results, B: CRX results).

GRX and all subspace methods perform quite bad, except for OSPRX that gives quite good results for targets T1, T2, and T4. Contrary to what we have seen in the other datasets, LRX does not perform particularly well in this dataset—despite good compliance with the multivariate normal distribution assumption locally near the targets, cf. Table 6 in the Appendix. The reason for this is probably that the remaining (non-target) parts of the scene are not well described by a local multivariate normal distribution, and hence we get lots of false alarms. This assumption is partly verified by comparing with the results of OSLO1. The target material of targets T2 and T3 in OSLO2 is the same as the material of the OSLO1 targets, and the local background is very similar (grass at two different places in Oslo), so a difference in performance between the images ought to be due to a difference in the number of false alarms.

As mentioned above, an urban scene presents many objects that may have an anomalous spectrum and that thus will be considered a false alarm in the above evaluation. It is therefore of interest to give an idea of these false alarms for each of the detectors. Figure 13 presents the results of the three best detectors (MMM, CRX, and TLES), superimposed on a grayscale image of the OSLO2 scene. The results shown are thresholded detection results with the threshold set to the lowest first detection level for the true targets.

The figure shows that target 1 is the most easy, and that it was detected completely by the three detectors. T2 and T3 have been completely detected by MMM, while the two other detectors (at the selected thresholds) detect only a part of the interior. On the contrary, T4 has been completely detected by TLES and CRX, while MMM only detects a part of its interior. The figure also shows that many of the "false alarms" are quite different for the three detectors. TLES detects parts of the vegetation. CRX detects a set of small objects next to the building on the lower left in the image. Some cars and small structures on roofs have been detected by a combination of detectors. This subjective evaluation shows that the three detectors that perform best according to the "objective" evaluation are quite complimentary. Examining further the properties of their results may lead to interesting ideas on fusion of anomaly detectors.

7. Conclusions

This paper evaluates the performance of anomaly detection methods in scenes with different backgrounds and types of targets: agricultural scenes with subpixel targets, an agricultural scene with some of the targets in shadow, and an urban scene. Three classes of anomaly detectors were considered besides the global RX: subspace methods, local methods, and segmentation-based anomaly detection (SBAD) methods.

For subpixel anomaly detection in scenes of low complexity (rural and non shadow), LRX gives the best results, followed by MMM. From the investigated global RX-based methods SSRX and OSPRX give the best results. For the SBAD and local methods and GRX, detection results are improved by data reduction, and (with minor exceptions) more by kurtosis dimension reduction than by spectral binning. The improvement of results after kurtosis-based data reduction for most of the detectors illustrates the potential of customizing the data reduction method.

For the rural scene with some of the targets in shadow the results show that it is important to account for the effects

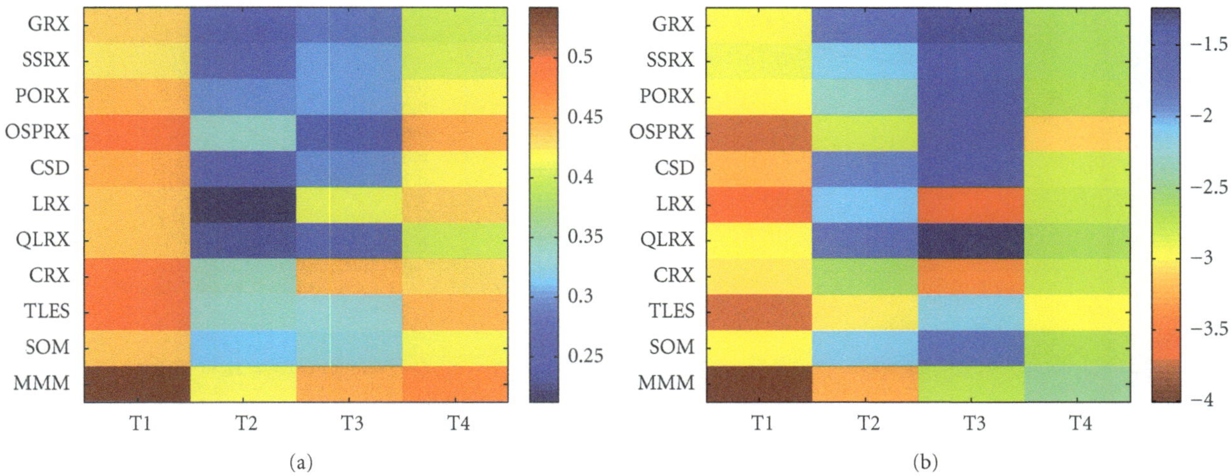

FIGURE 12: logAUC and logFARAt1stDet results per target for OSLO2 using the best data reduction method for each detector.

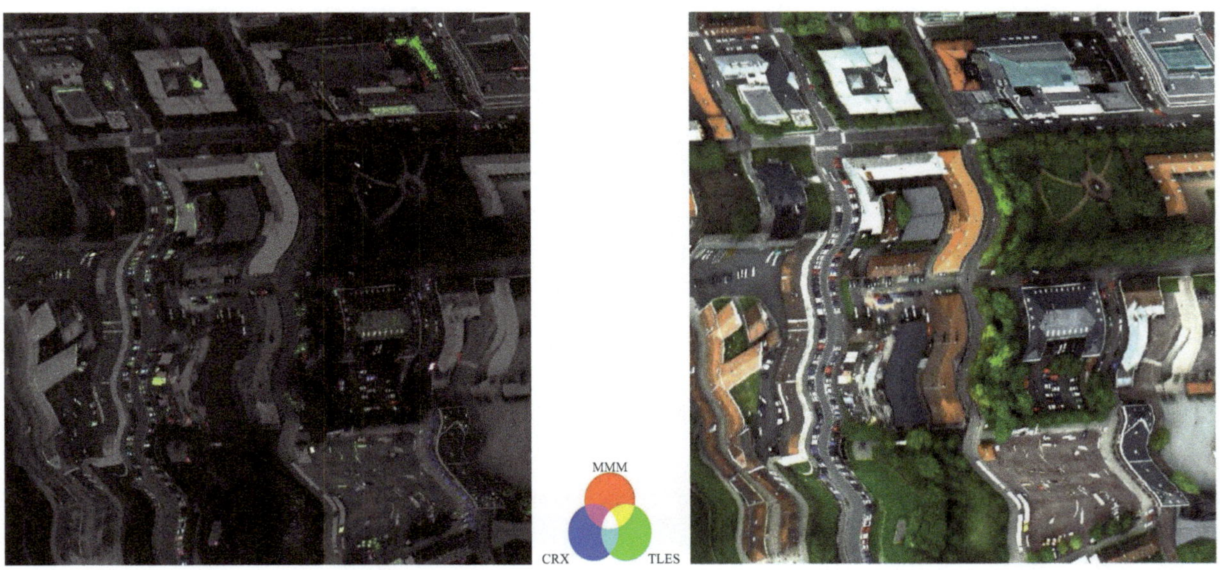

FIGURE 13: Left: color composite of the detection results for MMM, CRX, and TLES at lowest 1st detection threshold, superimposed on B/W image of OSLO2 (R: MMM results, G: CRX results, B: TLES results).

of shadow. In this paper this was done using a simplified approach consisting in square root transforming the data. After this transform MMM gives the best overall results, followed by LRX and CRX. For some targets LRX gives significantly better results than MMM. These three best detectors produce different false alarms while producing a common detection for all but one of the targets. They are thus complementary to each other and fusion of their results should be beneficial.

In the urban environment the SBAD methods perform best. The overall best result for the urban scene is obtained by MMM, TLES, and CRX. Of the globl RX-based methods OSPRX gives the best results in this dataset. Subjective evaluation of the detection results shows that the best performing detectors give complimentary results, and that "false alarms" are mainly due to objects with anomalous spectra in the scene such as cars and parts of buildings. Further investigation of this complementarity may lead to efficient detector fusion.

Appendix

Exploratory Data Analysis

The main aim of the exploratory data analysis is to investigate how well the different datacubes comply with the assumption of unimodal multivariate normality. If a distribution is multivariate normal, the square of the Mahalanobis distances of its samples follows a χ^2 distribution with degrees of freedom equal to the dimension of the multivariate variable [40]. The compliance can then be investigated visually using a Q-Q plot of the empirical cumulative distribution function (CDF) of the Mahalanobis distance and the CDF of the theoretical χ^2 distribution. Figure 14 shows the Q-Q plots

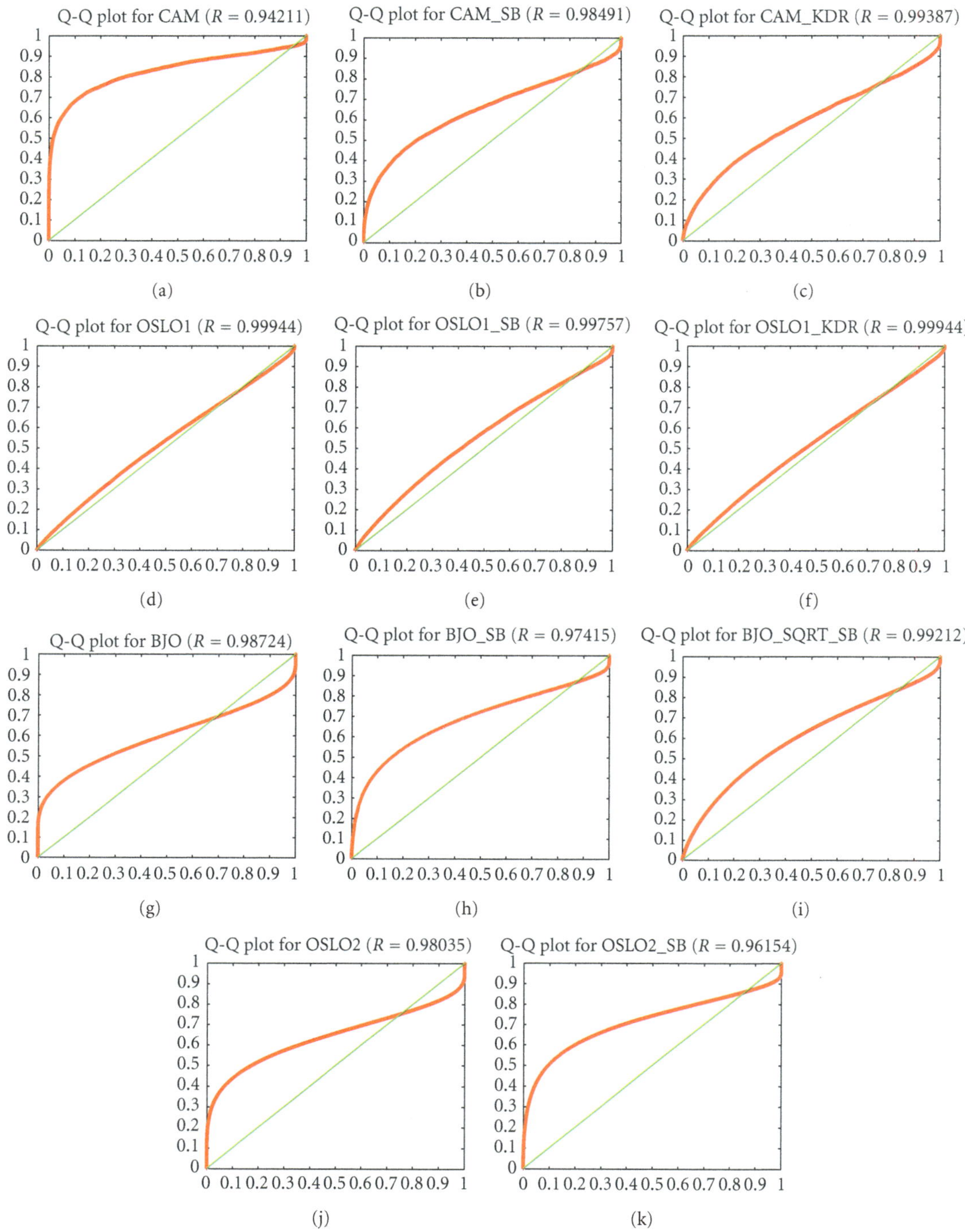

FIGURE 14: Q-Q plots for the different data cubes used in this paper.

for the four scenes and the different used preprocessing methods. Table 5 shows the correlation coefficient between the empirical and theoretical CDF of the Mahalanobis distance as well as the maximal deviation between the two (the Kolmogorov-Smirnov test statistic).

As can be expected the OSLO1 scene, consisting of a very homogeneous background, conforms best to the assumption of global multivariate normality. Both the original data and the data after kurtosis data reduction have a high correlation coefficient between the two CDFs and a low value for the KS-statistic. None of the other datasets comply with the global normality assumption. For the CAM scene the multivariate normality improves by preprocessing and the best normality is achieved after kurtosis data reduction when all kurtosis components are considered. When only the first five components are considered, the global normality

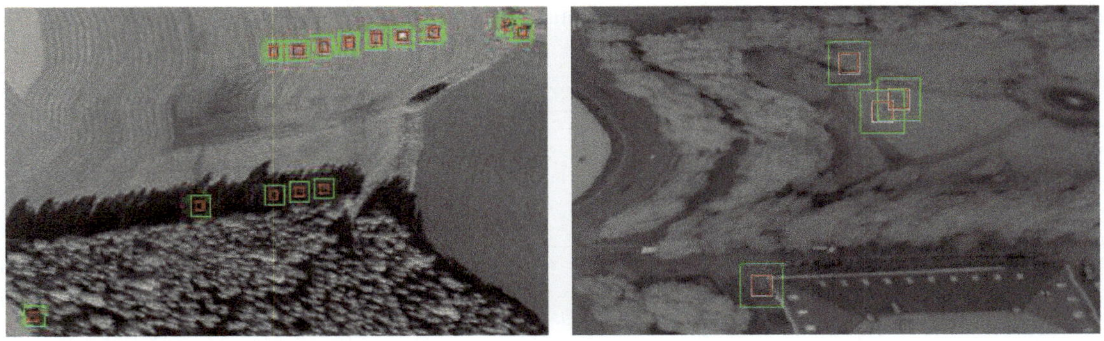

(a) (b)

FIGURE 15: Guard window (red) and outer window used for calculating local Σ centered at each target location for BJO (a) and OSLO2 (b).

TABLE 5: Correlation coefficient and Kolmogorov-Smirnov test statistic between empirical and theoretical CDF of the Mahalanobis distance and the condition number of global Σ.

Name	Preprocessing	Correlation	KS statistic	Condition number of Σ
CAM	None	0.937	0.579	$1.11e+10$
	SB	0.984	0.295	$7.23e+08$
	KDR	0.993	0.178	$2.91e+08$
	KDR (5 kcp)	0.892	0.540	$2.95e+00$
OSLO1	None	0.999	0.049	$9.11e+03$
	SB	0.997	0.093	$8.32e+03$
	KDR	0.999	0.049	$6.73e+00$
	KDR (5 kcp)	0.973	0.251	$4.91e+00$
BJO	None	0.986	0.278	$1.63e+05$
	SB	0.973	0.343	$1.03e+05$
	SQRT + SB	0.992	0.188	$1.06e+05$
OSLO2	None	0.979	0.336	$3.38e+04$
	SB	0.961	0.409	$2.27e+04$

assumption is not met. For BJO the combination of the square root transform and spectral binning improves the normality of the data.

Several of the investigated anomaly detection methods (the global RX-based methods and LRX) rely on the estimation and inversion of the spectral covariance matrix Σ. It is known that the sample covariance matrix in many cases needs to be regularized before inversion [41, 42]. The regularization makes the problem of finding the inverse mathematically stable, but if the initial matrix does not have a stable inverse, that is, is well-conditioned, the obtained inverse might not lead to a good detection result for the detector. In Table 5 the condition numbers of the covariance matrices of the complete images are also given. The condition number is the ratio between the highest and lowest singular value and it provides an indication of the accuracy of the results of matrix inversion. In CAM and OSLO1 the various preprocessing methods appear to decrease the condition number. This decrease is particularly significant for KDR in the OSLO1 scene. For KDR, contrary to the normality assumption, considering only the first five components reduces the condition number substantially.

The local anomaly detection methods estimate the characteristics of the background in a local window around the current pixel. For LRX the data in that local window are supposed to follow a multivariate normal distribution. In order to assess the validity of this assumption, the normality was checked in the neighborhood of each target in the different scenes. The neighborhood is defined in the same way as for the actual LRX detector. Table 6 shows the same estimators of normality as well as the condition number as in Table 5 but based on a local background estimation. As there are different targets in each of the scenes, the average and standard deviation over all targets is given. One can notice that for the BJO and OSLO2 scene the local normality assumption is much better met than the global one. For OSLO1 and BJO the condition number of local Σ is also better than that of the global Σ. The large standard deviation of the condition numbers are due to some targets for which local covariance matrix has a very high condition number. For CAM, this is the case for three targets and the median of the local condition number in that scene is 5.3×10^9 (no preprocessing). For the scenes with a limited number of targets this is explored in more detail in Table 7. The table shows the values obtained for the three estimators in the local window around each of the targets. Figure 15 shows the guard window (red) and the outer window (green) used for estimating local Σ superimposed on a grayscale representation of the two scenes. The green window is the outer window used when no data reduction is applied. For BJO and OSLO2, where the guard window is 15×15 and the number of bands $n = 80$, this means that the outer window has a size of 33×33. In OSLO2, LRX is applied after KDR and only 5 kurtosis components are kept. This results in an outer window of 17×17 represented by the white squares in Figure 15(b).

From Table 7 it appears that the normality assumption in the BJO scene is best obeyed for the local neighborhood of targets T8–T13. T3 has the most heterogeneous background, as can be also be seen in Figure 15(a). T1 and T2 deviate from the normality assumption because of target contamination: the outer windows overlap part of the adjacent target. It is

TABLE 6: Correlation coefficient and Kolmogorov-Smirnov test statistic between empirical and theoretical CDF of the Mahalanobis distance around each of the targets and the condition number of local Σ.

Name	Preprocessing	Correlation mean ± std	KS statistic mean ± std	Condition number of local Σ mean ± std
CAM	None	0.95 ± 0.03	0.48 ± 0.17	$1.15e+10 \pm 2.3e+10$
	SB	0.98 ± 0.01	0.28 ± 0.11	$3.97e+09 \pm 8.4e+09$
	KDR	0.992 ± 0.004	0.15 ± 0.04	$8.25e+08 \pm 2.6e+08$
	KDR (5 kcp)	0.991 ± 0.011	0.128 ± 0.07	$6.08e+01 \pm 1.04e+02$
OSLO1	None	0.999 ± 0.0003	0.04 ± 0.009	$6.33e+03 \pm 5.6e+03$
	SB	0.998 ± 0.0008	0.068 ± 0.01	$3.02e+03 \pm 2.5e+03$
	KDR	0.9995 ± 0.0003	0.04 ± 0.009	$3.90e+01 \pm 3.3e+01$
	KDR (5 kcp)	0.993 ± 0.003	0.11 ± 0.03	$1.20e+01 \pm 2.0e+01$
BJO	None	0.994 ± 0.02	0.076 ± 0.10	$7.34e+04 \pm 1.4e+05$
	SB	0.996 ± 0.006	0.089 ± 0.06	$5.38e+04 \pm 8.6e+04$
	SQRT + SB	0.996 ± 0.006	0.089 ± 0.06	$5.38e+04 \pm 8.6e+04$
OSLO2	None	0.996 ± 0.006	0.097 ± 0.06	$3.27e+04 \pm 1.0e+04$
	SB	0.996 ± 0.005	0.102 ± 0.05	$2.08e+04 \pm 1.8e+04$

TABLE 7: Correlation coefficient and Kolmogorov-Smirnov test statistic between empirical and theoretical CDF of the Mahalanobis distance calculated in a local window around each target and the condition number of local Σ for BJO and OSLO2, both without preprocessing.

TGT	Correlation	KS statistic	Condition number of local Σ
		BJO	
T1	0.9976	0.121	$1.746e+05$
T2	0.9968	0.104	$1.884e+05$
T3	0.9329	0.420	$5.405e+05$
T4	0.9991	0.055	$6.775e+03$
T5	0.9995	0.043	$6.864e+03$
T6	0.9994	0.052	$8.845e+03$
T7	0.9992	0.057	$6.654e+03$
T8	0.9999	0.027	$1.345e+04$
T9	0.9999	0.022	$1.264e+04$
T10	0.9998	0.033	$1.144e+04$
T11	0.9997	0.031	$1.134e+04$
T12	0.9997	0.032	$1.426e+04$
T13	0.9999	0.021	$1.123e+04$
		OSLO2	
T1	0.9995	0.052	$3.780e+04$
T2	0.9986	0.096	$4.352e+04$
T3	0.9991	0.052	$3.021e+04$
T4	0.9874	0.190	$1.935e+04$

well known that target contamination degrades the results of detectors [42, 43]. For the targets in the shadow at the edge of the forest (T4–T7) the normality assumption is reasonably well met and the condition number of the local covariance matrix is lower than for the other targets. The outer windows for these targets fall entirely in the shadow zone.

Table 7 shows that in OSLO2 the assumption of local normality is best met for targets T1 and T3 while for T4 the assumption is less valid. The corresponding figure shows that T4 has indeed the most heterogeneous local background. There is some target contamination between T2 and T3 when no data reduction is applied, while this is not the case when KDR is applied and only five kurtosis components are used.

Acknowledgments

The authors would like to acknowledge Salvatore Resta at the University of Pisa for providing the ground truth for the OSLO2 data set. The OSLO1, OSLO2, and BJO datasets were provided by the Norwegian Defence Research Establishment (FFI). The CAM scene was provided by ESA (DAISEX experiment) through a collaboration with ONERA. The MVSA matlab code, used in TLES, is kindly provided by Professor J. Bioucas on http://www.lx.it.pt/~bioucas/code.htm.

References

[1] D. W. J. Stein, S. G. Beaven, L. E. Hoff, E. M. Winter, A. P. Schaum, and A. D. Stocker, "Anomaly detection from hyperspectral imagery," *IEEE Signal Processing Magazine*, vol. 19, no. 1, pp. 58–69, 2002.

[2] S. Matteoli, M. Diani, and G. Corsini, "A tutorial overview of anomaly detection in hyperspectral images," *IEEE Aerospace and Electronic Systems Magazine*, vol. 25, no. 7, pp. 5–27, 2010.

[3] I. S. Reed and X. Yu, "Adaptive multiple-band CFAR detection of an optical pattern with unknown spectral distribution," *IEEE Transactions on Acoustics, Speech, and Signal Processing*, vol. 38, no. 10, pp. 1760–1770, 1990.

[4] C. I. Chang and S. S. Chiang, "Anomaly detection and classification for hyperspectral imagery," *IEEE Transactions on Geoscience and Remote Sensing*, vol. 40, no. 6, pp. 1314–1325, 2002.

[5] H. Kwon and N. M. Nasrabadi, "Kernel RX-algorithm: a nonlinear anomaly detector for hyperspectral imagery," *IEEE Transactions on Geoscience and Remote Sensing*, vol. 43, no. 2, pp. 388–397, 2005.

[6] A. Schaum, "Advanced methods of multivariate anomaly detection," in *Proceedings of the IEEE Aerospace Conference*, pp. 1–7, March 2007.

[7] A. P. Schaum, "Hyperspectral anomaly detection beyond RX," in *Proceedings of the SPIE Algorithms and Technologies for Multispectral, Hyperspectral and Ultraspectral Imagery XII*, vol. 6565, 2007.

[8] E. Lo and L. T. C. John Ingram, "Hyperspectral anomaly detection based on minimum generalized variance method," in *Proceedings of the Algorithms and Technologies for Multispectral, Hyperspectral, and Ultraspectral Imagery XIV*, vol. 6966, March 2008.

[9] C.E. Caefer, J. Silverman, O. Orthal, D. Antonelli, Y. Sharoni, and S.R. Rotman, "Improved covariance matrices for point target detection in hyperspectral data," *Optical Engineering*, vol. 47, no. 7, Article ID 076402, 2008.

[10] E. Lo and A. Schaum, "A hyperspectral anomaly detector based on partialling out a clutter subspace," in *Proceedings of the Algorithms and Technologies for Multispectral, Hyperspectral, and Ultraspectral Imagery XV*, vol. 7334, April 2009.

[11] E. Lo, "Maximized subspace model for hyperspectral anomaly detection," *Pattern Analysis and Applications*, pp. 1–11, 2011.

[12] E. Lo, "Variable subspace model for hyperspectral anomaly detection," *Pattern Analysis and Applications*, pp. 1–13, 2011.

[13] I. Kåsen, P. E. Goa, and T. Skauli, "Target detection in hyperspectral images based on multi-component statistical models for representation of background clutter," in *Proceedings of the SPIE on Electro-Optical and Infrared Systems: Technology and Applications*, vol. 5612, pp. 258–264, October 2004.

[14] D. Blumberg, E. Ohel, and S. Rotman, "Anomaly detection in noisy multi- and hyperspectral images of urban environments," in *Proceedings of the 3rd GRSS/ISPRS Symposium*, Tempe, Ariz, USA, March 2005.

[15] C. Willis, "Anomaly detection in hyperspectral imagery using statistical mixture models," in *Proceedings of the 2nd EMRS DTC Technical Conference*, Edinburgh, UK, 2005.

[16] O. Duran and M. Petrou, "A time-efficient method for anomaly detection in hyperspectral images," *IEEE Transactions on Geoscience and Remote Sensing*, vol. 45, no. 12, pp. 3894–3904, 2007.

[17] O. Duran and M. Petrou, "Spectral unmixing with negative and superunity abundances for subpixel anomaly detection," *IEEE Geoscience and Remote Sensing Letters*, vol. 6, no. 1, pp. 152–156, 2009.

[18] T. Veracini, S. Matteoli, M. Diani, and G. Corsini, "Fully unsupervised learning of Gaussian mixtures for anomaly detection in hyperspectral imagery," in *Proceedings of the 9th International Conference on Intelligent Systems Design and Applications (ISDA '09)*, pp. 596–601, December 2009.

[19] D. Borghys, E. Truyen, M. Shimoni, and C. Perneel, "Anomaly detection in hyperspectral images of complex scenes," in *Proceedings of the 29th Symposium of the European Association of Remote Sensing Laboratories*, Chania, Greece, June 2009.

[20] Y. Tarabalka, T. V. Haavardsholm, I. Kåsen, and T. Skauli, "Real-time anomaly detection in hyperspectral images using multivariate normal mixture models and GPU processing," *Journal of Real-Time Image Processing*, vol. 4, no. 3, pp. 287–300, 2009.

[21] D. Borghys, I. Kasen, V. Achard, and C. Perneel, "Comparative evaluation of hyperspectral anomaly detectors in different types of background," in *Proceedings of the SPIE Algorithms and Technologies for Multispectral, Hyperspectral, and Ultraspectral Imagery XVIII*, vol. 8390, Baltimore, April 2012.

[22] C. I. Chang, "Orthogonal subspace projection (OSP) revisited: a comprehensive study and analysis," *IEEE Transactions on Geoscience and Remote Sensing*, vol. 43, no. 3, pp. 502–518, 2005.

[23] P. Masson and W. Pieczynski, "SEM algorithm and unsupervised statistical segmentation of satellite images," *IEEE Transactions on Geoscience and Remote Sensing*, vol. 31, no. 3, pp. 618–633, 1993.

[24] J. Li and J. M. Bioucas-Dias, "Minimum volume simplex analysis: a fast algorithm to unmix hyperspectral data," in *Proceedings of the IEEE International Geoscience and Remote Sensing Symposium*, pp. III250–III253, July 2008.

[25] S. S. Shen and E. M. Bassett, "Information theory based band selection and utility evaluation for reflective spectral systems," in *Proceedings of the SPIE Algorithms and Technologies for Multispectral, Hyperspectral, and Ultraspectral Imagery VIII*, vol. 4725, pp. 18–29, April 2002.

[26] I. Kåsen, A. Rødningsby, T. V. Haavardsholm, and T. Skauli, "Band selection for hyperspectral target detection based on a multinormal mixture anomaly detection algorithm," in *Proceedings of the SPIE Algorithms and Technologies for Multispectral, Hyperspectral, and Ultraspectral Imagery XIV*, vol. 6966, March 2008.

[27] D. Peña, F. J. Prieto, and J. Viladomat, "Eigenvectors of a kurtosis matrix as interesting directions to reveal cluster structure," *Journal of Multivariate Analysis*, vol. 101, no. 9, pp. 1995–2007, 2010.

[28] E. A. Ashton, B. D. Wemett, R. A. Leathers, and T. V. Downes, "A novel method for illumination suppression in hyperspectral images," in *Proceedings of the SPIE Algorithms and Technologies for Multispectral, Hyperspectral, and Ultraspectral Imagery XIV*, vol. 6966, March 2008.

[29] R. Richter and A. Müller, "De-shadowing of satellite/airborne imagery," *International Journal of Remote Sensing*, vol. 26, no. 15, pp. 3137–3148, 2005.

[30] S. M. Adler-Golden, M. W. Matthew, G. P. Anderson, G. W. Felde, and J. A. Gardner, "An algorithm for de-shadowing spectral imagery," in *Proceedings of the 11th JPL Airborne Earth Science Workshop*, pp. 3–4, March 2002.

[31] M. Shimoni, G. Tolt, C. Perneel, and J. Ahlberg, "Detection of vehicles in shadow areas using combined hyperspectral and lidar data," in *Proceedings of the Geoscience and Remote Sensing Symposium (IGARSS)*, pp. 4427–4430, Vancouver, Canada, July 2011.

[32] B. D. Wemett, J. K. Riek, and R. A. Leathers, "Dynamic thresholding for hyperspectral shadow detection using Levenberg-Marquardt minimization on multiple gaussian illumination distributions," in *Proceedings of the SPIE Algorithms and Technologies for Multispectral, Hyperspectral, and Ultraspectral Imagery XV*, vol. 7334, April 2009.

[33] T. Skauli, "Sensor noise informed representation of hyperspectral data, with benefits for image storage and processing," *Optics Express*, vol. 19, no. 14, pp. 13031–13046, 2011.

[34] P. McCullagh and J. Nelder, *Generalized Linear Models*, chapter 6, CRC Press, 1999.

[35] J. Harsanyi, W. Farrand, and C. Chang, "Determining the number and identity of spectral endmembers: an integrated approach using neyman pearson eigenthresholding and iterative constrained rms error minimization," in *Proceedings of the 9th Thematic Conference on Geologic Remote Sensing*, February 1993.

[36] C. I. Chang and Q. Du, "Estimation of number of spectrally distinct signal sources in hyperspectral imagery," *IEEE Transactions on Geoscience and Remote Sensing*, vol. 42, no. 3, pp. 608–619, 2004.

[37] J. Bioucas-Diaz and M. Nascimiento, "Hyperspectral subspace identification," *IEEE Transactions on Geoscience and Remote Sensing*, vol. 46, no. 8, pp. 2435–2445, 2008.

[38] N. Acito, M. Diani, and G. Corsini, "A new algorithm for robust estimation of the signal subspace in hyperspectral images in the presence of rare signal components," *IEEE Transactions on Geoscience and Remote Sensing*, vol. 47, no. 11, pp. 3844–3856, 2009.

[39] N. Acito, "Hyperspectral signal subspace identification in the presence of rare signal components," *IEEE Transactions on Geoscience and Remote Sensing*, vol. 48, no. 4, pp. 1940–1954, 2010.

[40] B. Manly, Ed., *Multivariate Statistical Methods*, Chapman and Hall, Boca Raton, Fla, USA, 1995.

[41] O. Ledoit and M. Wolf, "A well-conditioned estimator for large-dimensional covariance matrices," *Journal of Multivariate Analysis*, vol. 88, no. 2, pp. 365–411, 2004.

[42] S. Matteoli, M. Diani, and G. Corsini, "Different approaches for improved covariance matrix estimation in hyperspectral anomaly detection," in *Proceedings of the Annual Meeting of the Italian National Telecommunications and Information Theory Group (GTTI '09)*, 2009.

[43] J. Theiler and B. R. Foy, "Effect of signal contamination in matched-filter detection of the signal on a cluttered background," *IEEE Geoscience and Remote Sensing Letters*, vol. 3, no. 1, pp. 98–102, 2006.

9

Resource Management in Satellite Communication Systems: Heuristic Schemes and Algorithms

Shahaf I. Wayer and Arie Reichman

School of Engineering, Ruppin Academic Center, Emek Hefer, Israel

Correspondence should be addressed to Arie Reichman, arier@ruppin.ac.il

Academic Editor: Yi Su

The high cost of frequency bandwidth in satellite communication emphasizes the need for good algorithms to cope with the resource allocation problem. In systems using DVB-S2 links, the optimization of resource allocation may be related to the classical multi-knapsack problem. Resource management should be carried out according to the requests of subscribers, their priority levels, and assured bandwidths. A satisfaction measure is defined to estimate the allocation processes. Heuristic algorithms together with some innovative scaling schemes are presented and compared using Monte Carlo simulation based on a traffic model introduced here.

1. Introduction

1.1. DVB-S2 in Satellite Communication. The most popular standard today in satellite communication is DVB-S2 [1], designed for transmission of digital television broadcasting. This relatively new standard is gradually replacing the classic DVB-S [2] in new systems as well as in existing systems. It is also used in new VSAT systems in forward link as well as in reverse link, in cases when the required data rates are higher than those supplied by DVB-RCS [3]. DVB-S2 is also used in satellite modems for point-to-point applications. In this paper we focus on a system that consists of DVB-S2 links with dynamic changes in the required parameters of the links due to changes in the traffic demands.

1.2. Resource Management. In order to save the operational cost of the satellite communication system, transmissions that share the total available frequency band have to be fairly divided according to the requirements of the subscribers [4]. For a star-type system with a central HUB that communicates with subscriber terminals, the control is conducted by the control center in the HUB according to subscriber needs. For other type of systems, like point-to-point, the control is conducted by a dedicated communication and control system that operates in addition to point-to-point links.

Regardless of the way the control is performed, a resource allocation algorithm has to be designed for optimal performances and best satisfaction.

In commercial systems run by a service provider, the Service Level Agreement (SLA), which is a set of parameters that defines the capabilities and the paid rights of every terminal in the system, connects the service provider and the subscribers. The Committed Information Rate (CIR) is the rate that is guaranteed by the system. Typically it is low enough so that the system can guarantee the user of this specific rate. The Peak Information Rate (PIR), on the other hand, is the maximum rate that may be allocated. When the user asks for a rate higher than CIR and lower or equal to PIR, the system will allocate this rate if available. When a user asks for a rate higher than PIR, the system will consider it as if the request was equal to PIR. The service provider has difficulty in providing the Committed Rate with certainty. In this research, we assumed that the bandwidth of a terminal is proportional to its rate. We use the term *assured bandwidth* as a quantity that the system can provide *most of the time* but not always.

The terminals have priority levels that determine their significance according to their payment to the service provider. In a situation of low resources, requests from high priority terminals may cause the system to decrease the rate of low priority terminals to their minimum.

In the course of the resource allocation cycle, subscribers issue requests for bandwidth. The bandwidths they own are removed from the spectrum list. The resource allocation process follows, and the bandwidths are reconfigured over the spectrum. This whole process of removing spectrum bands and reconfiguring the spectrum with the newly allocated bands creates gaps (*holes*) in the spectrum, resulting in a fragmented spectrum. The task of the resource allocation algorithms is to fit the requests into those *holes* in the spectrum.

In [5] it was shown that this task is equivalent to the multi-knapsack problem (MKP). Two heuristic algorithms were presented together with a single scaling scheme for the bandwidth requests. Scaling down the requests is necessary when their sum exceeds the available bandwidth budget, and utilizing the scaling scheme improves the algorithm results.

In this paper, we review the MKP problem and suggest an extension to it. We then present our idea of combining *heuristic algorithms* with scaling down schemes. Some innovative schemes are suggested to maximize a defined *satisfaction measure* and at the same time support the allocation of the *assured bandwidth* for each subscriber.

In Section 2, the system's quality of service parameters is presented and performance criteria are defined. The resource management process is presented in Section 3, including the mathematical background for the MKP problem with extension, together with a problem reformulation that may suggest a *scaling down* strategy as a natural way to solve the extended MKP problem.

Section 4 focuses on the algorithms used for the resource allocation, while Section 5 shows the significance of the scal-ing schemes. Several schemes are presented, based on the quality of service parameters. Then a simplified example is presented to illustrate the whole process. The techniques used to model the system dynamics are presented in Section 6, while Section 7 compares the performances obtained in simulation. Finally, we draw our conclusions in Section 8.

2. System Parameters and Performance Criteria

2.1. The Quality of Service Parameters. Parameters of quality of service determine the capability of providing better service to network traffic. The system we study assumes some quality of service characteristics for each subscriber. The algorithms presented here take into consideration two such parameters.

(a) Priority Level. A priority factor w_k is assigned to each subscriber type, such that higher level subscribers get priority in allocation of bandwidth following user request.

(b) Assured Level. Subscribers are guaranteed (with a high probability) to receive the bandwidth they request up to some assured amount Ω_k. Any bandwidth above this is optional, up to a maximal value determined by PIR.

These two parameters are used in the definition of the *satisfaction measure* function and play an important role in algorithm design and optimization of bandwidth utilization.

2.2. Disconnections. The fact that the DVB-S2 links were designed for fixed center frequency and symbol rate would cause disconnections while changing them. Therefore, allocating a new bandwidth to a user causes a disconnection to the subscriber. To take into account this fact, we use *effective allocated bandwidth* A related to the *allocated bandwidth* A_0 according to:

$$A = \begin{cases} \beta \cdot A_0, & \text{(if disconnected)}, \\ A_0, & \text{(otherwise)}, \end{cases} \tag{1}$$

where $\beta < 1$ is a constant expressing the effect of disconnection period, and the assumed dissatisfaction it causes. In this paper, we narrow ourselves to the case where users are not moved in the frequency range if they have no request for a change.

2.3. The Satisfaction Criterion. Resource allocation is conducted according to the requests of subscribers, their priority levels, and their assured bandwidths. For a subscriber with assured bandwidth Ω, a requested bandwidth R, and an effective allocated bandwidth A, and assuming that $R \geq A$, we suggest the satisfaction measure S that may be calculated for a factor $\alpha < 1$:

$$S(R, A, \Omega) = \begin{cases} \dfrac{\Omega + \alpha \cdot (A - \Omega)}{\Omega + \alpha \cdot (R - \Omega)}, & (R > A \geq \Omega), \\ \dfrac{A}{\Omega + \alpha \cdot (R - \Omega)}, & (R > \Omega \geq A), \\ \dfrac{A}{R}, & (\Omega > R \geq A). \end{cases} \tag{2}$$

The satisfaction function ranges from 0 to 1. The maximum satisfaction value $S = 1$ is reached when the allocated bandwidth is equal to the requested bandwidth. Also, one can see that when $A > \Omega$, the contribution of any increment ΔA to the function is smaller by a factor of $\alpha < 1$ than the contribution of the same increment ΔA when $A < \Omega$.

In order to reduce the number of variables involved in this formula, we express the satisfaction function in terms of the *normalized bandwidth variables*:

$$r = \frac{R}{\Omega}, \qquad a = \frac{A}{\Omega}, \tag{3}$$

such that the satisfaction measure may be defined as:

$$S(r, a) = \begin{cases} \dfrac{1 + \alpha \cdot (a - 1)}{1 + \alpha \cdot (r - 1)}, & (r > a \geq 1), \\ \dfrac{1}{1 + \alpha \cdot (r - 1)}, & (r > 1 \geq a), \\ \dfrac{a}{r}, & (1 > r \geq a). \end{cases} \tag{4}$$

This is a concave function increasing monotonically both in the r direction and in the a direction.

For a given set of users with normalized requests and allocated bandwidths r_i, a_i and priority factors w_i, the total (weighted) satisfaction $\overline{S}$ is defined as:

$$\overline{S} = \frac{\sum_i w_i \cdot S(r_i, a_i)}{\sum_i w_i}. \quad (5)$$

The total satisfaction function gets the maximum of 1 when all requests are fulfilled. This quantitative measure may be used to compare the performance of the various algorithms and schemes.

2.4. The Requests Scale Down Criterion. While resource allocation algorithms fit the bandwidth requests into the *holes* in the spectrum, some *holes* may become *under populated* and some *over-populated.*

The first situation (*under population*) happens when the sum of requests inserted into a *hole* does not fully cover it.

The other one (*over population*) happens when the sum of requests assigned into a *hole* is greater than the capacity of the *hole.* In this situation the requests assigned to that *hole* should be scaled down to fit into the *hole* capacity.

Let $\{R_j^{(k)}\}$ be the set of requests allocated into a *hole* $H^{(k)}$, Then the total amount δ of the requests bandwidth that is scaled down is:

$$\delta = \sum_k \max\left(\left[\sum_j R_j^{(k)}\right] - H^{(k)}, 0\right). \quad (6)$$

The amount δ is the *requests scale down* factor. It measures the effectiveness of the algorithms in finding a good arrangement of the requests into the gaps (*holes*) in spectrum. A lower value indicates better algorithm performance.

3. The Resource Allocation Process

The disconnections caused while changing frequency and symbol rates pose a serious constraint in the design of optimal allocation algorithms. Therefore, links of terminals that did not require any change should stay in the same location in the spectrum, while new requests should be allocated only in the free portions (*holes*) of the spectrum.

In this part, we will show that this problem is equivalent to the multi-knapsack problem (MKP) [6] known in computational complexity theory and then draw a scheme for a computational process to cope with it.

3.1. Resource Allocation as a Complexity Theory Problem. The connection between the MKP problem and satellite communication resource allocation was already reported by Birmani [7]. However, his work deals with *power allocation* and burst scheduling problems, which are *inherently quite different* in their nature than the *bandwidth allocation* problems in our concern.

The association between resource allocation in communication and the MKP problem was already observed by Rajkumar et al. [8]. Their research was on a QoS-based Resource Allocation Model (QRAM) for satisfying multiple quality-of-service dimensions in a resource-constrained environment. Using this model, available system resources can be apportioned across multiple applications, thus maximizing the net utility that accrues to the end users of those applications. They showed that the Q-RAM problem of finding the optimal resource allocation to satisfy multiple QoS dimensions is NP hard. There is a similarity between the resources, QoS, and utility factor in their research and the bandwidth, priority level, and satisfaction function in this research. However, in our work the satisfaction function is different than their utility factor and the methods we suggest are simpler to implement.

The task of the resource allocation algorithms is to find an optimal match between a set of N requests $\{R_i \mid 0 \le i < N\}$ and a set of M holes $\{H^{(j)} \mid 0 \le j < M\}$ in a spectrum such that *all* the N_j requests are inserted into those holes:

$$\sum_{i=0}^{N_j} R_{K_j(i)} \le H^{(j)}, \quad (7)$$

where $\{K_j\}$ are M disjoint subsets of indexes: the size of set K_j is N_j, such that their union covers the set $\{1, \ldots, N\}$:

$$\bigcap K_j = \phi, \qquad \bigcup K_j = \{i \mid i \le N\}, \quad (8)$$

where $K_j(i)$ is the ith element in the set K_j.

The algorithm theory classifies such a problem as a variation of the *multi-knapsack problem.* Given M containers (knapsacks) with capacities $H^{(j)}$ and N items with values R_i, the task is to split the set $\{R_i\}$ into M disjoint subsets, such that the sum of item values in each subset will not exceed the given capacities $H^{(j)}$ as in (7).

Although the set of inequalities in the problem definition resembles a typical linear programming formulation, it is not a linear programming problem. Instead, it belongs to a family of *subset sum* problems known as knapsack problems.

The knapsack problem is part of the family of combinatorial NP-complete problems [9], meaning that it is computationally difficult to solve in general (except for a $O(M^N)$ brute force solution). It is difficult enough to be chosen as a trapdoor cryptographic function (i.e., the Merkle-Hellman cryptosystem).

Yet the resource allocation problem does not fit the exact definition of a knapsack problem. There is one essential difference: when (7) cannot be fully satisfied, then the target of optimization should be replaced by:

$$\sum_{i=0}^{N_j} C_i^{(j)} \cdot R_{K_j(i)} \le H^{(j)}, \quad \left(0 < C_i^{(j)} \le 1\right), \quad (9)$$

such that the factors $C_i^{(j)}$ are maximal.

This is a "softer" version of the original problem (7). It extends the problem adding some degree of complexity (finding the factors $C_i^{(j)}$). On the other hand, it eases the task somewhat, adding several degrees of freedom so that a solution can be found, even in situations where the original problem is practically insolvable.

So, because this is an NP-Hard problem, we shall not try to solve it fully. Instead we may design some practical heuristic algorithms and reach suboptimal solutions [10].

3.2. The Resource Management Process. The resource allocation problem is too complex to be solved at once. Instead of solving (9) directly by a single algorithm, we suggest splitting it into three subproblems:

(1) find an initial approximation to the $C_i^{(j)}$ factors;

(2) solve the main resource allocation problem;

(3) find the final values of the $C_i^{(j)}$ factors.

Following this model, the entire resource management process consists of three stages, as illustrated in Figure 1.

(1) The prescaling Step. In order for the resource allocation algorithms to operate properly, subscriber requests for bandwidth *cannot* exceed the available bandwidth budget. However, this situation does occur frequently. To cope with this problem, we suggest the *pre-scaling* process, to scale down subscriber requests according to the priority factors w_i and the assured bandwidths Ω_j.

(2) The resource allocation algorithm finds the best fit between the requests and the *holes* in the spectrum.

(3) The postscaling step is performed in one of the following situations:

  (a) if the sum of the allocated requests assigned to a *hole* is bigger than the size of the *hole*, then the requests should be *downscaled*;

  (b) if the sum of the allocated requests assigned to a hole is smaller than the size of the *hole*, and the requests were downscaled (by means of pre-scaling stage), then they may be *upscaled*. The allocated bandwidths are scaled up such that their new value will not exceed the original requests, and in addition their sum will not exceed the size of the hole.

In either case, the scaling is conducted according to the priority factors w_i and the assured bandwidths Ω_i in a manner very similar to the prescale stage.

The feedback in Figure 1 marks the transition to the next phase of the algorithm, when new requests are generated.

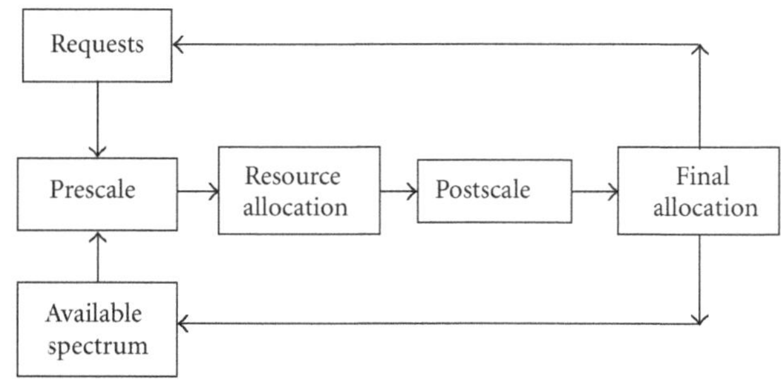

FIGURE 1: The resource allocation process flow chart.

4. The Resource Allocation Algorithms

Two heuristic algorithms are presented to solve the main resource allocation problem:

(i) the Fast MKP solver;

(ii) the IBF (Insert to Best Fit) MKP solver.

Both algorithms are classified as *greedy algorithms*. At each stage they take the locally optimal choice rather than aiming for the global optimum. Such algorithms are widely used to solve heuristically *dynamic programming* problems.

4.1. The Fast MKP Algorithm. When allocating a request to a hole, we define the residue remainder after the allocation, that is, the size of the hole less the size of the request. The *Fast MKP* algorithm adopts a strategy of *largest residues first*, that is, it fits the largest requests to *holes* with the largest residue in decreasing order.

In order to perform this efficiently, we maintain a binary tree of *hole* items (sorted by the residue size) to enable a *search and insert* operation in $O(\log(M))$ time steps [11].

The algorithm consists of the following steps.

(1) Make a sorted list of requests in *decreasing order* (largest to smallest) and a binary tree of residues associated with the *hole* items.

(2) Fit the *largest request* to the biggest residue *hole* item.

(3) Update the residue value corresponding to the *hole* item and insert it back into the tree according to the residue value.

(4) Discard the request from the list of requests.

(5) Repeat steps (2)–(4) until the request list is empty.

This algorithm is efficient with respect to run time and memory consumption.

4.2. The IBF MKP Algorithm. This algorithm adopts a strategy of *best fit residue*, that is, it fits requests in decreasing order into *holes* with the residue that fits best (if no residue fits, it inserts to the largest one—). A *best fit* for any R_j over a *decreasing* series of residues $\{Z_i\}$, is any Z_i, such that $Z_{i+1} < R_j \leq Z_i$.

(1) Start with a sorted list of requests in *decreasing order* (largest to smallest) and a binary tree of residues associated with the *hole* items.

(2) Insert the largest request into the *hole* item, with the smallest residue that is larger than the request.

(3) If there is no such residue, insert the request into the *hole* with the largest residue.

(4) Update the residue value corresponding to the *hole* item and insert it back into the tree according to the residue new value.

(5) Discard the request from the list of requests.

(6) Repeat steps (2)–(5) until the request list is empty.

The purpose of stage (3) in the algorithm is to ensure that *all* the requests are inserted into the *holes*, even when *holes* become *overpopulated*. This may involve temporarily assigning negative values to some residues. This is quite legitimate, because the residue values have instrumental meaning only. The *postscaling* stage described in next section would resolve this problem.

5. The Scaling Schemes

The algorithms presented above perform poorly under a shortage of bandwidth resources. We suggest activating some scaling schemes when the sum of the requests is greater than the sum of the holes. In those cases, such schemes can share *fairly* the available bandwidth resources, taking into consideration the *quality of service* parameters in order to resolve the users' diversity issues.

For a set of requests $\{R_i\}$ and a set of *holes* $\{H^{(j)}\}$ such that $R = \sum R_i$ and $H = \sum H_i$ we introduce *scaling schemes* to take place when $R > H$.

Introducing scaling stages into the allocation process is crucial. It is analogous to *adding filters* at the entrance and outlet of an electronic system in order to adjust the signal to the dynamic range of a system. The design of scaling schemes may dramatically influence the overall performance of the allocating process just like filters that affect system's behavior.

We present first the schemes for the *downscaling* performed in the prescaling step. In correspondence with the quality of service levels [12], we consider several levels of scaling schemes in accordance with the number of parameters taken into consideration:

(i) the basic scheme, which takes no parameters;

(ii) *the priority-oriented scheme*, which takes into consideration the priority factors w_i only;

(iii) advanced schemes, taking into consideration the assured bandwidth Ω_i as well as the priority factors w_i.

5.1. The Basic Scaling Scheme. In absence of any priority criteria for sharing the bandwidth, the requests are simply scaled down proportionally to their value:

$$R_k^{(\text{new})} = \frac{H}{R} \cdot R_k. \tag{10}$$

Such a scheme corresponds to a network system that supports *best-effort* subscribers only, with no quality of service parameters.

5.2. The Priority-Oriented Scaling Scheme. This scheme takes into consideration only the *priority factors* w_i. Such a scheme is applicable for network systems that support *at least* the *differentiated service* level.

The amount of bandwidth to be reduced is the difference between the sum of requests and the sum of holes:

$$D = R - H. \tag{11}$$

This amount should be shared proportionally to the requests R_j (larger requests are reduced more than small ones) and inversely proportional to the priority factors w_j (higher priority requests are reduced less than the lower ones). So, the request R_j should be reduced by amount d_j given by: $d_j = C \cdot (R_j/w_j)$ such that $D = \sum_j d_j$ is the total reduction.

Hence: $D = C \cdot (\sum_j R_j/w_j)$ where C is some proportion factor that can be written as:

$$C = \frac{D}{\sum_k R_k/w_k} = \frac{R-H}{\sum_k R_k/w_k}. \tag{12}$$

And finally, the requests are downscaled according to:

$$R_j^{(\text{new})} = \left(1 - \frac{C}{w_j}\right) \cdot R_j. \tag{13}$$

This scheme works unconditionally (but it does not take into consideration the assured bandwidth Ω).

5.3. Advanced Schemes. High-quality network systems should take into consideration the subscribers' *assured bandwidths* Ω_i, as well as their *priority factors* w_i. The constraint imposed by the assured bandwidth condition can be expressed as:

$$R_i^{(\text{new})} \geq \min(R_i, \Omega_i). \tag{14}$$

This condition could be met only when:

$$\sum H_i \geq \sum \min(R_i, \Omega_i). \tag{15}$$

When this condition is satisfied, we can allocate bandwidth taking into account both the priority factors and the assured bandwidths. Otherwise, the *priority-oriented scaling scheme* should be used.

Our strategy is to grant all the requests below the assured bandwidth and scale down *only* the others.

The two schemes we present here have the same "prolog", and differ only in the final stage.

(0) Start with a set of indexes $I = \{1, \dots, N\}$, where N is the number of requests.

(1) Split I into 2 subsets:

$$I^{(-)} = \{i \mid R_i \leq \Omega_i\} \text{ and}$$
$$I^{(+)} = \{i \mid R_i \leq \Omega_i\} \text{ (the complementary set)}.$$

(2) Then define

$$H^{(+)} = H - \sum_{i \in I^{(+)}} R_i, \qquad \Omega^{(+)} = \sum_{i \in I^{(+)}} \Omega_i. \tag{16}$$

(3) $R_i^{(\text{new})} = R_i$, $i \in I^{(-)}$ ("grant requests less than user's assured bandwidth").

(4) Scale the requests in $I^{(+)}$ using one of the schemes described next.

(i) Difference-Oriented Scaling. Using the *difference* $(R_i - \Omega_i)$ weighted by the priority factors w_i, we get:

$$R_i^{(\text{new})} = \Omega_i + C \cdot w_i \cdot (R_i - \Omega_i), \quad i \in I^{(+)}. \tag{17}$$

In similar arguments used in developing the *priority-oriented scaling scheme*, we find:

$$C = \frac{H^{(+)} - \Omega^{(+)}}{\sum_{i \in I^{(+)}} (R_i - \Omega_i) \cdot w_i}. \tag{18}$$

(Note that the denominator $(R_i - \Omega_i) \cdot w_i$ is always positive because $R_i > \Omega_i$ for all $i \in I^{(+)}$).

(ii) Ratio-Oriented Scaling. Using the *ratio* (R_i/Ω_i) weighted by the priority factors w_i, we get:

$$R_i^{(\text{new})} = \Omega_i + C \cdot w_i \cdot \left(\frac{R_i}{\Omega_i}\right), \quad i \in I^{(+)}, \tag{19}$$

where again by similar arguments we get:

$$C = \frac{H^{(+)} - \Omega^{(+)}}{\sum_{i \in I^{(+)}} (R_i/\Omega_i) \cdot w_i}. \tag{20}$$

After the prescaling procedure, the algorithms in Section 4 are applied. This prescaling is a necessary but not sufficient condition to be able to complete the procedure. However, the postscaling can be used to complete it.

5.4. The Postscaling Schemes. The postscaling stage makes the final correction to the allocated bandwidths. It may involve more downscaling to some of the requests. This happens when the allocating algorithm inserts to a hole a set of requests whose sum exceeds the hole size and assigns a negative value to the residue associated with the hole. In this case, the downscaling is performed in essentially same way as in the prescale step, except that the calculations are made separately for each *hole* j. Hence H_j is used instead of H and $\{R_i^{(j)}\}$, and the set of requests allocated to H_j would replace R_i.

There are situations when the allocating algorithm assigns to a hole requests whose sum is less than the *hole* size. In this case, if some of the requests were over-decreased in the prescaling stage, they should be now increased back (up to the original amount) to fit the hole size.

5.5. Concluding Remarks. We have presented here some scaling schemes to support *fair* sharing of available bandwidth resources in shortage conditions.

Naturally, the *advanced schemes* are expected to perform better than the *priority oriented scheme* because they take into consideration the assured bandwidths as well as the priority factors.

However, as we have pointed out, the other scaling schemes should not be abandoned. In some exceptional situations (heavy traffic periods for instance), when condition (15) is not satisfied, the *priority oriented* scheme (or the *Basic* one) should be used instead.

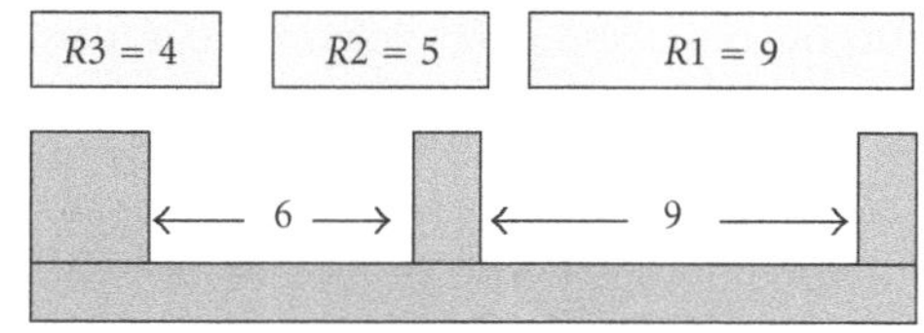

Figure 2: The initial state of *holes* and requests.

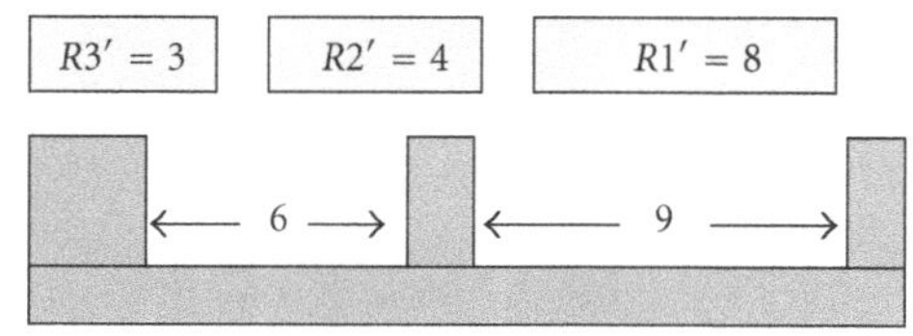

Figure 3: The state of *holes* and requests *after prescaling*.

5.6. An Illustrative Example for the Allocation Cycle. To illustrate the whole resource allocation cycle, consider the following simplified example. Consider three requests issued to the system:

$$R_1 = 9, \qquad R_2 = 5, \qquad R_3 = 4. \tag{21}$$

When there are two available "holes" in the spectrum:

$$H_1 = 9, \qquad H_2 = 6. \tag{22}$$

As shown in Figure 2.

The sum of the requests $\sum R_j = 18$ is greater than the available space in the spectrum $\sum H_i = 15$, meaning there is deviation of 3, and a scaling phase should take place to address it.

(i) Prescaling. For the sake of simplicity and in order to make it intuitive to the reader, tedious calculations involving QoS parameters considerations are dropped. Instead, we will focus on illustrating the concepts of the scaling stages.

To solve the deviation problem, the prescaling phase can reduce each request by 1. For instance,

$$R = \{9, 5, 4\} \longrightarrow \{8, 4, 3\}. \tag{23}$$

As shown in Figure 3.

This way $\sum R_j = 15$, which is exactly the amount of available space.

(ii) Allocation. The *residue* of H_i was denoted as Z_i. Initially set: $\{D\} = \{H\}$. Then:

(1) insert $R_1 = 8$ into H_1 (with $Z_1 = 9$) such that the residue is updated to $D_1 = 1$;

(2) insert $R_2 = 4$ into H_2 (with $Z_2 = 6$) which makes this residue become $Z_2 = 2$;

(3) finally, insert $R_3 = 3$ into H_2 (with $Z_2 = 2$). This makes the *residue* Z_2 receive a *negative value*: $Z_2 = -1$!

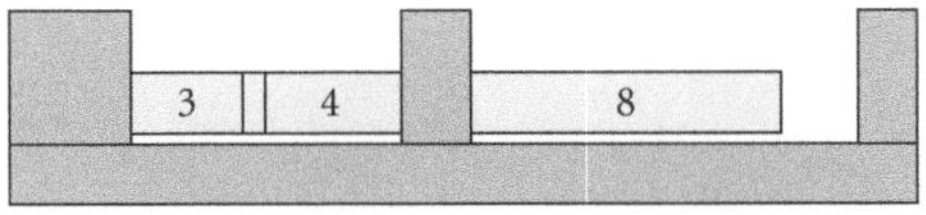

FIGURE 4: The requests inserted to the *holes by the allocation algorithm*. Notice the *overlapping* between R_3' and R_4'.

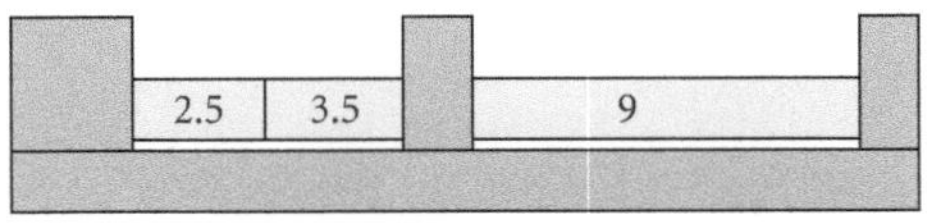

FIGURE 5: The final state *after the postscaling*.

This state of affairs is presented in Figure 4. The overpopulation situation in H_2 caused by the allocation algorithm is expressed by the *overlapping* R_2 and R_3 in the figure.

As stated in Section 4.2, assigning a negative value to some residues is legitimate as an intermediate stage in calculation, and the task of the *postscaling* phase is to fix this situation.

(iii) Postscaling. This phase fine tunes the bandwidths allocated. Recall that originally the request R_1 was 9, and then it was reduced to 8 and allocated to H_1, which has a capacity of 9. Obviously the original request can be granted. Thus the final allocation for R_1 should be $A_1 = 9$.

R_2 and R_3 were allocated to H_2 which has a capacity of 6, but $R_2 + R_3 = 7$, so they should be reduced again to: $A_2 = 3.5$ and $A_3 = 2.5$ such that $A_2 + A_3 = 6$ to fit the $H_2 = 6$ capacity.

So the final resource allocation yields:

$A_1 = 9$ inserted into $H_1 = 9$ and $\{A_2 = 3.5, A_3 = 2.5\}$ inserted into $H_2 = 6$. (See Figure 5).

6. Modeling the Resource Allocation Process

In order to simulate the system dynamics under the resource allocation process described above, we derived a model to produce the network traffic and the inputs to the process. We distinguish between the *demands* subscribers have and the *requests* they issue to the system. New demands for bandwidth emerge for each subscriber in random time steps. We assume that the demands ν_k of subscriber k are *gamma* distributed.

$$\nu_k \sim \Gamma(m, \lambda_k). \tag{24}$$

The gamma distribution [13] is a two-parameter family of continuous probability distributions. It has a shape parameter m, a scale parameter λ_k, and the average value:

$$E[\nu_k] = m \cdot \lambda_k. \tag{25}$$

The traffic models based on random variables with Gamma distribution are used in applications such as video and speech [14, 15]. Such a random variable has a probability distribution with a shape similar to a Gaussian random variable but has only positive values.

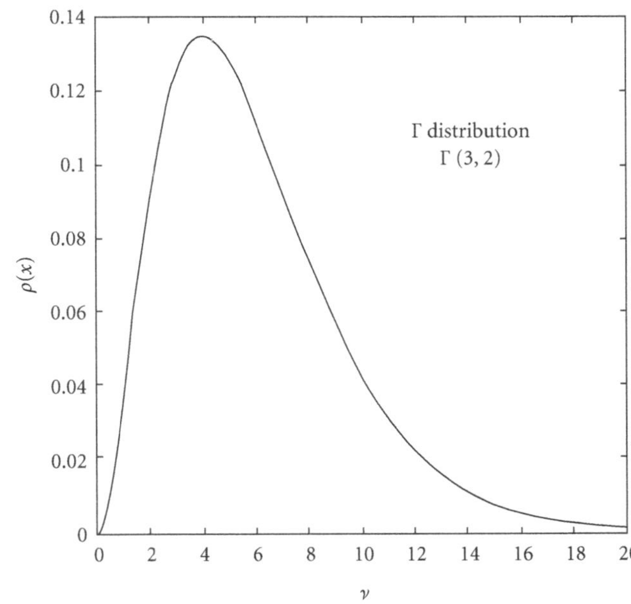

FIGURE 6: Gamma distribution: Γ(3,2).

The parameters are chosen such that the mean demand of subscriber k, ν_k, is proportional to Ω_k, the assured bandwidth of subscriber k, by a constant factor μ:

$$E[\nu_k] = \mu \cdot \Omega_k. \tag{26}$$

So comparing (25) and (26):

$$\lambda_k = \frac{\mu \cdot \Omega_k}{m}. \tag{27}$$

In the simulation, the values used are: $\mu = 2.0$ and $m = 3$, as shown in Figure 6.

We may assign a probability p for new demand to emerge such that in time step n we have demand $\nu_k^{(n)}$:

$$\nu_k^{(n)} = \begin{cases} \nu \sim \Gamma(m, \lambda_k), & \text{with probability } p, \\ 0, & \text{with probability } (1-p). \end{cases} \tag{28}$$

We assume that *very often user requests are not fully granted*. In this situation, a subscriber should request the remainder $R_k - A_k$ in the next time step, added to a new demand (if one emerges). This can be modeled by assigning a variable Q_k to each subscriber k to accumulate requests R_k, subtract the allocated bandwidth A_k, and add demands ν_k:

$$Q_k^{(n+1)} = Q_k^{(n)} + R_k^{(n)} - A_k^{(n)} + \nu_k^{(n+1)}. \tag{29}$$

Stability considerations suggest limiting the request amount a subscriber k may issue. In Section 1.1, we called this limit the Peak Information Rate (PIR). We set this limit to $\kappa \cdot \Omega_k$, where $\kappa > 1$ is a constant ($\kappa = 2$ in our simulation):

$$R_k^{(n+1)} = \min\left(Q_k^{(n+1)}, \kappa \cdot \Omega_k\right). \tag{30}$$

Table 1: The simulation configuration.

Parameter		Equation	Value
Satisfaction coefficient	α	(2)	0.75
Disconnection penalty factor	β	(1)	0.9
Mean demand to assured bandwidth ratio	μ	(18)	2.0
Demand update rate	p	(21)	0.25
Threshold for new request	γ	(24)	0.40
Request bound to assured bandwidth ratio	κ	(23)	2.0

Table 2: Comparison of the "requests scale-down" between the algorithms.

	Fast-MKP	IBF-MKP
$\delta = 0$	34.0%	43.2%
$0 < \delta \leq 250$	6.0%	9.8%
$250 < \delta \leq 500$	5.4%	8.8%
$500 < \delta \leq 1000$	11.0%	14.6%
$1000 < \delta \leq 2000$	19.8%	22.6%
$2000 < \delta \leq 4000$	23.0%	35.2%
$\delta > 4000$	6.0%	10.0%

Table 3: Comparison of the *requests scale-down* for different scaling schemes.

	Priority oriented	Difference oriented	Ratio oriented
$\delta = 0$	32.0%	44.6%	48.2%
$0 < \delta \leq 25$	5.8%	8.4%	7.2%
$25 < \delta \leq 50$	8.2%	11.4%	6.4%
$50 < \delta \leq 100$	10.4%	20.6%	8.0%
$100 < \delta \leq 200$	14.4%	24.2%	12.2%
$200 < \delta \leq 400$	17.8%	20.4%	13.8%
$\delta > 400$	11.4%	15.0%	7.8%

Finally, in order to avoid the penalty of unwanted disconnections associated with new bandwidth allocations, a restriction is superimposed on issuing new requests. Only changes exceeding some threshold can justify issuing new requests. To assure this, we set up a rule: *a new request shall be issued only when:*

$$\left| R_k^{(n+1)} - A_k^{(n)} \right| \geq \gamma \cdot A_k^{(n)}. \quad (31)$$

In the simulation, $\gamma = 0.4$ is used.

7. Comparison between Algorithms

Algorithms can be ranked either by their efficiency or by the quality of the results they produce. We shall examine and compare our algorithms in the following aspects:

(i) algorithm complexity (memory and time consumption);

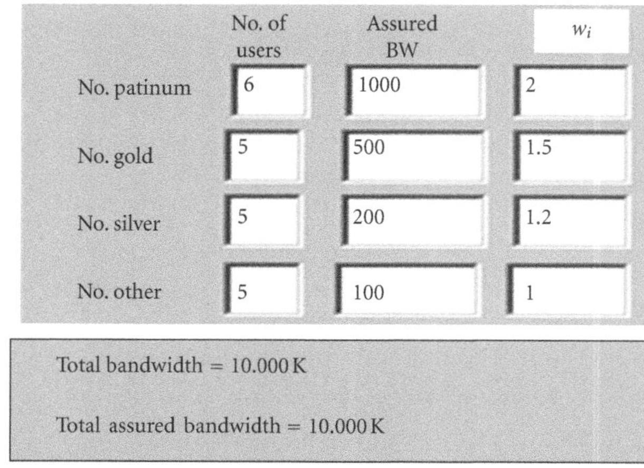

Figure 7: The system configuration in a simulation snapshot.

(ii) distribution of the satisfaction measure S;

(iii) distribution of the requests scale down δ.

We considered a typical system with 4 types of subscribers. Each type has its priority factor w_k and assured bandwidth Ω_k. The system setup is shown in a simulation snapshot presented in Figure 7.

The other simulation parameters are presented in Table 1.

7.1. The Complexity of the Algorithms. Let us consider the complexity of the algorithms (without the scaling part) for N new requests and M *holes* in the spectrum.

The time complexity of both Fast-MKP and IBF-MKP algorithm is bounded by $O(N \cdot \log(M))$ *integer operations*, as the *search and insert* operation in a binary tree is $O(\log(M))$.

The memory complexity of both algorithms is $O(N+M)$. However, we observed (experimentally) that the IBF-MKP algorithm runs slower than Fast-MKP by no more than 20% on the average. So, both algorithms are very efficient in runtime and memory consumption.

7.2. The Distributions of the Satisfaction Measure. The distributions for the satisfaction measure for the IBF-MKP and the Fast-MKP algorithms are presented by the histograms in Figures 8 and 9, obtained by simulation without the effect of the scaling schemes. For the criterion of satisfaction, the IBF-MKP algorithm achieves better results.

The effect of the scaling schemes on the satisfaction measure is shown in Figures 10, 11, and 12. The results are presented for the IBF-MKP algorithm only. The scaling schemes do improve results over those obtained without the scaling. The *ratio-oriented* scaling scheme yields better results than *Difference Oriented*, while the *Priority Oriented* scaling scheme is in third place.

7.3. The Distributions of the "Requests Scale-Down". The simulation records the amount of bandwidth scaled down in the resource allocation. Table 2 compares the distribution

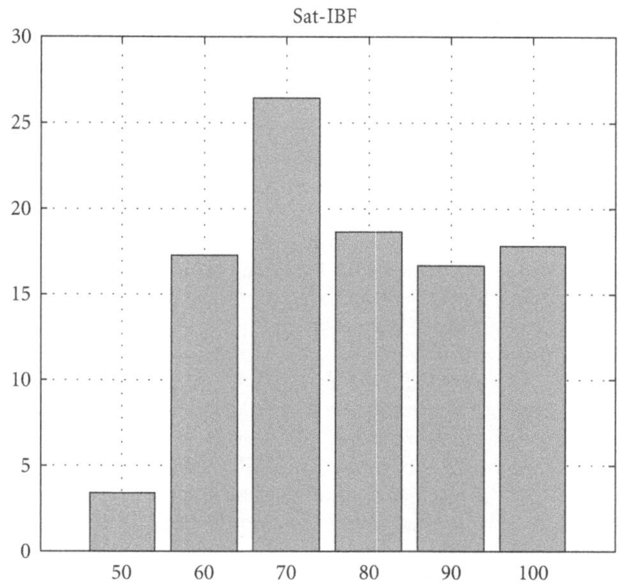

FIGURE 8: Satisfaction histogram IBF-MKP algorithm with *priority-oriented* scaling scheme.

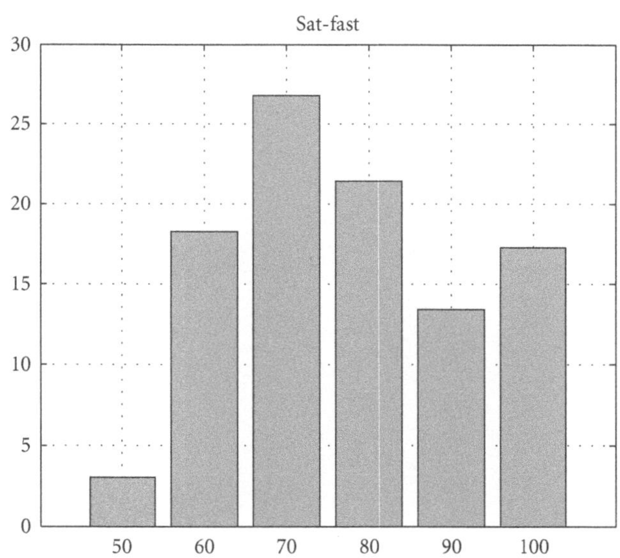

FIGURE 9: Satisfaction histogram for Fast-MKP Algorithm with priority-oriented scaling scheme.

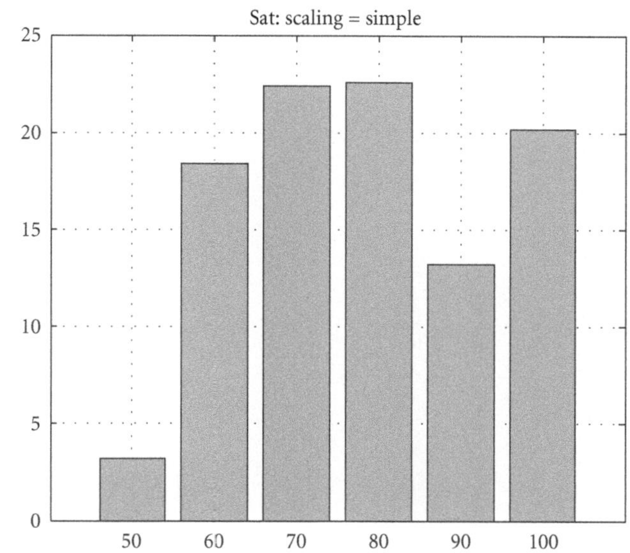

FIGURE 10: Satisfaction histogram for *priority-oriented* scaling scheme.

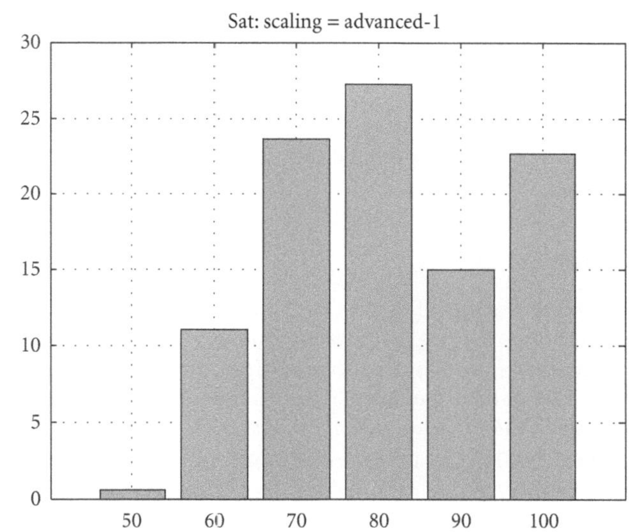

FIGURE 11: Satisfaction histogram for *difference-oriented* scaling scheme.

of this variable for the two algorithms (without utilizing the scaling schemes).

Clearly, the IBF-MKP algorithm achieves better results for the average amount of requests scale-down in the allocation process.

The effect of scaling schemes on the *requests scale-down* factor δ is compared in Table 3, which presents results where again the algorithm is the IBF-MKP.

As expected, scaling schemes improve the *scale-down* results. The *ratio-oriented* gives the best results, while the *difference-oriented* is somewhat less effective, and the *priority-oriented* offers the least effectiveness.

The difference is probably due to the ability to grant all requests below the assured bandwidth. This approach makes these advanced scaling schemes superior to other schemes.

8. Conclusion

In a rapidly evolving communications market with ever-growing demand for more bandwidth, a good resource management mechanism is essential for efficient digital communication of any kind. The algorithms suggested here together with the scaling schemes successfully perform this task with reasonable computational complexity, producing a high-quality resource allocation with remarkable performance.

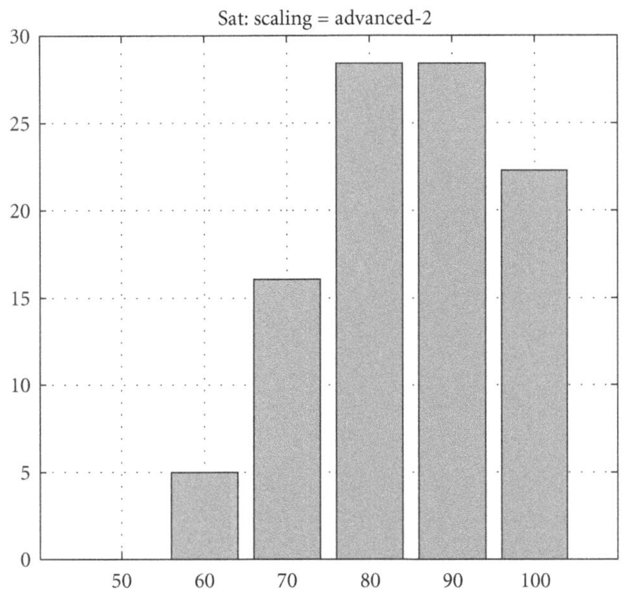

FIGURE 12: Satisfaction histogram for *ratio-oriented* scaling scheme.

Acknowledgments

The authors wish to acknowledge the cooperation of Ayecka Communication Systems Ltd. and the support of the MAGNETON program of the Israel Ministry of Industry, Trade and Labor.

References

[1] ETSI: EN 302 307, Digital Video Broadcasting (DVB), "Second generation framing structure, channel coding and modulation systems for Broadcasting, Interactive Services, News Gathering and other broadband satellite applications (DVB-S2)," *European Standard* (Telecommunications series), 2005.

[2] ETSI: EN 300 421, Digital Video Broadcasting (DVB), "Framing structure, channel coding and modulation for 11/12 GHz satellite services (DVB-S)," *European Standard* (Telecommunications series),1994.

[3] ETSI: EN 301 790, Digital Video Broadcasting (DVB), "Interaction channel for satellite distribution systems (DVB-RCS)," *European Standard* (Telecommunications series), 2005.

[4] G. Giambene, *Resource Management in Satellite Networks Optimization and Cross-Layer Design*, Springer, 2007.

[5] S. I. Wayer and A. Reichman, "Resource management in satellite communication systems—Heuristic algorithms," in *Proceedings of the IEEE 26th Convention of Electrical and Electronics Engineers in Israel (IEEEI '10)*, pp. 342–346, November 2010.

[6] D. Pisinger, "A Minimal Algorithm for the Bounded Knapsack Problem," *INFORMS Journal on Computing*, vol. 12, no. 1, pp. 75–82, 2000.

[7] V. Birmani, *Resource allocation for Ka-band broadband satellite systems [M.S. thesis]*, University of Maryland, 1999.

[8] R. Rajkumar, C. Lee, J. P. Lehoczky, and D. P. Siewiorek, "Practical solutions for QoS-based resource allocation problems," in *Proceedings of the 19th IEEE Real-Time Systems Symposium*, pp. 296–306, December 1998.

[9] S. Goddard, "P, NP, and NP-Complete," CSCE 310J lecture notes on "Data Structures & Algorithms", University of Nebraska-Lincoln, http://www.cse.unl.edu/~goddard/Courses/CSCE310J/Lectures/Lecture10-NPcomplete.pdf.

[10] M. Hifi, M. Michrafy, and A. Sbihi, "Heuristic algorithms for the multiple-choice multidimensional knapsack problem," *Journal of the Operational Research Society*, vol. 55, no. 12, pp. 1323–1332, 2004.

[11] T. H. Cormen, C. E. Leiserson, R. L. Rivest, and C. Stein, *Introduction to Algorithms*, MIT Press, 2nd edition, 2001.

[12] Quality of Service Networking, http://docwiki.cisco.com/wiki/Quality_of_Service_Networking.

[13] A. Papoulis, *Probability, Random Variables, and Stochastic Processes*, McGraw-Hill, New York, NY, USA, 2nd edition, 1984.

[14] M. Menth, A. Binzenhöfer, and S. Mühleck, "Source models for speech traffic revisited," *IEEE/ACM Transactions on Networking*, vol. 17, no. 4, pp. 1042–1051, 2009.

[15] O. Rose, "Simple and efficient models for variable bit rate MPEG video traffic," *Performance Evaluation*, vol. 30, no. 1-2, pp. 69–85, 1997.

10

Blind-Matched Filtering for Speech Enhancement with Distributed Microphones

Sebastian Stenzel and Jürgen Freudenberger

Institute for System Dynamics, HTWG Konstanz, Brauneggerstrasse 55, 78462 Konstanz, Germany

Correspondence should be addressed to Sebastian Stenzel, sebastian.stenzel@htwg-konstanz.de

Academic Editor: Sriram Srinivasan

A multichannel noise reduction and equalization approach for distributed microphones is presented. The speech enhancement is based on a blind-matched filtering algorithm that combines the microphone signals such that the output SNR is maximized. The algorithm is developed for spatially uncorrelated but nonuniform noise fields, that is, the noise signals at the different microphones are uncorrelated, but the noise power spectral densities can vary. However, no assumptions on the array geometry are made. The proposed method will be compared to the speech distortion-weighted multichannel Wiener filter (SDW-MWF). Similar to the SDW-MWF, the new algorithm requires only estimates of the input signal to noise ratios and the input cross-correlations. Hence, no explicit channel knowledge is necessary. A new version of the SDW-MWF for spatially uncorrelated noise is developed which has a reduced computational complexity, because matrix inversions can be omitted. The presented blind-matched filtering approach is similar to this SDW-MWF for spatially uncorrelated noise but additionally achieves some improvements in the speech quality due to a partial equalization of the acoustic system.

1. Introduction

In many speech communication systems, like hands-free car kits, teleconferencing systems, and speech recognition systems, the desired speech signal is linearly distorted by the room acoustics and also corrupted by undesired background noise. Therefore, efficient speech processing techniques are required to enhance the speech signal under the constraint of a small speech distortion. The use of multiple microphones can improve the performance compared to single microphone systems [1]. The most common way to place the microphones is beamformer arrays with designed array geometry. Beamforming algorithms exploit the spatial directivity effects by a proper combining, like the Frost beamformer [2] or the generalized sidelobe canceler (GSC) [3]. Usually, the microphones are located in close proximity and the same signal conditions at the microphone positions are assumed.

Alternatively, multimicrophone setups have been proposed that combine the processed signals of two or more distributed microphones. The microphones are positioned separately in order to ensure incoherent recording of noise [4–6]. Basically, all these approaches exploit the fact that speech components in the microphone signals are strongly correlated while the noise components are only weakly correlated if the distance between the microphones is sufficiently large.

For immersive communication, future communication devices must be able to collect the desired speech signal as naturally as possible. But the speech signal quality depends on the speaker's distance to the microphone (array). Therefore, we propose the use of a setup with distributed microphones, where the user can place the microphones arbitrarily. Hence, the array geometry is arbitrary and not a priori known.

In this paper, we discuss schemes for an optimal speech signal combining in real-world acoustic scenarios with distributed microphones. With distributed arrays, the transfer functions to the different microphones vary and these variations have to be taking into account providing an optimal signal combining. Often when the room acoustic is taken into account in a beamformer design, one microphone is taken as a reference channel, for example, the speech distortion weighted-multichannel Wiener filter (SDW-MWF) [7, 8] or the general transfer function GSC (TF-GSC) [9, 10]. For microphone arrays with close proximities and similar transfer functions, this, is a suitable solution. However, for distributed microphones, the a priori chosen reference channel is not necessarily the ideal choice. Moreover, possible equalization capabilities are often neglected.

The matched filter (MF) [11] and the special case of the MF, the minimum variance distortionless response (MVDR) beamformer, provide a signal combining that maximizes the signal-to-noise ratio (SNR) in the presence of additive noise. A direct implementation of matched filtering requires knowledge of the acoustic transfer functions. With perfect channel knowledge, the MVDR beamformer also provides perfect equalization. However, with speech applications, the acoustic transfer functions are unknown and we have no means to directly measure the room impulse responses. There exist several blind approaches to estimate the acoustic transfer functions (see, e.g., [12–14]) which were successfully applied to dereverberation. However, the proposed estimation methods are computationally demanding. In [15], an iterative procedure was proposed where the matched filter was utilized in combination with a least-mean-squares (LMSs) adaptive algorithm for blind identification of the required room impulse responses.

In general, a signal combining for distributed microphones is desirable which does not require explicit knowledge of the channel characteristics. In a previous work, we have developed a matched-filter approach under the assumption of a uniform incoherent noise field [16]. The optimal weighting of the matched filter can be estimated by an approximation of the input SNR values and a phase estimate. Similarly, a scaled version of the MVDR-beamformer coefficients can be found by maximizing the beamformer output SNR [17]. In the frequency domain, these coefficients can be obtained by estimating the dominating generalized eigenvector (GEV) of the noise covariance matrix and the input signal covariance matrix. For instance, an adaptive variant for estimating a GEV was proposed by Doclo and Moonen [18], and later by Warsitz and Haeb-Umbach [19]. Furthermore, it can be shown that the SDW-MWF also provides an optimal signal combining that maximizes the signal-to-noise ratio [20]. The SDW-MWF requires only estimates of the input and noise correlation matrices. Hence, no explicit channel knowledge is required. However, the SDW-MWF does not equalize the speech signal.

In this work, we consider speech enhancement with distributed microphones. In Section 3, we present some measurement results that motivate a distributed microphone array. In particular, we consider two different acoustic situations: a conference room where the noise level is typically low, but the speech signal is distorted due to reverberation, and a car environment where the reverberation time is low, but the strong background noise occurs.

The basic idea of the presented approach is to apply the well-known matched-filter technique for a blind equalization of the acoustic system in the presence of additive background noise. This concept is strongly related to the SDW-MWF. Therefore, we discuss different matched-filtering techniques and their relation to the multichannel Wiener filter in Section 4.

In many speech applications, a diffuse noise field can be assumed [21]. With a diffuse noise field, the correlation of the noise signals depends on the frequency and the distance between the microphones. Typically, for small microphone distances, the low-frequency band is highly correlated whereas the correlation is low for higher frequencies. With a larger microphone spacing, the noise correlation is further decreased and the noise components can be assumed to be spatially uncorrelated. In Sections 5 and 6, we demonstrate that this fact can be exploited to reduce the complexity of the SDW-MWF algorithm as well as to improve the equalization capabilities.

The calculation of the MWF requires the inversion of the correlation matrix of the input signals. This is a computationally demanding and also numerically sensitive task. In Section 5, we show that for a scenario with a single speech source and with spatially uncorrelated noise the matrix inversion can be omitted. Using the matrix inversion lemma [22] the equation of the MWF filter weights can be rewritten to an equation that only depends on the correlations of the input signals and the input noise power spectral densities at the different microphones.

In Section 7, we present a blind-matched filtering approach for speech recording in spatially uncorrelated noise where no assumption on the geometrical adjustment of the microphones is made. The approach presented in [16] is limited to uncorrelated noise signals where the noise power spectral densities are equal for all microphone inputs. In this work, we extend these results to situations where the noise signals are spatially uncorrelated, but the noise power spectral densities can vary. Furthermore, we show that combined with a single channel Wiener filter, this new structure is equivalent with the SDW-MWF with respect to noise suppression. However, the new approach provides a partial equalization of the acoustic transfer functions between the local speaker and the microphone positions.

Finally, we demonstrate in Section 8 that the presented filter structure can be utilized for blind system identification. For equal noise power spectral densities at all microphone inputs, the matched filter is equal to the vector of transfer function coefficients up to a common factor. Hence, by estimating the ideal matched filter, we estimate the linear acoustic system up to a common filter. Note that many known approaches for blind system identification can only infer the different channels up to a common filter [15]. Similarly, with the proposed system, all filters are biased. We derive the transfer function of the common filter and demonstrate that the biased acoustic transfer functions can be reliably estimated even in the presence of strong background noise.

2. Signal Model

In this section, we briefly introduce the notation. In general, we consider M microphones and assume that the acoustic system is linear and time invariant. Hence, the microphone signals $y_i(k)$ can be modeled by the convolution of the speech signal $x(k)$ with the impulse response $h_i(k)$ of the acoustic system plus an additive noise term $n_i(k)$. The M microphone signals $y_i(k)$ can be expressed in the frequency domain as

$$Y_i(\kappa,\nu) = H_i(\nu)X(\kappa,\nu) + N_i(\kappa,\nu), \quad (1)$$

where $Y_i(\kappa,\nu)$, $X(\kappa,\nu)$, and $N_i(\kappa,\nu)$ denote the corresponding short-time spectra and $H_i(\nu)$ the acoustic transfer functions. $S_i(\kappa,\nu) = H_i(\nu)X(\kappa,\nu)$ is the speech component of the ith microphone signal. The subsampled time index and the frequency bin index are denoted by κ and ν, respectively. In the following, the dependencies on κ and ν are often omitted for lucidity. Hence, we can define the M-dimensional vectors $\mathbf{S}$, $\mathbf{N}$, and $\mathbf{Y}$, in which the signals are stacked as follows:

$$\mathbf{S} = \begin{bmatrix} S_1 & S_2 & \cdots & S_M \end{bmatrix}^T,$$
$$\mathbf{N} = \begin{bmatrix} N_1 & N_2 & \cdots & N_M \end{bmatrix}^T, \quad (2)$$
$$\mathbf{Y} = \mathbf{S} + \mathbf{N}.$$

Note that T denotes the transpose of a vector or matrix, whereas the conjugate transpose is denoted by $\dagger$ and conjugation by $*$, respectively. $\mathbf{H}$ denotes the vector of channel coefficients:

$$\mathbf{H} = \begin{bmatrix} H_1 & H_2 & \cdots & H_M \end{bmatrix}^T. \quad (3)$$

In the following, we assume that the noise signals are zero-mean random processes with the variances $\sigma_{N_1}^2, \ldots, \sigma_{N_M}^2$. We denote the signal-to-noise ratio (SNR) at the microphone i by

$$\gamma_i = \frac{\sigma_X^2 |H_i|^2}{\sigma_{N_i}^2}, \quad (4)$$

where σ_X^2 is the speech power at the speaker's mouth.

3. Measurement and Simulation Setup

Throughout the paper, we will illustrate the proposed method with measurement and simulation results for two different acoustic situations: a conference room where the noise level is typically low, but the speech signal is distorted due to reverberation, and a car environment where the reverberation time is low, but the strong background noise may lead to very low input SNR values. In this section, we first present some measurement results that motivate a distributed microphone array. Then, we describe the setup for the simulations.

The basic idea of the presented approach is to apply the well-known matched filter technique for a blind equalization of the acoustic system in the presence of additive background noise. We first discuss some measurement results obtained in a conference room with a size of 4.7 × 4.8 × 3.0 m. For these measurements, we used three omnidirectional microphones which are placed on a table in the conference room as shown in Figure 1. The microphone distance was chosen 1.2 m for mic. 1 and mic. 3, and 1 m between the other microphone pairs (see Figure 1). This results in distances in the range from 0.5 m to 1.3 m between the local speakers and the microphones.

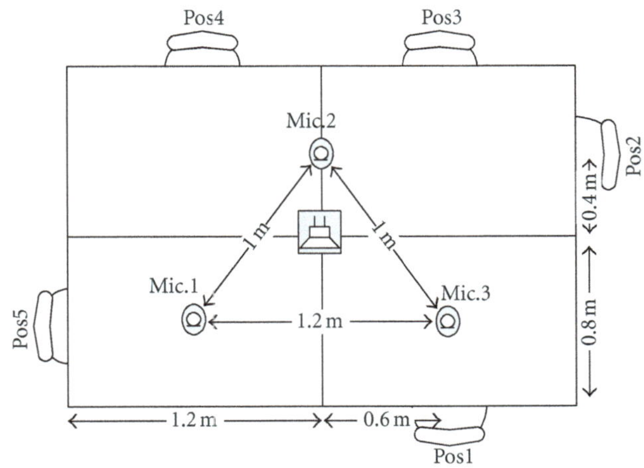

FIGURE 1: Measurement setup for the conference scenario.

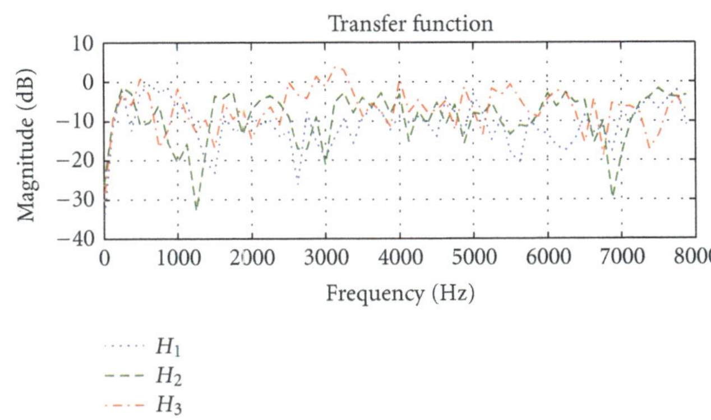

FIGURE 2: Magnitudes of the transfer functions for speaker three.

With an artificial head, we measured the room impulse responses for five local teleconference participants. For this scenario, Figure 2 shows the magnitudes of the acoustic transfer functions. The influence of the room acoustic is clearly visible. For some frequencies, the magnitudes of the acoustic transfer functions show differences of more than 20 dB. It can also be stated that the microphone with the best transfer function is not obvious, because for some frequencies $H_1(\nu)$, $H_2(\nu)$, and for others $H_3(\nu)$ has less attenuation.

Figure 3 depicts the SNR versus frequency for a situation with background noise which arises from a fan of a video projector. From this figure, we observe that the SNR values for frequencies above 1.5 kHz are quite distinct for these three microphone positions with differences of up to 10 dB depending on the particular frequency. Again, the best microphone position is not obvious in this case, because the SNR curves across several times.

Theoretically, if we assume spatially uncorrelated noise signals, a matched filter combining these input signals would

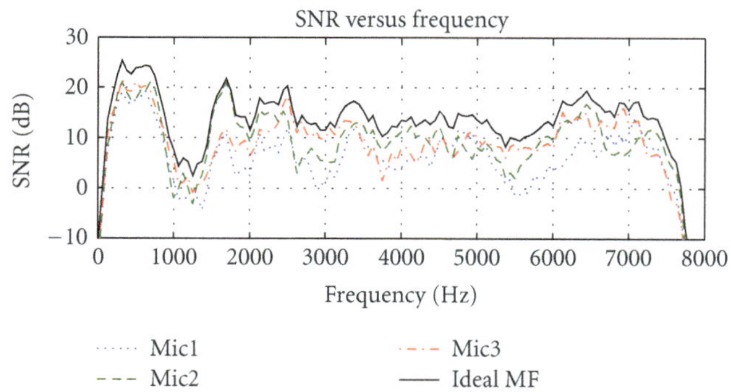

FIGURE 3: Input SNR for a conference scenario with background noise.

result in an output SNR equal to the sum of the input SNR values. With three inputs, a matched filter array achieves a maximum gain of 4.8 dB for equal input SNR values. In the case of varying input SNR values, the sum is dominated by the maximum value. Hence, for the curves in Figure 3, the output SNR would essentially be the envelope of the three curves. This is also shown in Figure 3, where the output SNR of an optimal signal combining is plotted (solid line).

For the simulation results presented throughout the paper the following setup was used: all processed signals are sampled at a sampling rate of $f_s = 16000$ Hz. For the short-time Fourier transform (STFT), we used a length of $L = 512$ and an overlap of $K = 384$ samples, while an overlap-add processing for the signal synthesis was performed. As clean speech signals we used two male and two female speech samples, each of a length of 8 seconds. Therefore, we took the German-speaking test samples from the recommendation P.501 of the International Telecommunication Union (ITU) [23]. To generate the speech signals s_i at the microphones, the clean speech was convolved with the corresponding room impulse responses. The reverberation time for the conference scenario was $T_{60} = 0.25$ s. We also show results for a conference scenario with $T_{60} = 0.5$ s, but for this scenario the impulse responses are generated using the image method [24]. Most presented algorithms require estimates of the noise power spectral density (PSD) and a voice activity detection (VAD), here we used the methods described in [16] throughout the paper.

For the measurements in the car environment, one microphone was installed close to the inside mirror, while the second microphone was mounted at the A-pillar (the A-pillar of a vehicle is the first pillar of the passenger compartment, usually surrounding the windscreen). This microphone setup leads to a distance of 0.6 m between the two microphones. We consider three different background noise situations for noise recordings: driving noise at 100 km/h and 140 km/h, and the noise arising from an electric fan (defroster, car in standstill). For a comparison with typical beamformer constellations, we installed a second pair of microphones at the inside mirror, such that the microphone distance between these two microphone was 0.04 m. This microphone setup is evaluated later in Section 5. Figure 4 shows the measured setup for the in-car environment. For all

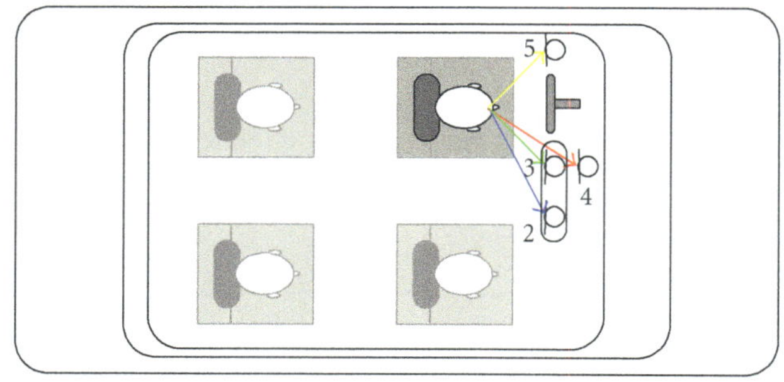

FIGURE 4: Illustration of the measured in-car scenario.

measurements in the car, we used cardioid microphones, which are often used for automotive applications. For the measurements of the room impulse responses also the artificial head was used. The reverberation time for this scenario was $T_{60} = 0.05$ s.

4. Optimal Signal Combining

In this section, we discuss combining schemes which are optimal in a certain manner. For such a combining of the microphone signals, each input Y_i is processed by a linear filter G_i before the signals are added. Stacking these filter functions into the vector $\mathbf{G}$, we have

$$\mathbf{G} = \begin{bmatrix} G_1 & G_2 & \cdots & G_M \end{bmatrix}^T. \tag{5}$$

Therefore, the processed signal at the output of the combining system can be expressed as follows:

$$\begin{aligned} \hat{X} &= \mathbf{G}^\dagger \mathbf{Y}, \\ \hat{X} &= \mathbf{G}^\dagger \mathbf{H} X + \mathbf{G}^\dagger \mathbf{N}. \end{aligned} \tag{6}$$

The SNR at the system output is defined as the ratio:

$$\begin{aligned} \gamma &= \frac{\mathbb{E}\left\{ |\mathbf{G}^\dagger \mathbf{H} X|^2 \right\}}{\mathbb{E}\left\{ |\mathbf{G}^\dagger \mathbf{N}|^2 \right\}}, \\ &= \frac{\sigma_X^2 \mathbf{G}^\dagger \mathbf{H}\mathbf{H}^\dagger \mathbf{G}}{\mathbf{G}^\dagger \mathbf{R}_N \mathbf{G}}, \end{aligned} \tag{7}$$

where

$$\mathbf{R}_N = \mathbb{E}\left\{ \mathbf{N}\mathbf{N}^\dagger \right\} \tag{8}$$

is the correlation matrix of the noise signals.

4.1. Maximization of the SNR. Our aim is now to find the filter functions $\mathbf{G}$, which are optimal in the sense of a maximal output SNR of the combining system. Hence, the maximization problem can be stated as

$$\mathbf{G}^{\mathrm{MF}} = \arg\max_{\mathbf{G}} \left\{ \frac{\sigma_X^2 \mathbf{G}^\dagger \mathbf{H}\mathbf{H}^\dagger \mathbf{G}}{\mathbf{G}^\dagger \mathbf{R}_N \mathbf{G}} \right\}. \tag{9}$$

This maximization problem leads to an eigenvalue problem with the matched filter (MF) solution [25]:

$$\mathbf{G}^{\mathrm{MF}} = c \cdot \mathbf{R}_N^{-1} \mathbf{H}, \tag{10}$$

where c is a nonzero constant value. Hence, one has to weight the input signals according to the acoustic transfer functions and the inverse of the noise correlation matrix. This weighting is also known as maximum SNR (MSNR) beamformer.

Applying a constant factor to $\mathbf{R}_N^{-1}\mathbf{H}$ does not affect the SNR at the filter output. Therefore, the matched filter can also be utilized for equalization to get a flat frequency response according to the source speech position ($\mathbf{G}^{\mathrm{MF}\dagger}\mathbf{H} = 1$). In this case, the following form is used:

$$\mathbf{G}^{\mathrm{MVDR}} = \frac{\mathbf{R}_N^{-1}\mathbf{H}}{\mathbf{H}^{\dagger}\mathbf{R}_N^{-1}\mathbf{H}}. \quad (11)$$

This algorithm is also called the minimum variance distortionless response (MVDR) beamformer and was described by Cox et al. [26]. For this technique, knowledge about the room impulse responses and the noise correlation matrix is needed. For an estimation of the noise power density and the cross-power density, several approaches exist in the literature [21, 27–30]. But the estimation of the room impulse responses is a blind estimation problem [12–14]. A reliable estimation of the room impulse responses in realtime is still an open issue. Often only a linear-phase compensation is done, by applying a sufficient time delay for the signals, thus the transfer functions are replaced by the steering vector:

$$\mathbf{d}(\nu) = \left[1, e^{-j\Delta_2(\nu)}, \ldots, e^{-j\Delta_M(\nu)}\right]^T, \quad (12)$$

where $\Delta_i(\nu)$ denotes the phase difference between the first and ith microphone. This corresponds to the classical Frost beamformer [2]. Thus, an estimate of the time delay of arrival (TDOA) is required. Note that for a known array geometry this information is equivalent to the direction of arrival (DOA).

4.2. Multichannel MMSE Criterion. Another related criterion is the minimization of the mean square error (MSE) between the output signal and a reference signal. This can also be used to find an optimal combining strategy for the microphone signals. To calculate the minimum mean squared error (MMSE) estimate of the clean speech signal X at the speaker's mouth, one has to minimize the following cost function:

$$\mathbf{G}^{\mathrm{MWF}} = \arg\min_{\mathbf{G}} \mathbb{E}\left\{\left|\mathbf{G}^{\dagger}\mathbf{Y} - X\right|^2\right\}. \quad (13)$$

By setting the complex derivative with respect to $\mathbf{G}^*$ to zero, one obtains the solution of this minimization problem as

$$\mathbf{G}^{\mathrm{MWF}} = \mathbf{R}_Y^{-1}\mathbf{R}_{Yx}, \quad (14)$$

where $\mathbf{R}_{Yx} = \mathbb{E}\{\mathbf{Y}X^*\}$ is the cross-correlation vector between the clean speech and the microphone input signal and $\mathbf{R}_Y = \mathbb{E}\{\mathbf{Y}\mathbf{Y}^{\dagger}\}$ is the correlation matrix of the microphone input signal, respectively. In the literature, this is often referred as the multichannel Wiener filter (MWF), which can be used for signal combining and noise reduction.

To overcome the problem of the required but unavailable cross-correlation vector $\mathbf{R}_{Yx}$ in the definition of the MWF, cf. (14), one can define an MWF that minimizes the mean squared error with respect to the speech signal of a reference microphone signal S_{ref}. In [7], the SDW-MWF was proposed, while a tradeoff parameter was introduced to the MWF. With this parameter, it is possible to adjust the noise reduction capabilities of the MWF with respect to the speech signal distortion at the output. Thus, the signal distortion is taken into account in the optimization. Therefore, the distortion is measured as the distance between the speech component of the output signal and the speech component of an input channel. This reference channel is selected arbitrarily in advance.

The error signal ε for the minimization is then defined as the difference between the output signal $\mathbf{G}^{\dagger}\mathbf{Y}$ and the speech component of the signal Y_{ref}:

$$\begin{aligned}\varepsilon &= \mathbf{G}^{\dagger}\mathbf{Y} - \mathbf{u}^T\mathbf{S}\\ &= \left(\mathbf{G}^{\dagger} - \mathbf{u}^T\right)\mathbf{S} + \mathbf{G}^{\dagger}\mathbf{N}\\ &= \varepsilon_s + \varepsilon_n.\end{aligned} \quad (15)$$

The column vector $\mathbf{u}$ selects the reference channel, that is, the corresponding entry is set to one and the others are set to zero. Using the two MSE cost functions:

$$\begin{aligned}J_n(\mathbf{G}) &= \mathbb{E}\left\{|\varepsilon_n|^2\right\},\\ J_s(\mathbf{G}) &= \mathbb{E}\left\{|\varepsilon_s|^2\right\},\end{aligned} \quad (16)$$

the unconstraint minimization criterion for the SDW-MWF is defined by

$$\mathbf{G}^{\mathrm{SDW}} = \arg\min_{\mathbf{G}} J_n(\mathbf{G}) + \frac{1}{\mu_W} J_s(\mathbf{G}), \quad (17)$$

where $1/\mu_W$ is a Lagrange multiplier. This results in the solution:

$$\mathbf{G}^{\mathrm{SDW}} = \left(\mathbf{R}_S + \mu_W\mathbf{R}_N\right)^{-1}\mathbf{R}_S\mathbf{u}, \quad (18)$$

where $\mathbf{R}_S$ is the speech correlation matrix and μ_W is a parameter which allows a trade-off between speech distortion and noise reduction (for details cf. [7]).

For further analyses, we assume that the single speaker speech signal is a zero-mean random proces with the PSD σ_X^2 and a time-invariant acoustic system. The correlation matrix of the speech signal can be written as

$$\begin{aligned}\mathbf{R}_S &= \mathbb{E}\left\{\mathbf{S}\mathbf{S}^{\dagger}\right\}\\ &= \sigma_X^2\mathbf{H}\mathbf{H}^{\dagger}.\end{aligned} \quad (19)$$

Using the matrix inversion lemma [22], the SDW-MWF can be decomposed as

$$\begin{aligned}\mathbf{G}^{\mathrm{SDW}} &= \frac{\sigma_X^2}{\sigma_X^2 + \mu_W\left(\mathbf{H}^{\dagger}\mathbf{R}_N^{-1}\mathbf{H}\right)^{-1}} \frac{\mathbf{R}_N^{-1}\mathbf{H}}{\mathbf{H}^{\dagger}\mathbf{R}_N^{-1}\mathbf{H}}\mathbf{H}^{\dagger}\mathbf{u},\\ \mathbf{G}^{\mathrm{SDW}} &= G^{\mathrm{WF}}\mathbf{G}^{\mathrm{MVDR}}H_{\mathrm{ref}}^*,\end{aligned} \quad (20)$$

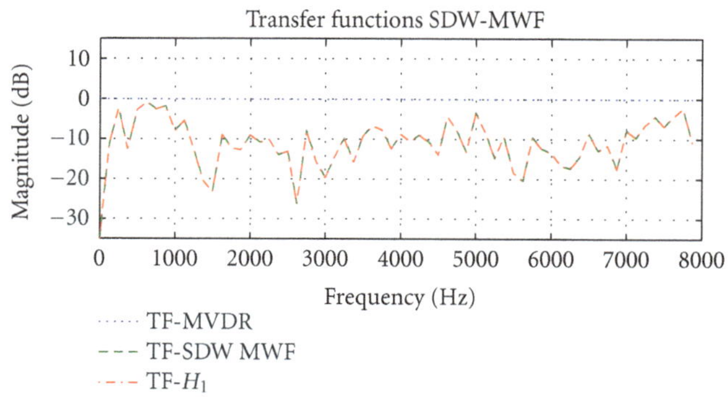

Figure 5: Comparison of the overall transfer function of the SDW-MWF and the ideal MVDR beamformer.

where $(\mathbf{H}^{\dagger}\mathbf{R}_N^{-1}\mathbf{H})^{-1}$ is the noise variance at the output of $\mathbf{G}^{\mathrm{MVDR}}$:

$$\left(\mathbf{H}^{\dagger}\mathbf{R}_N^{-1}\mathbf{H}\right)^{-1} = \mathbf{G}^{\mathrm{MVDR}\dagger}\mathbf{R}_N\mathbf{G}^{\mathrm{MVDR}}, \quad = \sigma^2{}_{N_{\mathrm{MVDR}}}. \tag{21}$$

Appendix A provides a derivation of this decomposition.

Thus the SDW-MWF is decomposed in an MVDR beamformer and a filter that is equal to the acoustic transfer function of the reference channel ($H_{\mathrm{ref}}^{*} = \mathbf{H}^{\dagger}\mathbf{u}$). Furthermore, the noise reduction is achieved by a single channel Wiener filter

$$G^{\mathrm{WF}} = \frac{\sigma_X^2}{\sigma_X^2 + \mu_W \sigma^2{}_{N_{\mathrm{MVDR}}}}, \tag{22}$$

where μ_W can be interpreted as a noise overestimation factor [31].

From this decomposition, it can be seen that the SDW-MWF provides an optimal signal combining with respect to the output SNR. Yet, it is not able to equalize the acoustic transfer functions. This is also obvious in Figure 5, where the overall system transfer function is depicted (dashed line). Here, the first microphone with the transfer function $H_1(\nu)$ was used as reference channel. For this plot, the Wiener filter part $G^{\mathrm{WF}}(\nu)$ of the transfer function was neglected. We observe that the overall transfer function of the SDW-MWF (dashed line) is equivalent to the transfer function of the reference channel (semidashed line). Note that we measured the transfer function between the speaker's mouth and the output of the SDW-MWF. Also, the flat transfer function of the MVDR beamformer is plotted for a comparison (dotted line).

5. The SDW-MWF for Spatially Uncorrelated Noise

The calculation of the MWF in (18) requires the inversion of the correlation matrix of the input signals. This is a computationally demanding and also numerically sensitive task. In this section, we show that for a scenario with a single speech source and with spatially uncorrelated noise, the matrix inversion can be omitted. Using the matrix inversion lemma [22], the equation of the MWF filter weights can be rewritten to an equation that only depends on the correlations of the input signals and the input noise PSDs at the different microphones.

Consider the speech-distortion-weighted multichannel Wiener filter according to (18). We can rewrite the inverse $(\mathbf{R}_S + \mu_W\mathbf{R}_N)^{-1}$ using the result in (A.1):

$$(\mu_W\mathbf{R}_N + \mathbf{R}_S)^{-1} = \frac{1}{\mu_W}\mathbf{R}_N^{-1} - \frac{\mathbf{R}_N^{-1}\mathbf{R}_S\mathbf{R}_N^{-1}}{\mu_W^2\left(1 + \mu_W^{-1}\sigma_X^2\mathbf{H}^{\dagger}\mathbf{R}_N^{-1}\mathbf{H}\right)}. \tag{23}$$

Furthermore, using (21), we have

$$\sigma_X^2\mathbf{H}^{\dagger}\mathbf{R}_N^{-1}\mathbf{H} = \frac{\sigma_X^2}{\sigma^2{}_{N_{\mathrm{MVDR}}}} = \gamma, \tag{24}$$

which is the signal-to-noise ratio at the output of the MVDR beamformer.

Using the inverse of $(\mu_W\mathbf{R}_N + \mathbf{R}_S)$ from (23) and the definition of the SDW-MWF in (18), we have

$$\begin{aligned}\mathbf{G}^{\mathrm{SDW}} &= \left(\frac{1}{\mu_W}\mathbf{R}_N^{-1} - \frac{\mathbf{R}_N^{-1}\mathbf{R}_S\mathbf{R}_N^{-1}}{\mu_W^2\left(1 + \mu_W^{-1}\sigma_X^2\mathbf{H}^{\dagger}\mathbf{R}_N^{-1}\mathbf{H}\right)}\right)\mathbf{R}_S\mathbf{u} \\ &= \left(1 - \frac{\gamma}{\mu_W + \gamma}\right)\frac{1}{\mu_W}\mathbf{R}_N^{-1}\mathbf{R}_S\mathbf{u} \\ &= \frac{\mathbf{R}_N^{-1}\mathbf{R}_S\mathbf{u}}{\mu_W + \gamma}.\end{aligned} \tag{25}$$

Because speech and noise are independent, we can estimate $\mathbf{R}_S$ by $\mathbf{R}_S = \mathbf{R}_Y - \mathbf{R}_N$. Therefore, we obtain

$$\begin{aligned}\mathbf{G}^{\mathrm{SDW}} &= \frac{\mathbf{R}_N^{-1}(\mathbf{R}_Y - \mathbf{R}_N)\mathbf{u}}{\mu_W + \gamma} \\ &= \frac{(\mathbf{R}_N^{-1}\mathbf{R}_Y - \mathbf{I})\mathbf{u}}{\mu_W + \gamma} \\ &= \frac{\mathbf{R}_N^{-1}\mathbf{R}_Y\mathbf{u} - \mathbf{u}}{\mu_W + \gamma}.\end{aligned} \tag{26}$$

Note that the column vector $\mathbf{u}$ selects the reference channel, that is, the corresponding entry is set to one and the others are set to zero.

Because we assume that the noise signals at the different microphones are uncorrelated, $\mathbf{R}_N$ is a diagonal matrix, and the elements of the main diagonal are the noise variances $\sigma_{N_1}^2, \ldots, \sigma_{N_M}^2$. Therefore, we obtain the inverse

$$\mathbf{R}_N^{-1} = \begin{pmatrix} \frac{1}{\sigma_{N_1}^2} & 0 & \cdots & 0 \\ & \ddots & & \\ 0 & 0 & \cdots & \frac{1}{\sigma_{N_M}^2} \end{pmatrix}. \tag{27}$$

Let ref be the index of the one in the vector $\mathbf{u}$. $\mathbf{R}_N^{-1}\mathbf{R}_Y\mathbf{u}$ results in the column vector:

$$\mathbf{w}_{\text{ref}} = \mathbf{R}_N^{-1}\mathbf{R}_Y\mathbf{u} = \left[\frac{\mathbb{E}\{Y_1 Y_{\text{ref}}^*\}}{\sigma_{N_1}^2}, \ldots, \frac{\mathbb{E}\{Y_M Y_{\text{ref}}^*\}}{\sigma_{N_M}^2}\right]^T. \quad (28)$$

Therefore, we obtain the following expression for the SDW-MWF for spatially uncorrelated noise signals:

$$\mathbf{G}^{\text{SDW}} = \frac{\mathbf{w}_{\text{ref}} - \mathbf{u}}{\mu_W + \gamma}. \quad (29)$$

This representation of the speech distortion weighted multichannel Wiener filter omits the inversion of the matrix $(\mathbf{R}_S + \mu_W\mathbf{R}_N)$. For spatially uncorrelated noise signals, the SNR γ can be calculated as the sum of the input signal-to-noise ratios:

$$\gamma = \sigma_X^2 \mathbf{H}^\dagger \mathbf{R}_N^{-1}\mathbf{H} = \sum_{i=1}^{M} \frac{\sigma_X^2 |H_i|^2}{\sigma_{N_i}^2}. \quad (30)$$

6. The SDW-MWF for Diffuse Noise

In the literature, many investigations on the spatial correlation properties of noise fields have been made. The assumption of spatially uncorrelated noise is rarely fulfilled in real-world scenarios, but it has been found, for example, by Martin and Vary in [21], that many noise fields can be assumed to be diffuse, like the noise in a car environment [32], office noise [21], or, for example, babble noise [33]. For diffuse noise, the spatial correlation depends on the intermicrophone distance and is dominant especially in the lower-frequency bands. Typically, the low-frequency band is highly correlated whereas the correlation is low for higher frequencies. This fact can be exploited by omitting the matrix inversion for higher frequencies.

To evaluate the correlation between the noise signals at different positions, the coherence function of the noise signals from different intermicrophone distances can be computed. The magnitude squared coherence (MSC) between two signals n_i and n_j is defined as follows:

$$C_{N_iN_j}(\nu) = \frac{\left|\sigma_{N_iN_j}(\nu)\right|^2}{\sigma_{N_i}^2(\nu)\sigma_{N_j}^2(\nu)}, \quad (31)$$

where $\sigma_{N_iN_j}(\nu)$ and $\sigma_{N_i}^2$ are the cross-power spectral density (CPSD) and the power spectral densities (PSDs) of the signals n_i and n_j, respectively. The values of the coherence function are between 0 and 1, where 0 means no correlation between the two signals at that frequency point. For highly correlated signals, the MSC will become close to 1 for all frequencies.

In [34], Armbrüster et al. have shown that the coherence of an ideal diffuse sound field recorded with omnidirectional microphones can be computed as follows:

$$C_{\text{theo}}(\nu) = \frac{\sin^2(2\pi\nu d_{\text{mic}} f_s/(Lc))}{(2\pi\nu d_{\text{mic}}/c)^2}, \quad (32)$$

where L denotes the length of the short-time Fourier transform (STFT) and f_s is the sampling frequency. The speed of sound is denoted by c, and d_{mic} represents the microphone distance. The zeros of the theoretical coherence function in (32) can by calculated by the following expression:

$$\nu_{\text{zero},m} = \frac{mLc}{2d_{\text{mic}} f_s}, \quad m = 1, 2, 3, \ldots. \quad (33)$$

In the following, we consider the coherence for the noise signals of the in-car scenario for the driving situation at 100 km/h (see Section 3). Figure 6(a) shows the coherence functions of the noise for the microphone pair with an intermicrophone spacing of $d_{\text{mic}} = 0.04$ m. Also, the theoretical coherence function computed according to (32) is shown. Obviously, there is a high correlation of the noise signals at frequencies below 2 kHz. Note that the coherence of the noise signals is closely approximated by the theoretical coherence function C_{theo}, although cardioid microphones were used for this measurement. In Figure 6(b), the coherence function of the noise for the microphone pair with a 0.6 m spacing is depicted. In this constellation, the noise signals at the two microphones are highly correlated for frequencies below 150 Hz only.

From (33) and Figure 6, it is obvious that the correlation of the diffuse noise signals depends on the intermicrophone distance. Therefore, the noise has only a high correlation at low frequencies and especially the high frequencies are only weakly correlated. Thus, the assumption of spatially uncorrelated noise is fulfilled for the higher frequency bands. Therefore, we propose to calculate the filter weights depending on the theoretical coherence $C_{\text{theo}}(\nu)$; for frequencies with a high coherence, we calculate the filter weights using the matrix inversion (see (18)), while for frequencies with a low coherence, we assume uncorrelated noise and thus the weights are computed according to (29). Hence, the filter function is calculated according to

$$\mathbf{G}^{\text{SDW}}(\nu) = \begin{cases} (\mathbf{R}_S(\nu) + \mu_W\mathbf{R}_N(\nu))^{-1}\mathbf{R}_S(\nu)\mathbf{u}, & C_{\text{theo}}(\nu) \geq C_{\text{lim}}, \\ \dfrac{\mathbf{w}_{\text{ref}}(\nu) - \mathbf{u}}{\mu_W + \gamma(\nu)}, & \text{otherwise}, \end{cases} \quad (34)$$

where C_{lim} is a parameter that allows a trade-off between accuracy and computing time.

The simulation results for the in-car microphone scenario with the two different microphone setups are given in Table 1. Each scenario (microphone setup and noise condition) was simulated twice. The first time it was simulated using the fullband matrix inversion according to (18) (denoted by *fullband MWF*). These results can be seen as an upper bound for the performance evaluation of the proposed method. The second time we used the proposed approach of the SDW-MWF with the partially inversion of the correlation matrix (*partial MWF*). Therefore, the inversion was omitted for all frequency bins with a theoretical coherence C_{theo} less than 0.7. This leads in our simulation setup to the threshold frequencies $f_{\text{lim}} = 1500$ Hz for the closed spaced microphone

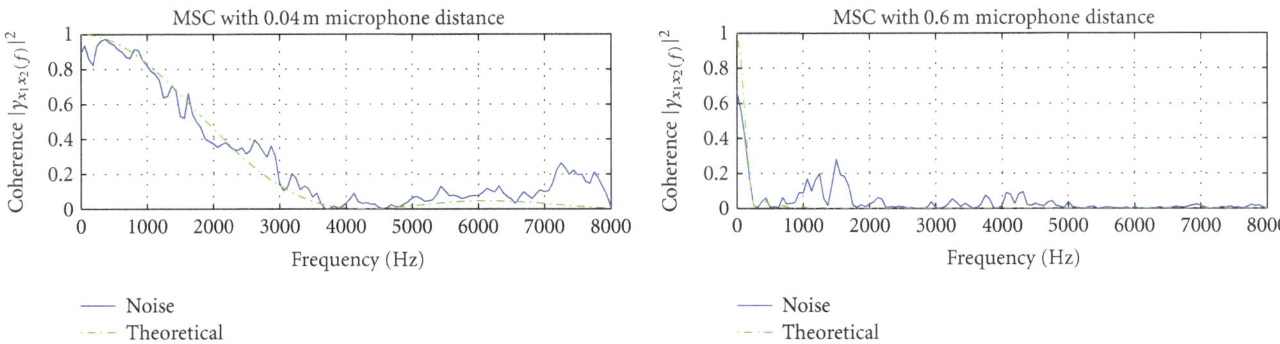

(a) MSC for intermicrophone spacing of 0.04 m

(b) MSC for intermicrophone spacing of 0.6 m

FIGURE 6: Comparison of the magnitude squared coherence.

TABLE 1: Comparison of the fullband (according to (18)) and partial MWF (according to (34)).

	d_{mic} [m]	0.04			0.6		
		100 km/h	140 km/h	defrost	100 km/h	140 km/h	defrost
	SSNR ch.1 [dB]	−2.7	−7.1	−0.8	−2.7	−7.7	−0.2
	SSNR ch.2 [dB]	−3.8	−8.2	−1.3	−1.1	−5.8	0.6
Fullband MWF	SSNR [dB]	4.4	−0.8	5.3	5.4	−0.1	7.6
	S-MOS	4.6	4.7	4.7	4.7	4.7	4.8
	N-MOS	3.2	2.6	3.0	3.1	2.6	3.0
	G-MOS	3.8	3.5	3.8	3.8	3.5	3.8
Partial MWF	SSNR [dB]	4.6	−0.6	4.9	5.4	−0.1	7.3
	S-MOS	4.6	4.7	4.7	4.6	4.7	4.7
	N-MOS	3.2	2.6	3.0	3.1	2.6	3.0
	G-MOS	3.8	3.5	3.8	3.7	3.5	3.7

pair and $f_{\text{lim}} = 100$ Hz for the setup with the microphone spacing of 0.6 m. As an objective evaluation criterion, we calculated the segmental signal-to-noise ratio (SSNR) of the output signal. Therefore, a voice activity detection according to the ITU P.56 was used [35]. Furthermore, we show results from an instrumental quality analysis in Table 1. The speech quality and noise reduction were evaluated according to the ETSI standard EG 202 396-3 [36]. This algorithm calculates three objective quality measures (according to the mean opinion score (MOS) scale): Speech-MOS (S-MOS), Noise-MOS (N-MOS), and Global-MOS (G-MOS). From these results, we observe that the *partial MWF* algorithm obtains nearly the same performance as the fullband SDW-MWF.

7. Matched Filtering for Spatially Uncorrelated Noise

We have seen in Section 4.2 that the SDW-MWF provides an optimal signal combining with respect to the output SNR, where the SDW-MWF does not require explicit channel knowledge to obtain this result. For spatially uncorrelated noise the SDW-MWF according to (29) requires only estimates of the input SNR values and the input cross-correlation with respect to the reference channel. However, in contrast to the MVDR beamformer, the SDW-MWF does not equalize the acoustic system.

In the following, we show that knowledge of the input SNR values and the input cross-correlation with respect to the reference channel is sufficient to provide at least a partial channel equalization. We consider the matched filter for spatially uncorrelated noise signals. If we assume that the noise signals at the different microphones are uncorrelated, $\mathbf{R}_N$ is a diagonal matrix, and the elements of the main diagonal are the noise variances $\sigma^2_{N_1}, \dots, \sigma^2_{N_M}$. Therefore, we obtain the inverse $\mathbf{R}_N^{-1}$ as in (27). In this case, the filter coefficients of the matched filter can be determined independently and we obtain

$$G_i^{\text{MF}} = \frac{H_i}{\sigma^2_{N_i}} \tag{35}$$

as ith coefficient of the matched filter according to (10) and

$$G_i^{\text{MVDR}} = \frac{H_i}{\sigma^2_{N_i}\left(|H_1|^2/\sigma^2_{N_1} + |H_2|^2/\sigma^2_{N_2} + \cdots\right)} \tag{36}$$

according to (11).

7.1. Filter Design. In [16], we have demonstrated that under the assumption of a uniform and spatially uncorrelated noise field, this optimal MF weighting can be obtained by the following filter:

$$G_i = \sqrt{\frac{\gamma_i}{\gamma}} = \sqrt{\frac{\gamma_i}{\gamma_1 + \gamma_2 + \cdots + \gamma_M}} \tag{37}$$

and an additional phase synchronization. γ denotes the sum of all input SNR values. Hence, this filter requires only estimates of the input SNRs. In the following, we extend this concept to nonuniform noise fields. In this case, the optimal weighting depends also on the noise power densities $\sigma^2_{N_i}$. Consider now the following filter:

$$G_i = \sqrt{\frac{\gamma_i \widetilde{\sigma}^2_N}{\sigma^2_{N_i}\gamma}} = \sqrt{\frac{\gamma_i \widetilde{\sigma}^2_N}{\sigma^2_{N_i}(\gamma_1 + \gamma_2 + \cdots + \gamma_M)}}, \tag{38}$$

where $\widetilde{\sigma}^2_N$ is the mean of the noise power spectral densities at the different microphones, defined by

$$\widetilde{\sigma}^2_N = \frac{1}{M}\sum_{i=1}^{M} \sigma^2_{N_i}. \tag{39}$$

This filter depends on the noise power density $\sigma^2_{N_i}$ and all input SNR values. Using (4), we obtain

$$\begin{aligned} G_i &= \sqrt{\frac{|H_i|^2 \widetilde{\sigma}^2_N}{\left(\sigma^2_{N_i}\right)^2 \left(|H_1|^2/\sigma^2_{N_1} + |H_2|^2/\sigma^2_{N_2} + \cdots\right)}} \\ &= \frac{|H_i|}{\sigma^2_{N_i}\sqrt{\left(1/\widetilde{\sigma}^2_N\right)\left(|H_1|^2/\sigma^2_{N_1} + |H_2|^2/\sigma^2_{N_2} + \cdots\right)}}. \end{aligned} \tag{40}$$

Note that the term $(1/\widetilde{\sigma}^2_N)(|H_1|^2/\sigma^2_{N_1} + |H_2|^2/\sigma^2_{N_2} + \cdots)$ is common to all filter coefficients. Hence, the filter according to (38) is proportional to the magnitude of the matched filter according to (35).

The proposed filter in (38) is real valued. To ensure cophasal signal combining, we require some additional system components for phase estimation.

7.2. Phase Estimation. For a coherent combining of the speech signals, we have to compensate the phase difference between the speech signals at each microphone. Therefore it is sufficient to estimate the phase differences to a reference microphone. Let $\phi_i(\nu)$ be the phase of the complex channel coefficient $H_i(\nu)$. We consider the phase differences to the a reference microphone $\Delta_i(\nu) = \phi_{\text{ref}}(\nu) - \phi_i(\nu)$, for all $i \neq \text{ref}$ and $\Delta_{\text{ref}}(\nu) = 0$. Cophasal addition is then achieved by

$$\hat{X} = \sum_{i=1}^{M} G_i e^{j\Delta_i} Y_i. \tag{41}$$

For multimicrophone systems with spatially separated microphones a reliable phase estimation is a challenging task. A coarse estimate of the phase difference can also be obtained from the time-shift τ_i between the speech components in the microphone signals, for example, using the generalized correlation method [37]. However, for distributed microphone

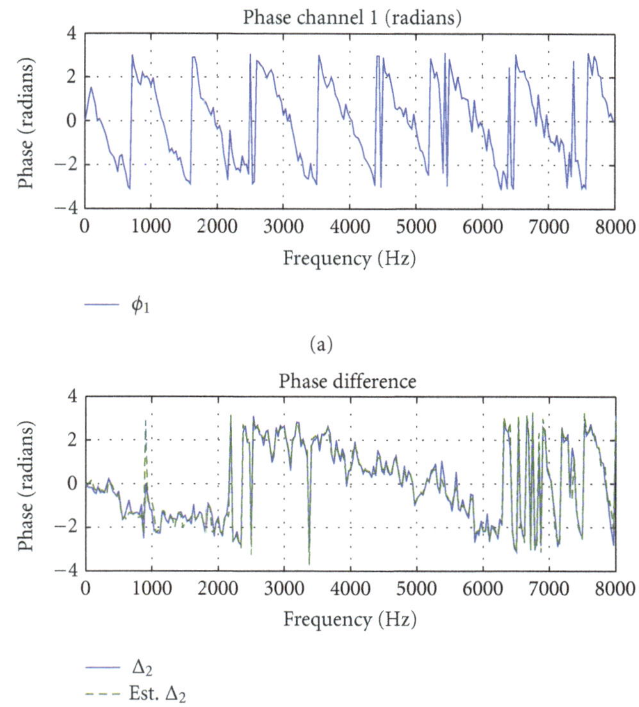

(b)

FIGURE 7: Actual phase of the reference channel (a) determined from the impulse response, actual phase difference and estimated phase difference for channel 2.

arrays in reverberant environments this phase compensation leads to a poor estimate of the actual phase differences. This can be observed in Figure 7 which depicts the phase $\phi_1(\nu)$ of the reference channel for the in car scenario with intermicrophone spacing of 0.6 m (see Section 3). In an anechoic environment, the phase of the reference channel as well as the phase difference Δ_2 for the second microphone would be linear functions of the frequency. Hence, we could expect ideal sawtooth functions if we consider the phase in the interval $[-\pi, \pi]$. From Figure 7, we observe that this is only a rough estimate of the actual phase values.

In order to ensure a cophasal addition of the signals, we employ a phase estimation similar to the approach presented in [16]. We use a frequency domain least-squares (FLMS) algorithm to estimate the required phase difference. Using Y_{ref} as reference signal, the filter $G_i^{\text{ANG}}(\kappa,\nu)$ is updated according to

$$G_i^{\text{ANG}}(\kappa+1,\nu) = G_i^{\text{ANG}}(\kappa,\nu) + \mu Y_i^*(\kappa,\nu)E_i(\kappa,\nu), \tag{42}$$

with

$$E_i(\kappa,\nu) = Y_{\text{ref}}(\kappa,\nu) - Y_i(\kappa,\nu)G_i^{\text{ANG}}(\kappa,\nu). \tag{43}$$

Note that the filter is only adapted if voice activity is detected, where we used the VAD method described in [16]. The FLMS algorithm minimizes the expected value:

$$\mathbb{E}\left\{\left|Y_{\text{ref}}(\kappa,\nu) - Y_i(\kappa,\nu)G_i^{\text{ANG}}(\kappa,\nu)\right|^2\right\}. \tag{44}$$

For stationary signals the adaptation converts to a filter transfer function:

$$G_i^{\mathrm{ANG}} = \frac{\mathbb{E}\{Y_i^* Y_{\mathrm{ref}}\}}{\mathbb{E}\left\{|Y_i|^2\right\}}, \tag{45}$$

where $\mathbb{E}\{Y_i^* Y_{\mathrm{ref}}\}$ is the cross-power spectrum of the two microphone signals and $\mathbb{E}\{|Y_i|^2\}$ is the power spectrum of the ith microphone signal. Assuming that the speech signal and the noise signals are spatially uncorrelated, (45) can be written as

$$G_i^{\mathrm{ANG}} = \frac{\mathbb{E}\left\{|X|^2\right\} H_i^* H_{\mathrm{ref}} + \mathbb{E}\left\{N_i^* N_j\right\}}{\mathbb{E}\left\{|X|^2\right\} |H_i|^2 + \mathbb{E}\left\{|N_i|^2\right\}}. \tag{46}$$

For frequencies where the noise components are uncorrelated, that is, $\mathbb{E}\{N_i^* N_1\} = 0$, this formula is reduced to

$$G_i^{\mathrm{ANG}} = \frac{\mathbb{E}\left\{|X|^2\right\} H_i^* H_{\mathrm{ref}}}{\mathbb{E}\left\{|X|^2\right\} |H_i|^2 + \mathbb{E}\left\{|N_i|^2\right\}}. \tag{47}$$

The phase of the filter G_i^{ANG} is determined by the two complex channel coefficients H_i and H_{ref} where the product $H_i^* H_{\mathrm{ref}}$ has the sought phase $\Delta_i(\nu) = \phi_{\mathrm{ref}}(\nu) - \phi_i(\nu)$. Hence, for the coherent signal combining, we use the phase of the filter G_i^{ANG}:

$$\hat{\Delta}_i(\kappa, \nu) = \arg\left(G_i^{\mathrm{ANG}}(\kappa, \nu)\right). \tag{48}$$

According to (28) and (29), the phase of the filter $\mathbf{G}^{\mathrm{SDW}}$ is determined by the cross-correlation of the input signals. Comparing (28) and (45), we note that the proposed approach leads to the same phase compensation as with the SDW-MWF. Note that the output signal of the SDW-MWF is computed as $\hat{X} = \mathbf{G}^{\mathrm{SDW}\dagger}\mathbf{Y}$.

With the estimated phase, we can now express the complex filter as

$$\widetilde{G}_i = \sqrt{\frac{\gamma_i \widetilde{\sigma}_N^2}{\sigma_{N_i}^2 \gamma}} e^{j\hat{\Delta}_i}. \tag{49}$$

Figure 7 presents simulation results for this phase estimation, where Δ_2 denotes the actual phase difference computed from the measured impulse responses and *est.* Δ_2 is the estimated phase difference. The presented results correspond to the driving situation with a car speed of 140 km/h and an intermicrophone distance of 0.6 m, as described in Section 3.

7.3. Residual Transfer Function. Next, we derive the residual transfer function of the proposed signal combining. Using (40), the complex filter transfer function can be expressed as

$$\widetilde{G}_i = \frac{|H_i| e^{j\hat{\Delta}_i}}{\sigma_{N_i}^2 \sqrt{\left(1/\widetilde{\sigma}_N^2\right)\left(|H_1|^2/\sigma_{N_1}^2 + |H_2|^2/\sigma_{N_2}^2 + \cdots\right)}}. \tag{50}$$

Assuming ideal knowledge of the SNR values and a perfect phase estimation, we can derive the overall transfer function. Comparing the MVDR beamformer in (36) with (50), we observe that the proposed system has a resulting transfer function:

$$\widetilde{H} = e^{j\phi_{\mathrm{ref}}} \sqrt{\widetilde{\sigma}_N^2 \left(\frac{|H_1|^2}{\sigma_{N_1}^2} + \frac{|H_2|^2}{\sigma_{N_2}^2} + \cdots\right)}. \tag{51}$$

That is,

$$\widetilde{G}_i = G_i^{\mathrm{MVDR}} \widetilde{H}. \tag{52}$$

Hence, the proposed system does not provide perfect equalization. However, the filter provides partial dereverberation, where the dips of the acoustic transfer functions are smoothed if the dips occur not in all transfer functions. Moreover, if the noise is uniform and stationary, the number of channels M is sufficiently high and in case of spatially uncorrelated channel coefficients, the sum $(|H_1|^2/\sigma_{N_1}^2 + |H_2|^2/\sigma_{N_2}^2 + \cdots)$ tends to a constant value independent of the frequency (cf. [15]).

7.4. Noise Reduction. As shown by the decomposition of the speech-distortion-weighted multichannel Wiener filter in (20), the noise reduction of the MWF is achieved by a single-channel Wiener filter. Therefore, we combine the proposed matched filter approach with a single channel Wiener filter. In the reminder of this section, we discuss the integration of the Wiener postfilter to the blind-matched filter approach. The considered system is shown in Figure 8.

The single channel Wiener filter in (20) can be rewritten to an equation which only depends on the output SNR γ:

$$G^{\mathrm{WF}} = \frac{\gamma}{\gamma + \mu_W}. \tag{53}$$

It is possible to integrate the Wiener filter function from (53) in the filter functions of the proposed blind-matched filter (38). This leads to the filter function G_i^{MFWF}, which consists of a blind matched filter (MF) with a single-channel Wiener postfilter (WF):

$$\begin{aligned} G_i^{\mathrm{MFWF}} &= G^{\mathrm{WF}} \widetilde{G}_i \\ &= \frac{\gamma}{\gamma + \mu_W} \sqrt{\frac{\gamma_i \widetilde{\sigma}_N^2}{\sigma_{N_i}^2 \gamma}} e^{j\hat{\Delta}_i} \\ &= \sqrt{\frac{\gamma_i \gamma \widetilde{\sigma}_N^2}{\sigma_{N_i}^2}} \frac{e^{j\hat{\Delta}_i}}{\gamma + \mu_W}, \end{aligned} \tag{54}$$

thus the MSE with respect to the speech component of the combined signal is minimized.

7.5. Simulation Results. In this section, we present some simulation results for the proposed combining system with additional noise suppression. Therefore, we used the simulation setup described in Section 3. Table 2 presents the results for the simulated in-car environment using the configuration with the intermicrophone distance of 0.6 m. As an objective

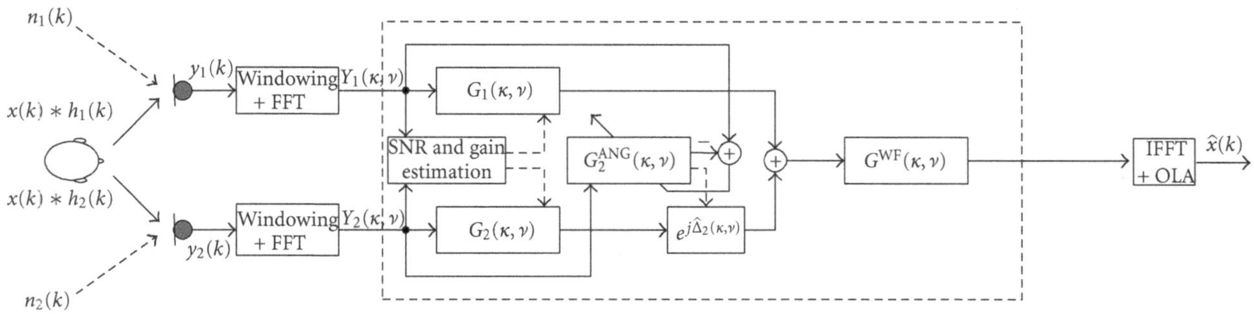

Figure 8: System structure of the blind-matched filtering with noise reduction for two channels.

Table 2: Simulation results for the proposed system in the car environment, cf. results on the right of Table 1.

d_{mic} [m]		0.6	
	100 km/h	140 km/h	defrost
SSNR ch.1 [dB]	−2.7	−7.7	−0.2
SSNR ch.2 [dB]	−1.1	−5.8	−0.6
SSNR [dB]	10.3	4.8	12.1
S-MOS	4.6	4.5	4.7
N-MOS	3.8	3.2	3.8
G-MOS	4.0	3.7	4.0

Table 3: Simulation results for the system in the conference environment with with $T_{60} = 0.25$ s.

Speaker pos.	1	2	3	4	5
SSNR ch.1 [dB]	8.8	8.3	8.5	10.1	13.2
SSNR ch.2 [dB]	8.3	10.1	10.2	9.9	7.8
SSNR ch.3 [dB]	13.6	10.1	9.9	8.0	9.3
$\mathbf{G}^{MFWF}$ SSNR [dB]	21.3	19.8	19.8	19.6	20.8
S-MOS	4.7	4.7	4.7	4.8	4.8
N-MOS	4.6	4.5	4.5	4.6	4.6
G-MOS	4.3	4.3	4.3	4.4	4.4
$\mathbf{G}^{SDW}$ SSNR [dB]	16.8	15.9	15.7	16.6	18.8
S-MOS	4.8	4.8	4.8	4.8	4.8
N-MOS	3.9	3.9	4.9	4.1	4.4
G-MOS	4.2	4.1	4.1	4.2	4.4

evaluation criterion, we calculated also the segmental SNR of the output signal. Also the input signal-to-noise ratios are shown in Table 2. Furthermore, we show results from an instrumental quality analysis in Table 1. Comparing these values with the results shown in Table 1, we observe that the proposed algorithm outperforms the SDW-MWF for this scenario. For this simulation, a higher noise overestimation factor μ_W can be used, while the S-MOS results are nearly the same in comparison to the results of SDW-MWF. This is because the proposed combining system partial equalizes the acoustic system and, therefore, the speech signal components are equalized with respect to the speech signal X at the speaker's mouth. Thus, the system can achieve a higher output SNR at the same level of speech distortion.

For the conference scenario, Table 3 shows the results of the performed simulations. It can be seen that the output SNR of the system as well as the MOS values is nearly the same for all speaker positions. This is a result of the proposed combining scheme.

The effect of the partial equalization is obvious by a comparison of the individual acoustic transfer functions with the overall system transfer function $\widetilde{H}(\nu)$. This is shown in Figure 9(a) for the speaker at position three, where the transfer functions between the speaker's mouth and the microphones are plotted as dashed and dotted lines. The overall system transfer function (including the system and the acoustic signal path) is plotted as solid line. It can be seen that the deep dips of the individual transfer functions are equalized. The overall transfer function follows the envelope of all transfer functions (which may include also the microphone characteristic). In Figure 9(b), the transfer functions of the SDW-MWF without the Wiener filter part and of the MVDR beamformer are plotted in comparison with the proposed blind-matched filter approach.

To show the applicability of the proposed system also in more reverberant environments, we used a simulated conference scenario with a reverberation time $T_{60} = 0.5$ s. For the generation of the room impulse responses, we used the image method described by Allen and Berkley in [24]. The results are presented in Table 4, again the output SNRs for the different speaker positions are in the same range. Also, simulation results for the SDW-MWF are given for a comparison of these two techniques.

8. Blind System Identification

In order to demonstrate that the filter $\widetilde{G}_i$ approximates the matched filter, we show that the structure in Figure 10 can be used for blind system identification. The SNR values for speech signals are fast time-varying. Hence, we use again FLMS filters G_i^{LMS} to estimate the average filter transfer functions. Note that if we have equal noise power spectral densities at all microphone inputs, the matched filter $\mathbf{G}^{MF}$ is equal to the vector of transfer function coefficients $\mathbf{H}$ up to a common factor. This factor can vary with the frequency. Hence, by estimating the ideal matched filter, we estimate the linear acoustic system up to a common filter. Furthermore, note that many known approaches for blind

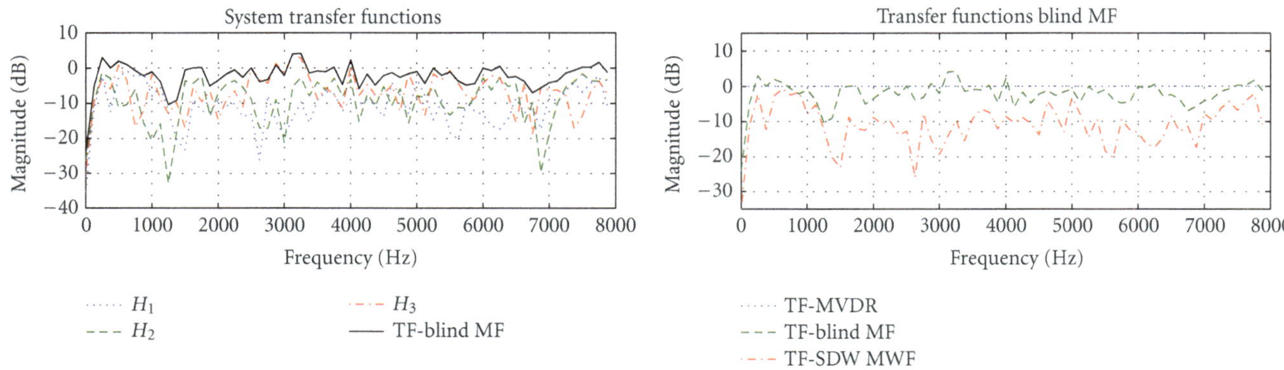

(a) Comparison of the blind MF with the transfer functions to the microphones

(b) Comparison of the blind MF with the ideal MVDR beamformer and the SDW-MWF

FIGURE 9: Comparison of the system transfer functions.

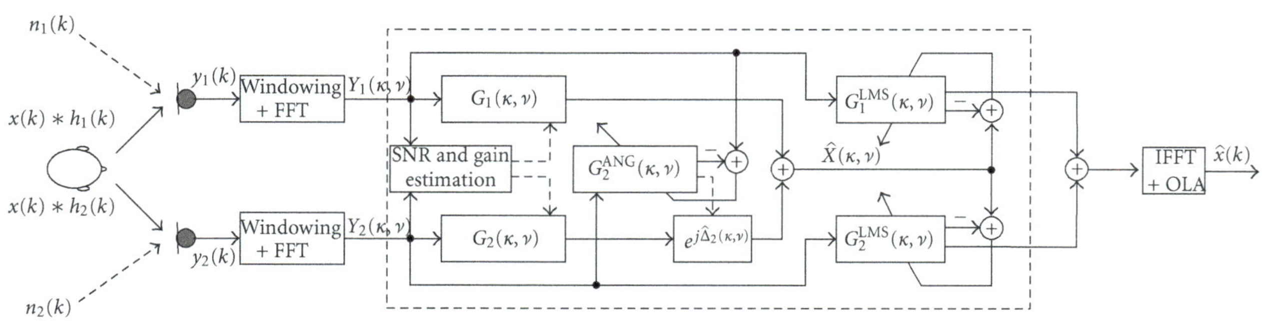

FIGURE 10: Basic system structure for the system identification approach for two channels.

TABLE 4: Simulation results for the system in a simulated conference environment with $T_{60} = 0.5$ s.

Speaker pos.	1	2	3	4	5
SSNR ch.1 [dB]	9.8	8.8	9.7	9.1	12.0
SSNR ch.2 [dB]	10.1	11.2	12.2	12.2	11.2
SSNR ch.3 [dB]	12.4	9.9	8.3	8.8	8.8
$\mathbf{G}^{\text{MFWF}}$ SSNR [dB]	21.5	21.0	21.5	20.8	21.6
S-MOS	4.7	4.7	4.7	4.7	4.7
N-MOS	4.5	4.4	4.5	4.5	4.6
G-MOS	4.3	4.3	4.3	4.3	4.3
$\mathbf{G}^{\text{SDW}}$ SSNR [dB]	17.8	17.0	17.1	16.9	19.0
S-MOS	4.8	4.7	4.7	4.8	4.8
N-MOS	3.8	3.7	3.7	3.7	4.1
G-MOS	4.1	4.0	4.1	4.1	4.2

system identification can only infer the M different channels up to a common filter [15]. Similarly, with the proposed system, all filters G_i are biased by a common factor. For equal noise power spectral densities, this common filter has the transfer function:

$$\widetilde{H} = e^{j\phi_{\text{ref}}}\sqrt{|H_1|^2 + |H_2|^2 + \cdots + |H_M|^2} \qquad (55)$$

and the LMS filters should converge to

$$G_i^{\text{LMS}} = \frac{H_i^*}{\widetilde{H}}. \qquad (56)$$

For simulations, we use the two microphone in-car setup with an intermicrophone distance of 0.6 m. We consider the driving situation with a car speed of 140 km/h. The magnitude of the actual transfer functions H_i and the magnitude of the corrected filter transfer function $G_i^{\text{LMS}}\widetilde{H}$ are depicted in Figure 11. We observe that the transfer functions are well approximated. As a quality measure, we use the distance

$$D_i = 10\log_{10}\left(\frac{\sum_\nu \left| G_i^{\text{LMS}}\widetilde{H} - H_i^* \right|^2}{\sum_\nu |H_i|^2}\right) \qquad (57)$$

and obtain values of $D_1 = -16.4$ dB and $D_2 = -11$ dB after 5 seconds of speech activity. For a driving situation with a car speed of 100 km/h, we obtain $D_1 = -17.9$ dB and $D_2 = -11.9$ dB, respectively.

9. Conclusions

In this paper, we have presented a speech enhancement system with distributed microphones, where the array geometry is arbitrary and not a priori known. The system is based

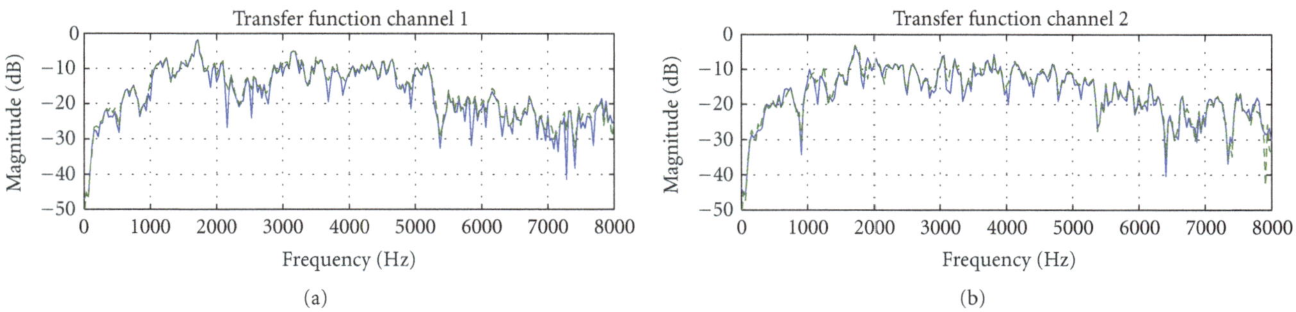

FIGURE 11: Estimated (dashed line) and actual transfer functions for the two channels.

on a blind-matched filtering approach where the filter coefficients depend only on the input signal-to-noise ratios and the correlation between the input signals. For spatially uncorrelated but not necessarily uniform noise, the system provides an optimal signal combining that maximizes the output SNR.

Moreover, the presented approach achieves a partial equalization of the acoustic system up to a common filter. To demonstrate that the ideal filter coefficients can be reliably estimated, we have presented an application for blind system identification. The system is able to identify the M different channels up to a common filter. The presented simulation results indicate that this identification is robust against background noise. To provide a perfect equalization, the remaining filter ambiguity needs to be resolved separately. However, the presented system could also be combined with other speech dereverberation algorithms, for example, the single channel reverberation suppression algorithms presented in [38, 39]. The system assumes a single speech source, but a situation with more than one active speaker cannot be avoided in real conference scenarios, so further investigations are needed to evaluate the concept for such scenarios.

Appendix

A. Decomposition of the SDW-MWF

Let $\mathbf{A}$ and $\mathbf{B}$ be two square matrices, where $\mathbf{A}$ has full rank and $\mathbf{B}$ has rank one. In this case, we can rewrite the inverse of the sum $\mathbf{A}+\mathbf{B}$ using of the matrix inversion lemma [22]:

$$(\mathbf{A}+\mathbf{B})^{-1} = \mathbf{A}^{-1} - \frac{1}{1+g}\mathbf{A}^{-1}\mathbf{B}\mathbf{A}^{-1}, \tag{A.1}$$

where g is the trace of $\mathbf{A}^{-1}\mathbf{B}$.

Now, consider the term $(\mathbf{R}_S + \mu_W \mathbf{R}_N)^{-1}$ in the definition of the SDW-MWF in (18). $\mathbf{R}_S = \sigma_X^2 \mathbf{H}\mathbf{H}^\dagger$ has rank one due to the structure of the matrix $\mathbf{H}\mathbf{H}^\dagger$ (see, e.g., [40]), and if we assume that the noise variances at all microphones are nonzero, $\mathbf{R}_N$ will have full rank. Using (A.1) the inverse $(\mu_W \mathbf{R}_N + \mathbf{R}_S)^{-1}$ in (18) can be rewritten as

$$(\mu_W \mathbf{R}_N + \mathbf{R}_S)^{-1} = \frac{1}{\mu_W}\mathbf{R}_N^{-1} - \frac{\mathbf{R}_N^{-1}\mathbf{R}_S\mathbf{R}_N^{-1}}{\mu_W^2\left(1+\mu_W^{-1}\sigma_x^2\mathbf{H}^\dagger\mathbf{R}_N^{-1}\mathbf{H}\right)}. \tag{A.2}$$

Hence, we can decompose the SDW-MWF as follows [20]:

$$\begin{aligned}
\mathbf{G}^{\mathrm{SDW}} &= \left(\frac{1}{\mu_W}\mathbf{R}_N^{-1} - \frac{\mathbf{R}_N^{-1}\mathbf{R}_S\mathbf{R}_N^{-1}}{\mu_W^2\left(1+\mu_W^{-1}\sigma_X^2\mathbf{H}^\dagger\mathbf{R}_N^{-1}\mathbf{H}\right)}\right)\mathbf{R}_S\mathbf{u}\\
&= \frac{1}{\mu_W}\mathbf{R}_N^{-1}\left(\mathbf{I} - \frac{\sigma_X^2\mathbf{H}\mathbf{H}^\dagger\mathbf{R}_N^{-1}}{\mu_W+\sigma_X^2\mathbf{H}^\dagger\mathbf{R}_N^{-1}\mathbf{H}}\right)\sigma_X^2\mathbf{H}\mathbf{H}^\dagger\mathbf{u}\\
&= \frac{1}{\mu_W}\mathbf{R}_N^{-1}\left(\mathbf{H} - \frac{\sigma_X^2\mathbf{H}\mathbf{H}^\dagger\mathbf{R}_N^{-1}\mathbf{H}}{\mu_W+\sigma_X^2\mathbf{H}^\dagger\mathbf{R}_N^{-1}\mathbf{H}}\right)\sigma_X^2\mathbf{H}^\dagger\mathbf{u}\\
&= \frac{1}{\mu_W}\mathbf{R}_N^{-1}\left(1 - \frac{\sigma_X^2\mathbf{H}^\dagger\mathbf{R}_N^{-1}\mathbf{H}}{\mu_W+\sigma_X^2\mathbf{H}^\dagger\mathbf{R}_N^{-1}\mathbf{H}}\right)\sigma_X^2\mathbf{H}\mathbf{H}^\dagger\mathbf{u}\\
&= \left(1 - \frac{\sigma_X^2\mathbf{H}^\dagger\mathbf{R}_N^{-1}\mathbf{H}}{\mu_W+\sigma_X^2\mathbf{H}^\dagger\mathbf{R}_N^{-1}\mathbf{H}}\right)\frac{1}{\mu_W}\mathbf{R}_N^{-1}\sigma_X^2\mathbf{H}\mathbf{H}^\dagger\mathbf{u}\\
&= \frac{\sigma_X^2}{\mu_W+\sigma_X^2\mathbf{H}^\dagger\mathbf{R}_N^{-1}\mathbf{H}}\mathbf{R}_N^{-1}\mathbf{H}\mathbf{H}^\dagger\mathbf{u}\\
&= \frac{\sigma_X^2}{\sigma_X^2+\mu_W\left(\mathbf{H}^\dagger\mathbf{R}_N^{-1}\mathbf{H}\right)^{-1}}\frac{\mathbf{R}_N^{-1}\mathbf{H}}{\mathbf{H}^\dagger\mathbf{R}_N^{-1}\mathbf{H}}\mathbf{H}^\dagger\mathbf{u}\\
&= \frac{\sigma_X^2}{\sigma_X^2+\mu_W\sigma^2{}_{N_{\mathrm{MVDR}}}}\frac{\mathbf{R}_N^{-1}\mathbf{H}}{\mathbf{H}^\dagger\mathbf{R}_N^{-1}\mathbf{H}}\mathbf{H}^\dagger\mathbf{u}\\
&= \frac{\gamma}{\gamma+\mu_W}\frac{\mathbf{R}_N^{-1}\mathbf{H}}{\mathbf{H}^\dagger\mathbf{R}_N^{-1}\mathbf{H}}\mathbf{H}^\dagger\mathbf{u}\\
&= G^{\mathrm{WF}}\mathbf{G}^{\mathrm{MVDR}}H_{\mathrm{ref}}^*,
\end{aligned} \tag{A.3}$$

where $\sigma^2{}_{N_{\mathrm{MVDR}}}$ is the noise variance at the output of $\mathbf{G}^{\mathrm{MVDR}}$

$$\begin{aligned}
\sigma^2{}_{N_{\mathrm{MVDR}}} &= \mathbf{G}^{\mathrm{MVDR}\dagger}\mathbf{R}_N\mathbf{G}^{\mathrm{MVDR}}\\
&= \frac{\mathbf{H}^\dagger\mathbf{R}_N^{-1}\mathbf{H}}{\left(\mathbf{H}^\dagger\mathbf{R}_N^{-1}\mathbf{H}\right)^2}\\
&= \left(\mathbf{H}^\dagger\mathbf{R}_N^{-1}\mathbf{H}\right)^{-1}.
\end{aligned} \tag{A.4}$$

Acknowledgments

The Research for this article was supported by DFG (FR 2673/2-1). Last but not the least, the authors would like to thank the reviewers for their constructive comments and suggestions which greatly improve the quality of this manuscript.

References

[1] E. Hänsler and G. Schmidt, *Speech and Audio Processing in Adverse Environments: Signals and Communication Technologie*, Springer, 2008.

[2] O. L. Frost, "An algorithm for linearly constrained adaptive array processing," *Proceedings of IEEE*, vol. 60, no. 8, pp. 926–935, 1972.

[3] L. J. Griffiths and C. W. Jim, "An alternative approach to linearly constrained adaptive beamforming," *IEEE Transactions on Antennas and Propagation*, vol. 30, no. 1, pp. 27–34, 1982.

[4] A. Guérin, R. Le Bouquin-Jeannès, and G. Faucon, "A two-sensor noise reduction system: applications for hands-free car kit," *Eurasip Journal on Applied Signal Processing*, vol. 2003, no. 11, pp. 1125–1134, 2003.

[5] T. Gerkmann and R. Martin, "Soft decision combining for dual channel noise reduction," in *Proceedings of the 9th International Conference on Spoken Language Processing (INTERSPEECH '06-ICSLP*, pp. 2134–2137, Pittsburgh, Pa, USA, September 2006.

[6] J. Freudenberger, S. Stenzel, and B. Venditti, "Spectral combining for microphone diversity systems," in *Proceedings of the European Signal Processing Conference (EUSIPCO '09)*, pp. 854–858, Glasgow, Scotland, UK, 2009.

[7] S. Doclo, A. Spriet, M. Moonen, and J. Wouters, "Speech distortion weighted multichannel wiener filtering techniques for noise reduction," in *Speech Enhancement*, chapter 9, Springer, 2005.

[8] S. Doclo, A. Spriet, J. Wouters, and M. Moonen, "Frequency-domain criterion for the speech distortion weighted multichannel Wiener filter for robust noise reduction," *Speech Communication*, vol. 49, no. 7-8, pp. 636–656, 2007.

[9] S. Gannot, D. Burshtein, and E. Weinstein, "Signal enhancement using beamforming and nonstationarity with applications to speech," *IEEE Transactions on Signal Processing*, vol. 49, no. 8, pp. 1614–1626, 2001.

[10] S. Gannot and I. Cohen, "Speech enhancement based on the general transfer function GSC and postfiltering," *IEEE Transactions on Speech and Audio Processing*, vol. 12, no. 6, pp. 561–571, 2004.

[11] S. Affes and Y. Grenier, "A signal subspace tracking algorithm for microphone array processing of speech," *IEEE Transactions on Speech and Audio Processing*, vol. 5, no. 5, pp. 425–437, 1997.

[12] S. Gannot and M. Moonen, "Subspace methods for multimicrophone speech dereverberation," *Eurasip Journal on Applied Signal Processing*, vol. 2003, no. 11, pp. 1074–1090, 2003.

[13] M. Delcroix, T. Hikichi, and M. Miyoshi, "Dereverberation and denoising using multichannel linear prediction," *IEEE Transactions on Audio, Speech and Language Processing*, vol. 15, no. 6, pp. 1791–1801, 2007.

[14] I. Ram, E. Habets, Y. Avargel, and I. Cohen, "Multi-microphone speech dereverberation using LIME and least squares filtering," in *Proceedings of the European Signal Processing Conference (EUSIPCO '08)*, Lausanne, Switzerland, Augast 2008.

[15] D. Schmid and G. Enzner, "Robust subsystems for iterative multichannel blind system identification and equalization," in *Proceedings of IEEE Pacific Rim Conference on Communications, Computers and Signal Processing (PACRIM '09)*, pp. 889–893, August 2009.

[16] J. Freudenberger, S. Stenzel, and B. Venditti, "Microphone diversity combining for In-car applications," *Eurasip Journal on Advances in Signal Processing*, vol. 2010, Article ID 509541, 2010.

[17] S. Applebaum, "Adaptive arrays," *IEEE Transactions on Antennas and Propagation*, vol. 24, no. 5, pp. 585–598, 1976.

[18] S. Doclo and M. Moonen, "Robust time-delay estimation in highly adverse acoustic environments," in *Proceedings of IEEE Workshop on Applications of Signal Processing to Audio and Acoustics*, pp. 59–62, October 2001.

[19] E. Warsitz and R. Haeb-Umbach, "Blind acoustic beamforming based on generalized eigenvalue decomposition," *IEEE Transactions on Audio, Speech and Language Processing*, vol. 15, no. 5, pp. 1529–1539, 2007.

[20] S. Doclo, S. Gannot, M. Moonen, and A. Spriet, "Acoustic beamforming for hearing aid applications," in *Handbook on Array Processing and Sensor Networks*, chapter 9, Wiley & Sons, 2009.

[21] R. Martin and P. Vary, "Combined acoustic echo cancellation, dereverberation and noise reduction: a two microphone approach," *Annales Des Télécommunications*, vol. 49, no. 7-8, pp. 429–438, 1994.

[22] K. S. Miller, "On the inverse of the sum of matrices," *Mathematics Magazine*, vol. 54, no. 2, pp. 67–72, 1981.

[23] ITU-T, *Test Signals for Use in Telephonometry*, Recommendation ITU-T P.501, International Telecommunication Union, Geneva, Switzerland, 2007.

[24] J. B. Allen and D. A. Berkley, "Image method for efficiently simulating small-room acoustics," *Journal Acoustic Society of America*, vol. 65, no. 4, pp. 943–950, 1979.

[25] B. Holter and G. E. Oien, "The optimal weights of a maximum ratio combiner using an eigenfilter approach," in *Proceedings of the 5th IEEE Nordic Signal Processing Symposium (NORSIG '02)*, Hurtigruten, Norway, 2002.

[26] H. Cox, R. Zeskind, and M. Owen, "Robust adaptive beamforming," *IEEE Transactions on Acoustics, Speech and Signal Processing*, vol. 35, no. 10, pp. 1365–1376, 1987.

[27] G. Doblinger, "Computationally efficient speech enhancement by spectral minima tracking in subbands," in *Proceedings of the European Conference on Speech Communication and Technology (EUROSPEECH '95)*, vol. 2, pp. 1513–1516, Madrid, Spain, 1995.

[28] I. Cohen, "On speech enhancement under signal presence uncertainty," in *Proceedings of the Interntional Conference on Acoustics, Speech, and Signal Processing (ICASSP '01)*, vol. 1, pp. 661–664, May 2001.

[29] M. Rahmani, A. Akbari, and B. Ayad, "An iterative noise cross-PSD estimation for two-microphone speech enhancement," *Applied Acoustics*, vol. 70, no. 3, pp. 514–521, 2009.

[30] R. Hendriks and T. Gerkmann, "Noise correlation matrix estimation for multi-microphone speech enhancement," *IEEE Transactions on Audio, Speech, and Language Processing*, vol. 20, no. 1, p. 223, 2011.

[31] M. Berouti, R. Schwartz, and J. Makhoul, "Enhancement of speech corrupted by acoustic noise," in *Proceedings of the International Conference on Acoustics, Speech, and Signal Processing (ICASSP '79)*, pp. 208–211, 1979.

[32] N. Dal Degan and C. Prati, "Acoustic noise analysis and speech enhancement techniques for mobile radio applications," *Signal Processing*, vol. 15, no. 1, pp. 43–56, 1988.

[33] M. Doerbecker, "Sind kohärenzbasierte Störgeräuschreduktionsverfahren für elektrische Hörhilfen geeignet?—Modelle zur Beschreibung der Kohärenzeigenschaften kopfbezogener Mikrofonsignale ," in *ITG-Fachtagung Sprachkommunikation und 9. Konferenz Elektronische Sprachsignalverarbeitung*, pp. 53–56, Dresden, Germany, 1998.

[34] W. Armbrüster, R. Czarnach, and P. Vary, "Adaptive noise cancellation with reference input," in *Signal Processing III*, pp. 391–394, Elsevier, 1986.

[35] ITU-T, *Objective Measurement of Active Speech Level*, Recommendation ITU-T P.56, International Telecommunication Union, Geneva, Switzerland, 1996.

[36] J. Reimes, H. W. Gierlich, F. Kettler, and S. Poschen, "Ein neues Verfahren zur objektiven Beurteilung der Sprachqualität bei Hintergrundger äuschen : 3QUEST," in *8. ITG-Fachtagung Sprachkommunikation*, 2008.

[37] C. H. Knapp and G. C. Carter, "The generalized correlation method for estimation of time delay," *IEEE Transactions on Acoustics, Speech, and Signal Processing*, vol. 24, no. 4, pp. 320–327, 1976.

[38] E. A. P. Habets, "Single-channel speech dereverberation based on spectral subtraction," in *Proceedings of the 15th Annual Workshop on Circuits*, vol. 10, pp. 250–254, 2004.

[39] H. W. Löllmann and P. Vary, "Low Delay Noise Reduction and Dereverberation for Hearing Aids," *Eurasip Journal on Advances in Signal Processing*, vol. 2009, Article ID 437807, 2009.

[40] G. Strang, *Linear Algebra and Its Applications*, Harcourt Brace Jovanovich, Lisbon, Portugal, 1988.

A Secure Image Encryption Algorithm Based on Rubik's Cube Principle

Khaled Loukhaoukha, Jean-Yves Chouinard, and Abdellah Berdai

Department of Electrical and Computer Engineering, Laval University, QC, Canada G1K 7P4

Correspondence should be addressed to Khaled Loukhaoukha, khaled.loukhaoukha.1@ulaval.ca

Academic Editor: Fouad Khelifi

In the past few years, several encryption algorithms based on chaotic systems have been proposed as means to protect digital images against cryptographic attacks. These encryption algorithms typically use relatively small key spaces and thus offer limited security, especially if they are one-dimensional. In this paper, we proposed a novel image encryption algorithm based on Rubik's cube principle. The original image is scrambled using the principle of Rubik's cube. Then, XOR operator is applied to rows and columns of the scrambled image using two secret keys. Finally, the experimental results and security analysis show that the proposed image encryption scheme not only can achieve good encryption and perfect hiding ability but also can resist exhaustive attack, statistical attack, and differential attack.

1. Introduction

The end of the 20th century was marked by an extraordinary technical revolution from analog to numerical as documents and equipments became increasingly used in various domains. However, the advantages of the digital revolution were not achieved without drawbacks such as illegal copying and distribution of digital multimedia documents. To meet this challenge, researchers were motivated more than ever to protect multimedia documents with new and efficient document protection techniques. In this context, different techniques have been introduced such as encryption and digital watermarking. The first one consists in transforming multimedia documents using an algorithm to make it unreadable to anyone except for the legitimate users. The second one consists of embedding digital watermarks into multimedia documents to guarantee the ownership and the integrity of the digital multimedia contents.

The protection of images is of particular interest in this paper. Traditional image encryption algorithms such as private key encryption standards (DES and AES), public key standards such as Rivest Shamir Adleman (RSA), and the family of elliptic-curve-based encryption (ECC), as well as the international data encryption algorithm (IDEA), may not be the most desirable candidates for image encryption, especially for fast and real-time communication applications. In recent years, several encryption schemes have been proposed [1–12]. These encryption schemes can be classified into different categories such as value transformation [1–4], pixels position permutation [5–8], and chaotic systems [9–12].

In the first category, Liu et al. [1] presented an image encryption scheme based on iterative random phase encoding in gyrator transform domains. A two-dimensional chaotic mapping is employed to generate many random data for iterative random phase encoding. In [2], a color image encryption method using discrete fractional random transform (DFRNT) and the Arnold transform (AT) in the intensity-hue-saturation (IHS) color space has been proposed. Each color space component is then encrypted independently with different approaches. In [3], an image encryption algorithm based on Arnold transform and gyrator transform has been proposed. The amplitude and phase of the gyrator transform are separated into several subimages, which are scrambled using the Arnold transform. The parameters of gyrator transforms and separating scheme serve as the key of the encryption method. Tao et al. [4] proposed an image encryption algorithm based on fractional

Fourier transform (FRFT) and which can be applied to the double or more image encryptions. The encrypted image is obtained by the summation of different orders of inverse discrete fractional Fourier transform (IDFRFT) of the interpolated subimages. The whole transform orders of the utilized FRFT are used as the secret keys for the decryption of each subimage.

In the second category, Zunino [5] use Peano-Hilbert curves as pixels position permutation to destroy the spatial autocorrelation of an image. Zhang and Liu [6] proposed image encryption scheme based on permutation-diffusion architecture and skew tent map system. In the proposed scheme, the P-box is chosen as the same size of original image, which shuffles the positions of pixels totally. To enhance the security, the keystream in the diffusion step depends on both the key and original image. Zhao and Chen [7] proposed to used ergodic matrices for scrambling and encryption of digital images. The authors analyzed the isomorphism relationship between ergodic matrices and permutation. Zhu et al. [8] proposed an innovative permutation method to confuse and diffuse the gray-scale image at the bitlevel, which changes the position of the pixel and modifies its value. This algorithm uses also the Arnold cat map to permute the bits and the logistic map to further encrypt the permuted image.

In the third category, Huang and Nien [9] proposed a novel pixel shuffling method for color image encryption which used chaotic sequences generated by chaotic systems as encryption codes. In [10], the two-dimensional chaotic cat map has been generalized to three-dimensional one and then was used to design a fast and secure symmetric image encryption scheme. This scheme employs the 3D cat map to shuffle the positions and the values of image pixels. Wang et al. [11] presented an image encryption algorithm based on simple Perceptron and using a high-dimensional chaotic system in order to produce three sets of pseudorandom sequence. Then to generate weight of each neuron of perceptron as well as a set of input signal, a nonlinear strategy is adopted. Recently, a new image encryption algorithm combining permutation and diffusion was proposed by Wang et al. [12]. The original image is partitioned into blocks and a spatiotemporal chaotic system is then employed to generate the pseudorandom sequence used for diffusing and shuffling these blocks.

The security of image encryption has been extensively studied. Almost some encryption schemes based on permutation had already been found insecure against the ciphertext-only and known/chosen-plaintext attacks, due to the high information redundancy, and it is quite understandable since the secret permutations can be recovered by comparing the plaintexts and the permuted ciphertexts. Generally, chaos-based image encryption algorithms are used more often than others but require high computational cost. Moreover, a chaos system is defined on real numbers while the cryptosystems are defined on finite sets of integers. One-dimensional chaotic cryptosystems are limited by their small key spaces and weak security in [1, 13].

In this paper, we present a novel image encryption algorithm based on the principle of Rubik's cube. First-, in order to scramble the pixels of gray-scale original image the principle of Rubik's cube is deployed which only changes the position of the pixels. Using two random secret keys, the bitwise XOR is applied into the odd rows and columns. Then, the bitwise XOR is also applied to even rows and columns using the flipped secret keys. These steps can be repeated while the number of iteration is not reached. Numerical simulation has been performed to test the validity and the security of the proposed encryption algorithm.

The remaining of this paper is organized as follows. Section 2 describes the proposed image encryption algorithm based on Rubik's cube principle. Experimental results and security analysis are presented in Section 3. Finally, we conclude in Section 4.

2. Rubik's Cube Image Encryption

In this section, the proposed encryption algorithm based on Rubik's cube principle is described along with the decryption algorithm.

2.1. Rubik's Cube Based Encryption Algorithm. Let I_o represent an α-bit gray scale image of the size $M \times N$. Here, I_o represent the pixels values matrix of image I_o. The steps of encryption algorithm are as follows:

(1) Generate randomly two vectors K_R and K_C of length M and N, respectively. Element $K_R(i)$ and $K_C(j)$ Each take a random value of the set $\mathcal{A} = \{0, 1, 2, \dots, 2^{\alpha} - 1\}$. Note that both K_R and K_C must not have constant values.

(2) Determine the number of iterations, $\mathrm{ITER}_{\max}$, and initialize the counter ITER at 0.

(3) Increment the counter by one: ITER = ITER + 1.

(4) For each row i of image I_o,

(a) compute the sum of all elements in the row i, this sum is denoted by $\alpha(i)$

$$\alpha(i) = \sum_{j=1}^{N} \mathrm{I}_o(i,j), \quad i = 1, 2, \dots, M, \tag{1}$$

(b) compute modulo 2 of $\alpha(i)$, denoted by $M_{\alpha(i)}$,

(c) row i is left, or right, circular-shifted by $K_R(i)$ positions (image pixels are moved $K_R(i)$ positions to the left or right direction, and the first pixel moves in last pixel.), according to the following:

$$\begin{aligned} &\text{if } M_{\alpha(i)} = 0 \longrightarrow \text{right circular shift} \\ &\qquad \text{else} \longrightarrow \text{left circular shift.} \end{aligned} \tag{2}$$

(5) For each column j of image I_o,

(a) compute the sum of all elements in the column j, this sum is denoted by $\beta(j)$,

$$\beta(j) = \sum_{i=1}^{M} \mathsf{I}_o(i, j), \quad j = 1, 2, \ldots, N, \tag{3}$$

(b) compute modulo 2 of $\beta(j)$, denoted by $M_{\beta(j)}$.

(c) column j is down, or up, circular-shifted by $K_C(i)$ positions, according to the following:

$$\begin{aligned} &\text{if } M_{\beta(j)} = 0 \longrightarrow \text{up circular shift} \\ &\quad \text{else} \longrightarrow \text{down circular shift.} \end{aligned} \tag{4}$$

Steps 4 and 5 above will create a scrambled image, denoted by I_{SCR}.

(6) Using vector K_C, the bitwise XOR operator is applied to each row of scrambled image I_{SCR} using the following expressions:

$$\begin{aligned} \mathsf{I}_1(2i-1, j) &= \mathsf{I}_{\text{SCR}}(2i-1, j) \oplus K_C(j), \\ \mathsf{I}_1(2i, j) &= \mathsf{I}_{\text{SCR}}(2i, j) \oplus \text{rot}\,180(K_C(j)), \end{aligned} \tag{5}$$

where $\oplus$ and rot $180(K_C)$ represent the bitwise XOR operator and the flipping of vector K_C from left to right, respectively.

(7) Using vector K_R, the bitwise XOR operator is applied to each column of image I_1 using the following formulas:

$$\begin{aligned} \mathsf{I}_{\text{ENC}}(i, 2j-1) &= \mathsf{I}_1(i, 2j-1) \oplus K_R(j), \\ \mathsf{I}_{\text{ENC}}(i, 2j) &= \mathsf{I}_1(i, 2j) \oplus \text{rot}\,180(K_R(j)). \end{aligned} \tag{6}$$

with rot $180(K_R)$ indicating the left to right flip of vector K_R.

(8) If ITER $=$ $\text{ITER}_{\max}$, then encrypted image I_{ENC} is created and encryption process is done; otherwise, the algorithm branches to step 3.

Vectors K_R, K_C and the max iteration number $\text{ITER}_{\max}$ are considered as secret keys in the proposed encryption algorithm. However, to obtain a fast encryption algorithm it is preferable to set $\text{ITER}_{\max} = 1$ (single iteration). Conversely, if $\text{ITER}_{\text{MAX}} > 1$, then the algorithm is more secure because the key space is larger than for $\text{ITER}_{\text{MAX}} = 1$. Nevertheless, in the simulations presented in Section 3, the number of iterations $\text{ITER}_{\max}$ was set to one.

2.2. Rubik's Cube Decryption Algorithm. The decrypted image, I_o, is recovered from the encrypted image, I_{ENC}, and the secret keys, K_R, K_C, and $\text{ITER}_{\max}$ as follows in the following.

(1) Initialize ITER $= 0$.

(2) Increment the counter by one: ITER $=$ ITER $+ 1$.

(3) The bitwise XOR operation is applied on vector K_R and each column of the encrypted image I_{ENC} as follows:

$$\begin{aligned} \mathsf{I}_1(i, 2j-1) &= \mathsf{I}_{\text{ENC}}(i, 2j-1) \oplus K_R(j), \\ \mathsf{I}_1(i, 2j) &= \mathsf{I}_{\text{ENC}}(i, 2j) \oplus \text{rot}\,180(K_R(j)), \end{aligned} \tag{7}$$

(4) Then, using the K_C vector, the bitwise XOR operator is applied to each row of image I_1:

$$\begin{aligned} \mathsf{I}_{\text{SCR}}(2i-1, j) &= \mathsf{I}_1(2i-1, j) \oplus K_C(j), \\ \mathsf{I}_{\text{SCR}}(2i, j) &= \mathsf{I}_1(2i, j) \oplus \text{rot}\,180(K_C(j)). \end{aligned} \tag{8}$$

(5) For each column j of the scrambled image I_{SCR},

(a) compute the sum of all elements in that column j, denoted as $\beta_{\text{SCR}}(j)$:

$$\beta_{\text{SCR}}(j) = \sum_{i=1}^{M} \mathsf{I}_{\text{SCR}}(i, j), \quad j = 1, 2, \ldots, N, \tag{9}$$

(b) compute modulo 2 of $\beta_{\text{SCR}}(j)$, denoted by $M_{\beta_{\text{SCR}}(j)}$,

(c) column j is down, or up, circular-shifted by $K_C(i)$ positions according to the following:

$$\begin{aligned} &\text{if } M_{\beta_{\text{SCR}}(j)} = 0 \longrightarrow \text{up circular shift} \\ &\quad \text{else} \longrightarrow \text{down circular shift.} \end{aligned} \tag{10}$$

(6) For each row i of scrambled image I_{SCR},

(a) compute the sum of all elements in row i, this sum is denoted by $\alpha_{\text{SCR}}(i)$:

$$\alpha_{\text{SCR}}(i) = \sum_{j=1}^{N} \mathsf{I}_{\text{SCR}}(i, j), \quad i = 1, 2, \ldots, M, \tag{11}$$

(b) compute modulo 2 of $\alpha_{\text{SCR}}(j)$, denoted by $M_{\alpha_{\text{SCR}}(j)}$,

(c) row i is then left, or right, circular-shifted by $K_R(i)$ according to the following:

$$\begin{aligned} &\text{if } M_{\alpha_{\text{SCR}}(j)} = 0 \longrightarrow \text{right circular shift} \\ &\quad \text{else} \longrightarrow \text{left circular shift.} \end{aligned} \tag{12}$$

(7) If ITER $=$ $\text{ITER}_{\max}$, then image I_{ENC} is decrypted and the decryption process is done; otherwise, the algorithm branches back to step 2.

TABLE 1: Difference measures between original and encrypted images.

Image	NPCR (in %)	UACI (in %)
Lena	99.5850	28.6210
Black	99.6078	50.1931
Baboon	99.6094	27.4092
Checkerboard	99.6201	50.0233

TABLE 2: Number of pixel change rate (NPCR) between images encrypted with keys K_1 and K_2 (with 1 bit difference).

Image	Mean (%)	Standard deviation (%)
Lena	99.6111	0.0251
Black	99.6064	0.0238
Baboon	99.6085	0.0253
Checkerboard	99.6113	0.0274

3. Experimental Results

In this section, we present the tests that were conducted to assess the efficiency and security of the proposed image encryption algorithm. These tests involve visual testing and security analysis.

3.1. Visual Testing. For visual testing, four gray-scale images of size 256 × 256 pixels were used. Figure 1 depicts these test images—`lena`, `black`, `baboon` and `checkerboard`—as well as the images encrypted using the proposed Rubik's cube algorithm. From this figure, one can see that there is no perceptual similarity between original images and their encrypted counterparts.

The encrypted image should greatly differ from its original form. In general, two difference measures are used to quantify this requirement. The first measure is the number of pixels change rate (NPCR), which indicate the percentage of different pixels between two images. The second one is the unified average changing intensity (UACI), which measures the average intensity of differences in pixels between two images [10]. Let $\mathsf{I}_o(i,j)$ and $\mathsf{I}_{\text{ENC}}(i,j)$ be the pixels values of original and encrypted images, I_o and I_{ENC}, at the ith pixel row and jth pixel column, respectively. Equations (13) and (15) give the mathematical expressions of the NPCR and UACI measures:

$$\text{NPCR} = \frac{\sum_{i=1}^{M}\sum_{j=1}^{N}\mathsf{D}(i,j)}{M\times N}\times 100\%, \quad (13)$$

$$\text{with}: \mathsf{D}(i,j) = \begin{cases} 0 & \text{if } \mathsf{I}_o(i,j) = \mathsf{I}_{\text{ENC}}(i,j), \\ 1 & \text{otherwise.} \end{cases} \quad (14)$$

$$\text{UACI} = \left[\sum_{i=1}^{M}\sum_{j=1}^{N}\frac{|\mathsf{I}_o(i,j) - \mathsf{I}_{\text{ENC}}(i,j)|}{255}\right]\times\frac{100\%}{M\times N}. \quad (15)$$

To approach the performances of an ideal image encryption algorithm, NPCR values must be as large as possible and UACI values must be around 33%. Table 1 gives the NPCR and UACI values for the original images and their encrypted versions. The values are very close to unity for the NPCR measure. The UACI values are also appropriate. The high percentage values of the NPCR measure indicate that the pixels positions have been randomly changed.

Furthermore, the UACI values show that almost all pixel gray-scale values of encrypted image have been changed from their values in the original images, making the original and encrypted image pixels more difficult to discriminate. It iss also observed that the UACI values for the case of the `black` and the `checkerboard` images are larger than those of the other images. This is due to the fact that all the pixels values of the `black` and `checkerboard` images are at the extremity of pixels values range, that is, 0 for the black pixels and 255 for the white pixels, leading to a large absolute difference between the original and encrypted images.

3.2. Security Analysis. Security is a major issue in cryptology. A good image encryption scheme should resist various attacks such as known plain text attack, cipher-text-only attack, statistical analysis attack, and brute-force attacks. In this section, a security analysis on the proposed image encryption algorithm is done. The security assessment has been done on key space analysis and statistical analysis.

3.2.1. Key Space Analysis. A secure image encryption scheme must have a large key space in order to make brute-force attack practically (computationally) infeasible. In theory, the proposed algorithm can accommodate an infinite key space. However, the encryption key used in our scheme is composed of the $(K_R, K_C, \text{ITER}_{\max})$ triplet. For an α-bit gray-scale image I_o of size $M \times N$ pixels, the vectors K_R and K_C can take $2^{M\cdot\alpha}$ and $2^{N\cdot\alpha}$ possible values, respectively. If we consider that both vectors must not have constant values, and the key space size is $2^{\alpha\cdot(M+N)} \times \text{ITER}_{\max} - 2^{2\alpha}$ keys, one can see that the size of the key space can be expanded when the number of iteration $\text{ITER}_{\max}$ is increased. For instance, for an 8-bits, scale gray image of size 256 × 256 pixels and $\text{ITER}_{\max} = 1$. The key space size is equal to $2^{4096} - 2^{16} \approx 10^{1233}$; this key space is large enough to resist exhaustive attack and it is larger than the key space size of the image encryption algorithms proposed in [1, 9, 10, 14–16].

3.2.2. Key Sensibility. Encryption algorithms should also have high sensibility to encryption key: this means that any small change in the key should lead to a significant change in the encrypted, or decrypted, image. We performed two tests to illustrate the key sensibility of our scheme. The first one shows the impact of a key change in the image encryption process. Here, the original image, I_o, is encrypted using the key $K_1 = (K_R, K_C, \text{ITER}_{\max})$, where K_R, K_C, and ITER_{MAX} are randomly generated. Then, the same image, that is, I_o, is encrypted using another key K_2 which differs only from the first key, K_1, in the least significant bit, that is, $K_2 = (K_R, K_C, \text{ITER}_{\max} + 1)$. This experiment is repeated 100 times using different key pairs K_1 and K_2 (still only differing by the least significant bit). Table 2 represents the mean and the standard deviation of the NPCR values between the

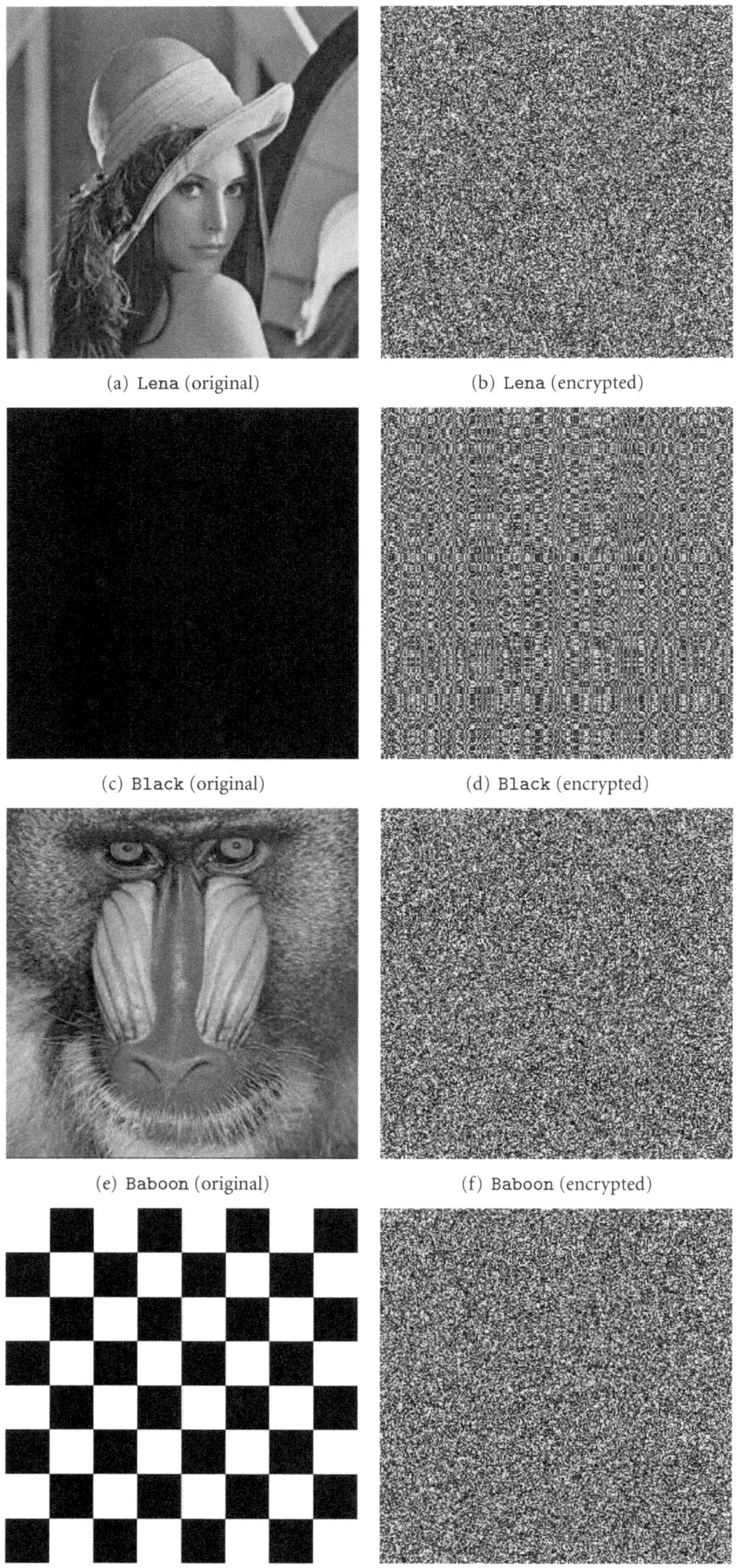

(a) Lena (original) (b) Lena (encrypted)

(c) Black (original) (d) Black (encrypted)

(e) Baboon (original) (f) Baboon (encrypted)

(g) Checkerboard (original) (h) Checkerboard (encrypted)

FIGURE 1: Original and encrypted images.

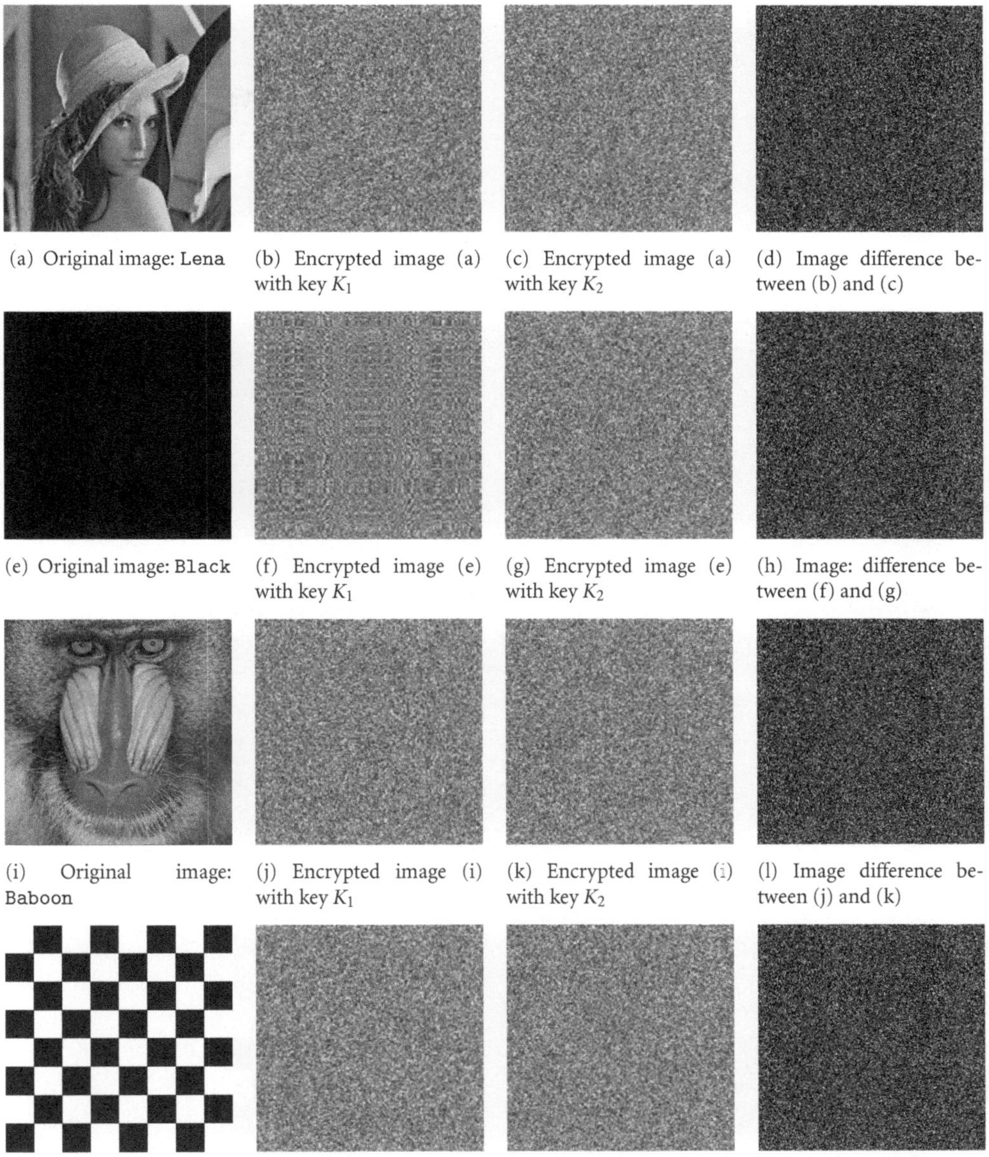

(a) Original image: Lena (b) Encrypted image (a) with key K_1 (c) Encrypted image (a) with key K_2 (d) Image difference between (b) and (c)

(e) Original image: Black (f) Encrypted image (e) with key K_1 (g) Encrypted image (e) with key K_2 (h) Image: difference between (f) and (g)

(i) Original image: Baboon (j) Encrypted image (i) with key K_1 (k) Encrypted image (i) with key K_2 (l) Image difference between (j) and (k)

(m) Original image: Checkerboard (n) Encrypted image (m) with key K_1 (o) Encrypted image (m) with key K_2 (p) Image: difference between (n) and (o)

FIGURE 2: Key sensibility for image encryption using the proposed algorithm.

encrypted image with key K_1 and the encrypted image using the key K_2 using 100 different key pairs. One can see from Table 2 that the mean values of NPCR are close to 100%, which means that image encrypted by the key K_1 differs significantly from the one encrypted by key K_2. Moreover, standard deviation values are very small: this indicates that the NPCR values are clustered closely around the mean.

Figure 2 represents the original images, their encrypted images using two different keys K_1 and K_2 and the image difference between the encrypted images, respectively. As mentioned earlier, keys K_1 and K_2 differ only by one bit.

The second test consists of measuring the key sensibility in the image decryption process. Let original image I_o be encrypted using the key $K_1 = (K_R, K_C, \text{ITER}_{\text{MAX}})$, where K_R, K_C, and ITER_{MAX} are again randomly generated, to give the encrypted image I_{ENC}. This image is decrypted separately using the keys K_1 and K_2; these keys always differ by only one bit in the least significant bit location. Figure 3 illustrates the original image, the encrypted image I_{ENC} with key K_1, the decrypted image of I_{ENC} using correct key K_1, and the decrypted image of I_{ENC} using the wrong key K_2. It is clear from this figure that decryption using a wrong key does not succeed.

3.3. Statistical Analysis. In a paper published in 1949 [17], Shannon stated that "*It is possible to solve many kinds of ciphers by statistical analysis.*" Consequently, he suggested two methods based on confusion and diffusion in order to counteract powerful attacks based on statistical analysis. In

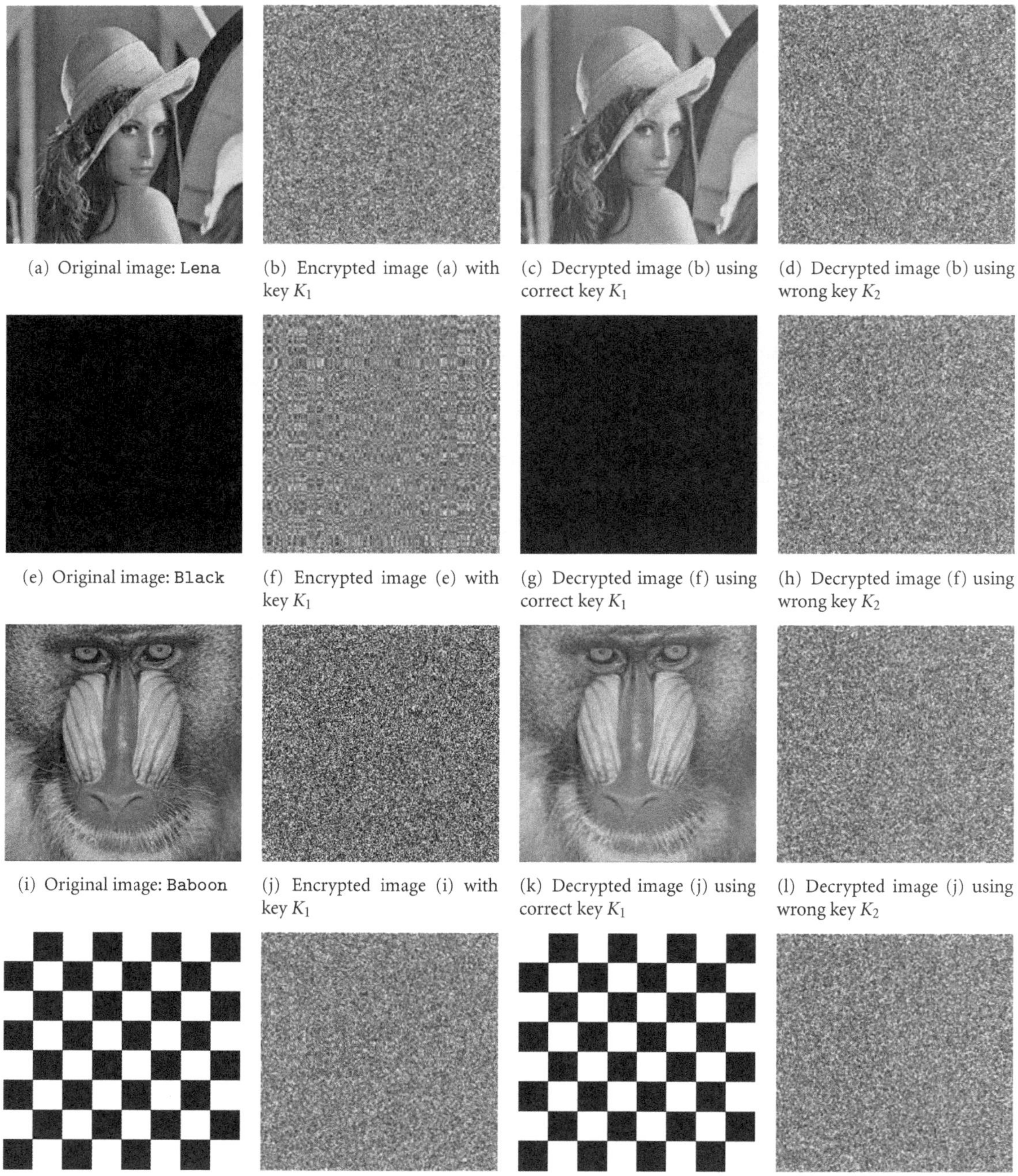

(a) Original image: `Lena` (b) Encrypted image (a) with key K_1 (c) Decrypted image (b) using correct key K_1 (d) Decrypted image (b) using wrong key K_2

(e) Original image: `Black` (f) Encrypted image (e) with key K_1 (g) Decrypted image (f) using correct key K_1 (h) Decrypted image (f) using wrong key K_2

(i) Original image: `Baboon` (j) Encrypted image (i) with key K_1 (k) Decrypted image (j) using correct key K_1 (l) Decrypted image (j) using wrong key K_2

(m) Original image: `Checkerboard` (n) Encrypted image (m) with key K_1 (o) Decrypted image (n) using correct key K_1 (p) Decrypted image (n) using wrong key K_2

FIGURE 3: Key sensibility for image decryption using the proposed algorithm.

the present paper, statistical analysis has been performed to demonstrate the superior confusion and diffusion properties of the proposed encryption algorithm against statistical attacks. This is done by performing two series of tests: histograms analysis of the encrypted images and the correlations computation of the adjacent pixels in encrypted images.

Figure 4 represents the histograms of the original and the encrypted images illustrated previously in Figure 1. One can see those the histograms of the encrypted images are almost uniform and are significantly different from that of the four original images. For instance, the histogram of original image `Checkerboard` shows as expected only two values: 0 and 255; however, the histogram of the encrypted `Checkerboard` image is fairly uniform. Therefore, the proposed image encryption algorithm responds well to the diffusion properties: it does not provide information that can be exploited for attacks based on statistical analysis of the encrypted image.

The other statistical test consists of computing the correlation between adjacent pixels [10]. It is obvious

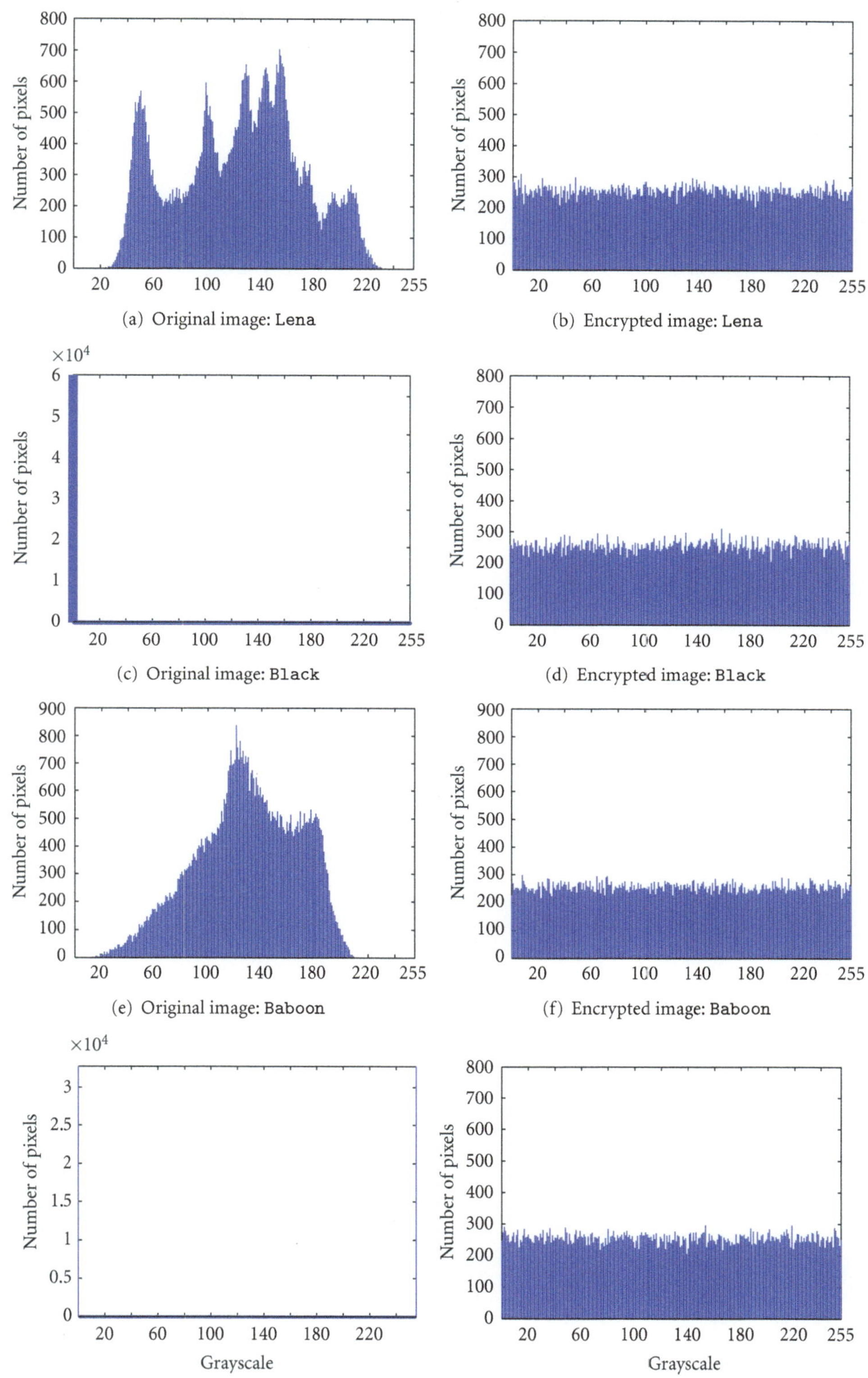

(a) Original image: `Lena`
(b) Encrypted image: `Lena`
(c) Original image: `Black`
(d) Encrypted image: `Black`
(e) Original image: `Baboon`
(f) Encrypted image: `Baboon`
(g) Original image: `Checkerboard`
(h) Encrypted image: `Checkerboard`

FIGURE 4: Histograms of original and encrypted images.

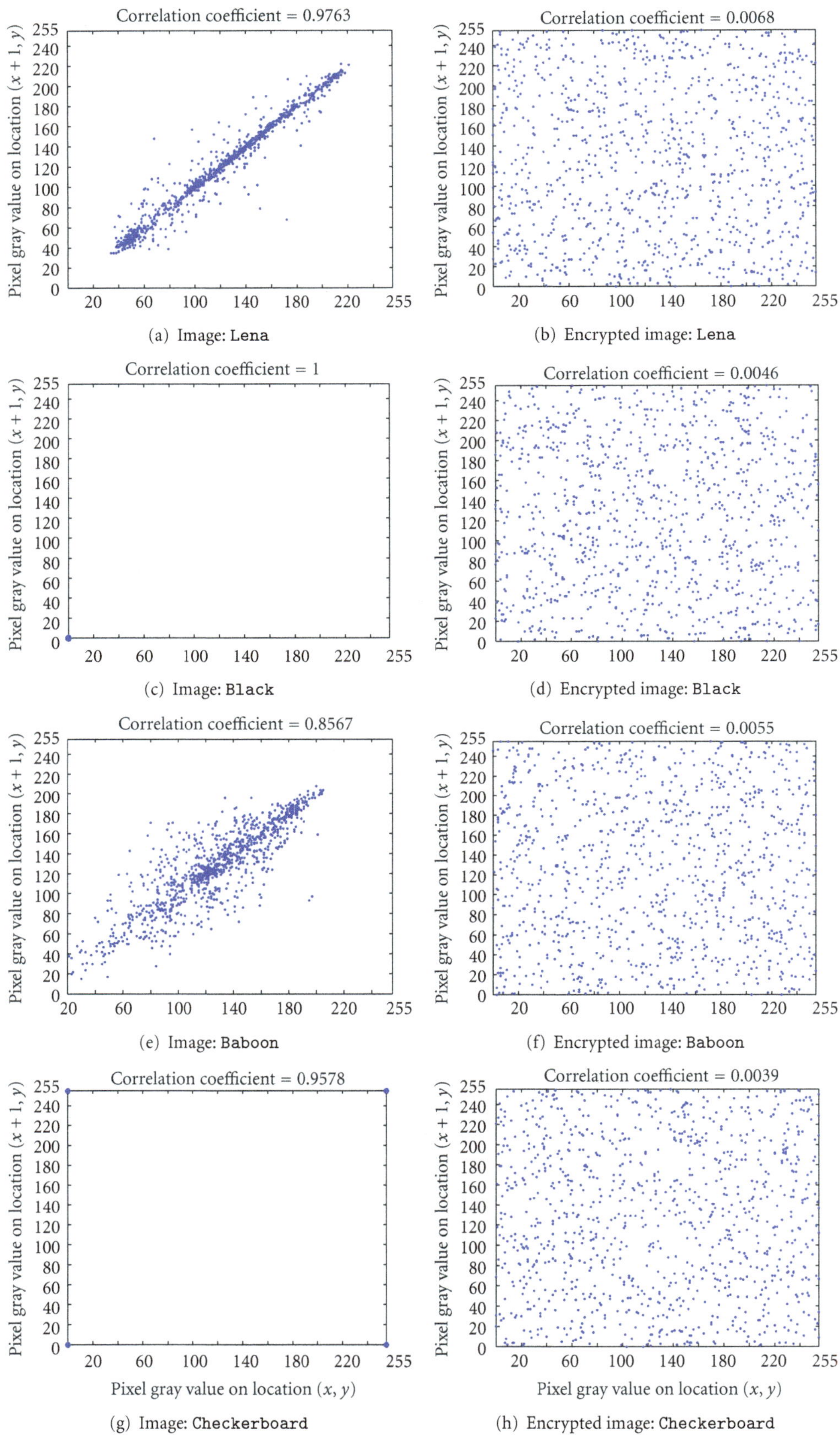

(a) Image: `Lena` (b) Encrypted image: `Lena`

(c) Image: `Black` (d) Encrypted image: `Black`

(e) Image: `Baboon` (f) Encrypted image: `Baboon`

(g) Image: `Checkerboard` (h) Encrypted image: `Checkerboard`

FIGURE 5: Correlation distribution of the pairs horizontal to adjacent pixels.

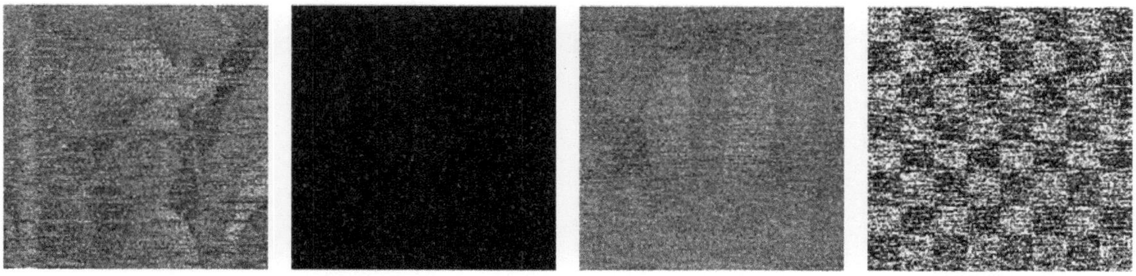

(a) Decrypted image: Lena (b) Decrypted image: Black (c) Decrypted image: Baboon (d) Decrypted image: Checkerboard

FIGURE 6: Attack by salt & pepper noise.

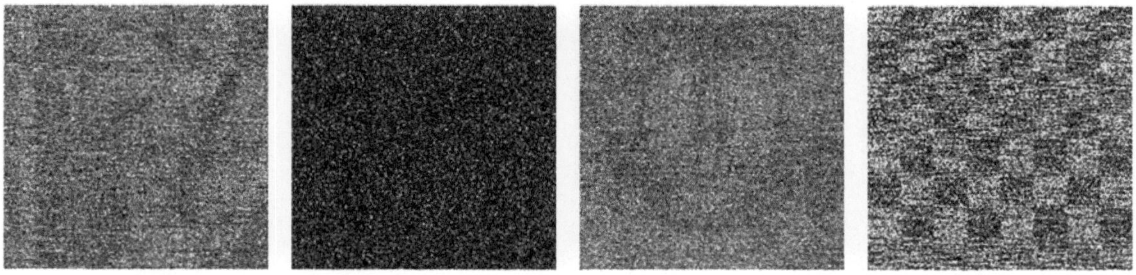

(a) Decrypted image: Lena (b) Decrypted image: Black (c) Decrypted image: Baboon (d) Decrypted image: Checkerboard

FIGURE 7: Attack by speckle noise.

that an arbitrarily chosen pixel in an image is generally strongly correlated with adjacent pixels, and its in either horizontal, vertical or diagonal directions. However, a secure image encryption algorithm must produce an encrypted image having low correlation between adjacent pixels. This correlation test consists of randomly selecting N pairs of adjacent pixels (vertical, horizontal, and diagonal) from the original and the encrypted images separately. Then, the correlation coefficient of each pair is calculated using (19)

$$E(x) = \frac{1}{N}\sum_{i=1}^{N} x_i, \quad (16)$$

$$D(x) = \frac{1}{N}\sum_{i=1}^{N} (x_i - E(x))^2, \quad (17)$$

$$\text{cov}(x, y) = \frac{1}{N}\sum_{i=1}^{N} (x_i - E(x))(y_i - E(y)), \quad (18)$$

$$\gamma_{xy} = \frac{\text{cov}(x, y)}{\sqrt{D(x)}\sqrt{D(y)}} \quad \text{with } D(x) \neq 0,\ D(y) \neq 0, \quad (19)$$

where x_i and y_i are the grayscale values of two adjacent pixels, N is the number of pairs (x_i, y_i), and $E(x)$ and $E(y)$, are respectively, the mean values of x_i and y_i.

Table 3 gives the correlation coefficient values of adjacent pixels in the horizontal, vertical, and diagonal directions of the original images and their encrypted versions. It is clear that for the original images, the coefficient correlation values are very high (close to one) contrary to those observed for the encrypted images. This confirms that adjacent pixels in the original images are strongly correlated. However, for the encrypted images, those values are close to zero, which means that the adjacent pixels (horizontal, vertical and diagonal directions) are very weakly correlated. Figure 5 illustrates the correlation distributions of the horizontal adjacent pixels of the original images and the corresponding encrypted images using the proposed algorithm. One can see from Figure 5 that adjacent pixels in encrypted images are indeed very weakly correlated.

TABLE 3: Correlation coefficients between adjacent pairs of pixels for original and encrypted images.

Correlation	Horizontal	Vertical	Diagonal
Original image: lena	0.9763	0.9453	0.9282
Encrypted image: lena	0.0068	0.0091	0.0063
Original image: black	1.0000	1.0000	1.0000
Encrypted image: black	0.0046	0.0055	0.0056
Original image: baboon	0.8567	0.8772	0.7880
Encrypted image: baboon	0.0055	0.0078	0.0042
Original image: checkerboard	0.9578	0.9521	0.9118
Encrypted image: checkerboard	0.0039	0.0090	0.0045

3.4. Entropy Analysis. The concept of entropy analysis for image encryption algorithm was introduced by Edward [18].

(a) Encrypted `lena` image (b) Encrypted `black` image (c) Encrypted `baboon` image (d) Encrypted `checkerboard` image

(e) Image (a) decrypted (f) Image (b) decrypted (g) Image (c) decrypted (h) Image (d) decrypted

FIGURE 8: Encrypted images under cropping attack in center (cropping over 1/8 of the entire image area).

TABLE 4: Comparison of entropy values of original images and their encrypted version.

Image	Original	Encrypted
`Lena`	7.4318 Sh	7.9968 Sh
`Black`	0 Sh	7.9966 Sh
`Baboon`	7.2279 Sh	7.9974 Sh
`Checkerboard`	1 Sh	7.9972 Sh

TABLE 5: Comparison of entropy values of encrypted images under different image encryption algorithms.

Algorithm	Entropy (Sh)
Proposed algorithm	7.9968
Baptista [19]	7.9260
Wong et al. [20]	7.9690
Xiang et al. [21]	7.9950
Lin and Wang [22]	7.9890

For gray-scale images of 256 levels, if each level of gray is assumed to be equiprobable, then the entropy of this image will be theoretically equal to 8 Sh (or bits). Ideally, an algorithm for encryption of images should give an encrypted image having equiprobable gray levels. Table 4 gives the entropy values of the four original images and those of their encrypted versions.

From those entropy values, we note that the entropy values of original images are far from ideal value of entropy since information sources are highly redundant and thus rarely generate uniformly distributed random messages. On the other hand, the entropy values of the encrypted images are very close to the ideal value of 8 Sh, which means that the proposed encryption algorithm is highly robust against entropy attacks.

Table 5 gives a comparison of the entropy values for encrypted image `Lena` with various image encryption algorithms.

3.5. Analysis against Attacks. An attacker who intercepts encrypted image can easily modify it, while the legitimate user can receive it and decrypt it successfully. This is the principle of attacks in image encryption; these attacks can include additive noise, filtering, rotation and cropping, and so forth.

3.5.1. Additive Noise. To verify the performance of the proposed encrypted algorithm against additive noise attacks, we considered two types of noises: salt and pepper noise and speckle noise. An additive noise attack consists in adding random noise to the intercepted encrypted image. Then, the noisy encrypted image will be decrypted. To measure the robustness of the proposed image encryption algorithm against this attack, mean squared error (MSE) measures are used.

Table 6 gives the MSE values between original images and their decrypted ones under the salt and pepper noise with different noise density values and the speckle noise with different variances.

Figures 6 and 7 illustrate the decrypted images: their encrypted version has been attacked separately by salt & pepper noise with 0.05 density and by speckle noise of variance 0.05. From these results, we can conclude that random noise attacks seriously affect decrypted images.

TABLE 6: MSE between original images and the decrypted versions under different noise attacks.

Image	Salt and pepper noise		Speckle noise	
	Density	MSE	Variance	MSE
Lena	0.05	3.42×10^3	0.05	5.30×10^3
	0.1	3.68×10^3	0.1	5.79×10^3
Black	0.05	1.06×10^3	0.05	8.11×10^3
	0.1	2.20×10^3	0.1	1.02×10^4
Baboon	0.05	2.38×10^3	0.05	3.97×10^3
	0.1	2.49×10^3	0.1	4.50×10^3
Checkerboard	0.05	2.44×10^4	0.05	2.12×10^4
	0.1	2.50×10^4	0.1	2.19×10^4

TABLE 7: MSE between original images and the decrypted versions under cropping attack.

Image	Center cropping	Sides cropping
Lena	2.96×10^3	2.90×10^3
Black	2.72×10^3	2.69×10^3
Baboon	2.05×10^3	2.02×10^3
Checkerboard	1.81×10^4	1.64×10^4

TABLE 8: Speed test results of the proposed algorithm with image Lena using a 2.7 GHz personal computer.

Image size	Encryption time	Decryption time
64×64	0.03 s	0.03 s
128×128	0.04 s	0.04 s
256×256	0.12 s	0.12 s
512×512	0.66 s	0.66 s
1024×1024	5.40 s	5.40 s

3.5.2. Analysis against Cropping Attacks. The cropping attacks consist of modifying the intercepted encrypted image by deleting one or several areas of the image. Table 7 gives the MSE values between original images and the decrypted and cropped images either in their center or on the image sides with parameter values equal to 1/8. This one indicates the fraction of the encrypted image that has been cropped. Figure 8 represents the encrypted images cropped in the center and their decrypted versions. From these results, we can conclude that the proposed image encryption algorithm resists to this attack lightly.

3.6. Speed Test. Apart from security considerations, another important consideration in the design of image encryption techniques is the actual algorithm execution speed, particularly for real-time applications. The proposed encryption algorithm is indeed very fast compared to other algorithms [10]. Our experimental results show that the average speed for encryption and for decryption is of around 0.9 Mb/s (megabits per second). The peak speed can reach up to 2.2 Mb/s on personnel computer equipped with an AMD Athlon processor with clock speed of 2.70 GHz, 1 GB (gigabytes) of RAM memory and 160 GB hard-disk capacity.

Table 8 illustrates the performance of the proposed algorithm using original image Lena with different sizes ranging from 64×64 to 1024×1024 pixels. The proposed algorithm was written using the MATLAB software platform.

4. Conclusion

In this paper, a novel image encryption algorithm is proposed. This algorithm is based on the principle of Rubik's cube to permute image pixels. To confuse the relationship between original and encrypted images, the XOR operator is applied to odd rows and columns of image using a key. The same key is flipped and applied to even rows and columns of image. Experimental tests have been carried out with detailed numerical analysis which demonstrates the robustness of the proposed algorithm against several types of attacks such as statistical and differential attacks (visual testing). Moreover, performance assessment tests demonstrate that the proposed image encryption algorithm is highly secure. It is also capable of fast encryption/decryption which is suitable for real-time Internet encryption and transmission applications.

References

[1] Z. Liu, L. Xu, C. Lin, J. Dai, and S. Liu, "Image encryption scheme by using iterative random phase encoding in gyrator transform domains," *Optics and Lasers in Engineering*, vol. 49, no. 4, pp. 542–546, 2011.

[2] Q. Guo, Z. Liu, and S. Liu, "Color image encryption by using Arnold and discrete fractional random transforms in IHS space," *Optics and Lasers in Engineering*, vol. 48, no. 12, pp. 1174–1181, 2010.

[3] Z. Liu, H. Chen, T. Liu et al., "Image encryption by using gyrator transform and Arnold transform," *Journal of Electronic Imaging*, vol. 2, no. 4, pp. 345–351, 1993.

[4] R. Tao, X. Y. Meng, and Y. Wang, "Image encryption with multiorders of fractional fourier transforms," *IEEE Transactions on Information Forensics and Security*, vol. 5, no. 4, pp. 734–738, 2010.

[5] R. Zunino, "Fractal circuit layout for spatial decorrelation of images," *Electronics Letters*, vol. 34, no. 20, pp. 1929–1930, 1998.

[6] G. Zhang and Q. Liu, "A novel image encryption method based on total shuffling scheme," *Optics Communications*, vol. 284, no. 12, pp. 2775–2780, 2011.

[7] X.-Y. Zhao and G. Chen, "Ergodic matrix in image encryption," in *Proceedings of the 2nd International Conference on Image and Graphics*, vol. 4875, pp. 394–401, August 2002.
[8] Z.-L. Zhu, W. Zhang, K.-W. Wong, and H. Yu, "A chaos-based symmetric image encryption scheme using a bit-level permutation," *Information Sciences*, vol. 181, no. 6, pp. 1171–1186, 2011.
[9] C. K. Huang and H. H. Nien, "Multi chaotic systems based pixel shuffle for image encryption," *Optics Communications*, vol. 282, no. 11, pp. 2123–2127, 2009.
[10] G. Chen, Y. Mao, and C. K. Chui, "A symmetric image encryption scheme based on 3D chaotic cat maps," *Chaos, Solitons and Fractals*, vol. 21, no. 3, pp. 749–761, 2004.
[11] X. Y. Wang, L. Yang, R. Liu, and A. Kadir, "A chaotic image encryption algorithm based on perceptron model," *Nonlinear Dynamics*, vol. 62, no. 3, pp. 615–621, 2010.
[12] Y. Wang, K. W. Wong, X. Liao, and G. Chen, "A new chaos-based fast image encryption algorithm," *Applied Soft Computing Journal*, vol. 11, no. 1, pp. 514–522, 2011.
[13] S. Li, C. Li, G. Chen, N. G. Bourbakis, and K. T. Lo, "A general quantitative cryptanalysis of permutation-only multimedia ciphers against plaintext attacks," *Signal Processing: Image Communication*, vol. 23, no. 3, pp. 212–223, 2008.
[14] C. K. Huang, H. H. Nien, S. K. Changchien, and H. W. Shieh, "Image encryption with chaotic random codes by grey relational grade and Taguchi method," *Optics Communications*, vol. 280, no. 2, pp. 300–310, 2007.
[15] S. Mazloom and A. M. Eftekhari-Moghadam, "Color image encryption based on Coupled Nonlinear Chaotic Map," *Chaos, Solitons and Fractals*, vol. 42, no. 3, pp. 1745–1754, 2009.
[16] Y. Tang, Z. Wang, and J. A. Fang, "Image encryption using chaotic coupled map lattices with time-varying delays," *Communications in Nonlinear Science and Numerical Simulation*, vol. 15, no. 9, pp. 2456–2468, 2010.
[17] C. E. Shannon, "Communication theory of secrecy systems," *Bell System Technical Journal*, vol. 28, no. 4, pp. 656–715, 1949.
[18] O. Edward, *Chaos in Dynamical Systems*, Cambridge University Press, Cambridge, UK, 2nd edition, 2003.
[19] M. S. Baptista, "Cryptography with chaos," *Physics Letters, Section A*, vol. 240, no. 1-2, pp. 50–54, 1998.
[20] K. W. Wong, S. W. Ho, and C. K. Yung, "A chaotic cryptography scheme for generating short ciphertext," *Physics Letters, Section A*, vol. 310, no. 1, pp. 67–73, 2003.
[21] T. Xiang, X. Liao, G. Tang, Y. Chen, and K. W. Wong, "A novel block cryptosystem based on iterating a chaotic map," *Physics Letters, Section A*, vol. 349, no. 1–4, pp. 109–115, 2006.
[22] Z. Lin and H. Wang, "Efficient image encryption using a chaos-based PWL memristor," *IETE Technical Review*, vol. 27, no. 4, pp. 318–325, 2010.

12

Analysis of DC Electrical Conductivity Models of Carbon Nanotube-Polymer Composites with Potential Application to Nanometric Electronic Devices

Rafael Vargas-Bernal,[1] Gabriel Herrera-Pérez,[1] Ma. Elena Calixto-Olalde,[1] and Margarita Tecpoyotl-Torres[2]

[1] *Departamento de Ingeniería en Materiales, Instituto Tecnológico Superior de Irapuato, Carretera Irapuato Silao Km. 12.5, 36821 Irapuato, Guanajuato, Mexico*
[2] *Centro de Investigación en Ingeniería y Ciencias Aplicadas (CIICAp), Universidad Autónoma del Estado de Morelos, Avenida Universidad No. 1001, Col. Chamilpa, 62209 Cuernavaca, Morelos, Mexico*

Correspondence should be addressed to Rafael Vargas-Bernal; rvargasbernal@gmail.com

Academic Editor: Esteban Tlelo-Cuautle

The design of nanometric electronic devices requires novel materials for improving their electrical performance from stages of design until their fabrication. Until now, several DC electrical conductivity models for composite materials have been proposed. However, these models must be valued to identify main design parameters that more efficiently control the electrical properties of the materials to be developed. In this paper, four different models used for modeling DC electrical conductivity of carbon nanotube-polymer composites are studied with the aim of obtaining a complete list of design parameters that allow guarantying to the designer an increase in electrical properties of the composite by means of carbon nanotubes.

1. Introduction

In the nanometer era, VLSI circuits must be simulated by novel models that allow us to predict and estimate the electrical behavior waited from their phase of design and during their performance. With the introduction of the nanomaterials as an alternative to silicon, numerous research groups around the world have been developing models to simulate electrical properties of such materials. Within this class of nanomaterials, carbon nanotube-polymer nanocomposites are offering improved electrical properties thanks to inherent electrical properties of the carbon nanotubes either semiconducting or conducting. Previous studies of these materials have predicted successful technological applications [1–3]. However, their design must overcome important barriers involved with the control of phenomena between their interfaces and fabrication methods with the aim of achieving repetitive results.

Electrical properties of materials are determined through four fundamental parameters called dielectric constant, tangent of dielectric loss angle, dielectric breakdown, and electrical conductivity. Most polymers exhibit a highly insulating dielectric behavior and a very high electrical resistivity. Thus, it is required adding an electrically conductive filler to polymers for changing their electrical properties. The resulting nanocomposite of such combination can give place to a semiconducting or conducting behavior. Electrical fillers can be ceramic materials, metals, and/or carbon nanotubes. In particular, carbon nanotubes form composite materials which can be applied in gas sensors, biosensors, electromagnetic shielding, antistatic coatings, high-strength low-density corrosion-resistant components, and lightweight energy storage. The use of these nanocomposites will be extended to VLSI circuit design when a complete domain of electrical properties and fabrication methods can be achieved. In addition, fabrication of these materials must

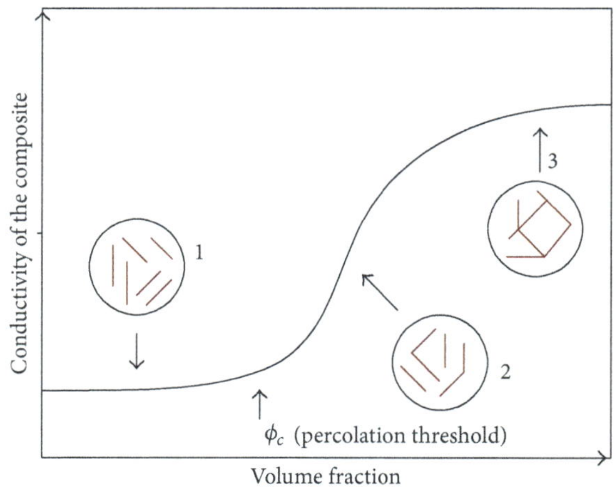

FIGURE 1: Dependence of electrical conductivity on filler volume fraction.

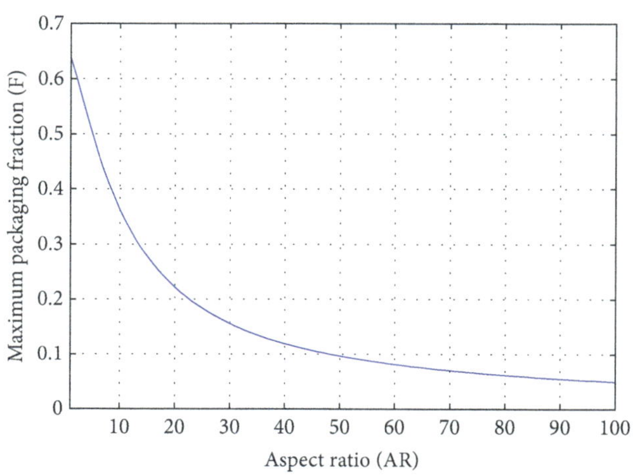

FIGURE 2: Maximum packaging fraction of the filler used in the composite versus aspect ratio of the filler involved.

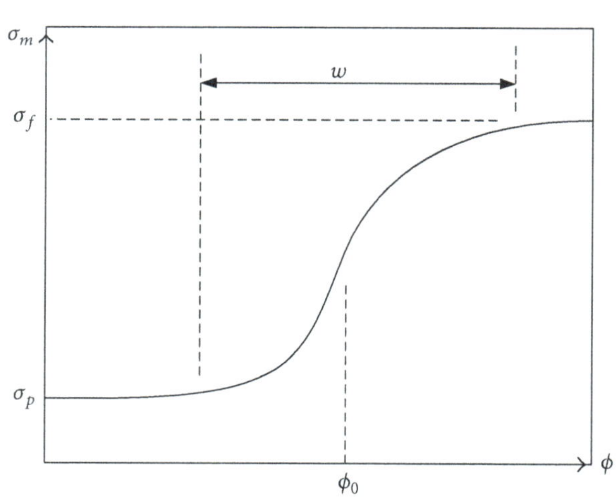

FIGURE 3: Sigmoidal function model for DC electrical conductivity for nanocomposites.

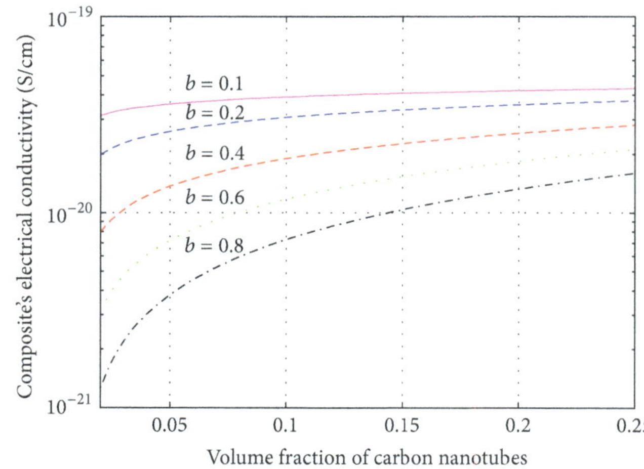

FIGURE 4: DC electrical conductivity of the nanocomposite when the critical exponent of the percolation is varied (Kirkpatrick's model).

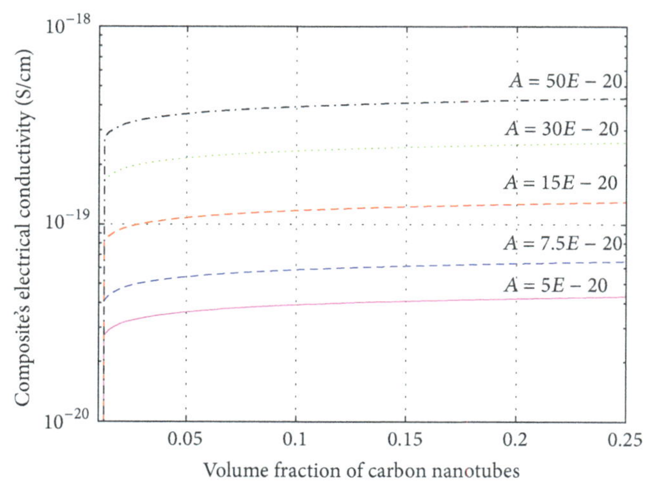

FIGURE 5: DC electrical conductivity of the nanocomposite at varying the conductivity of the filler (Kirkpatrick's model).

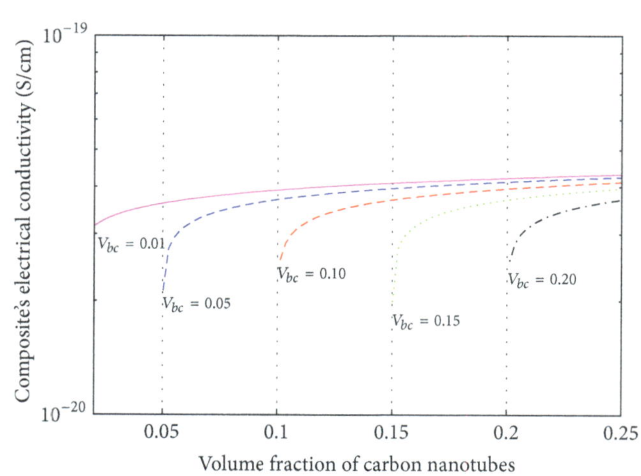

FIGURE 6: DC electrical conductivity of the nanocomposite when the percolation threshold of the filler is varied (Kirkpatrick's model).

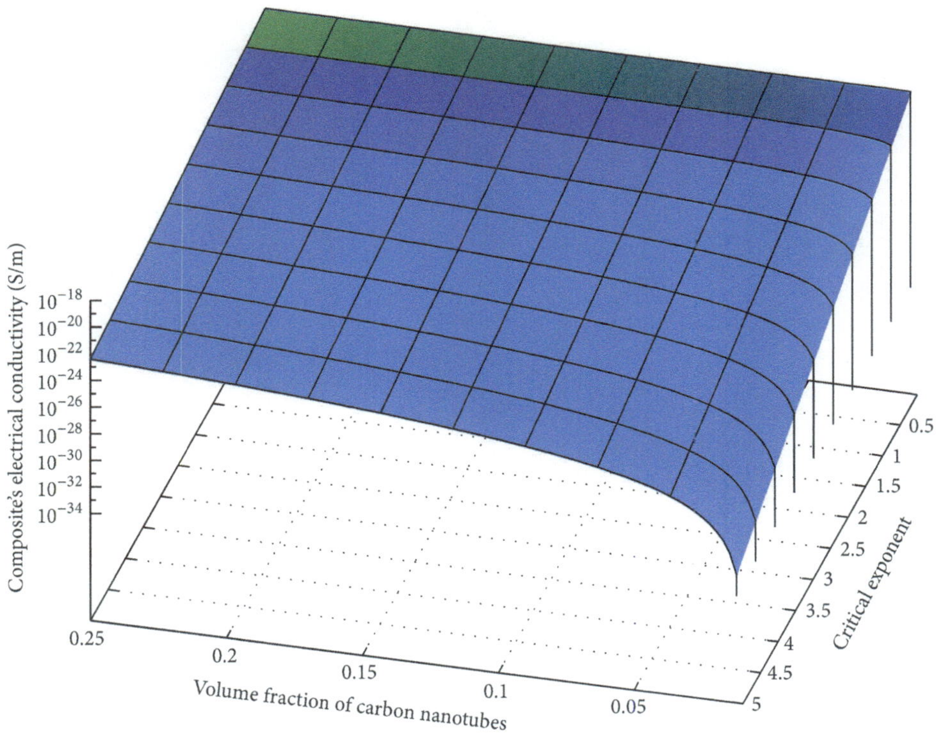

FIGURE 7: DC electrical conductivity of the nanocomposite at varying both the critical exponent of the percolation and the volume fraction of the filler (Kirkpatrick's model).

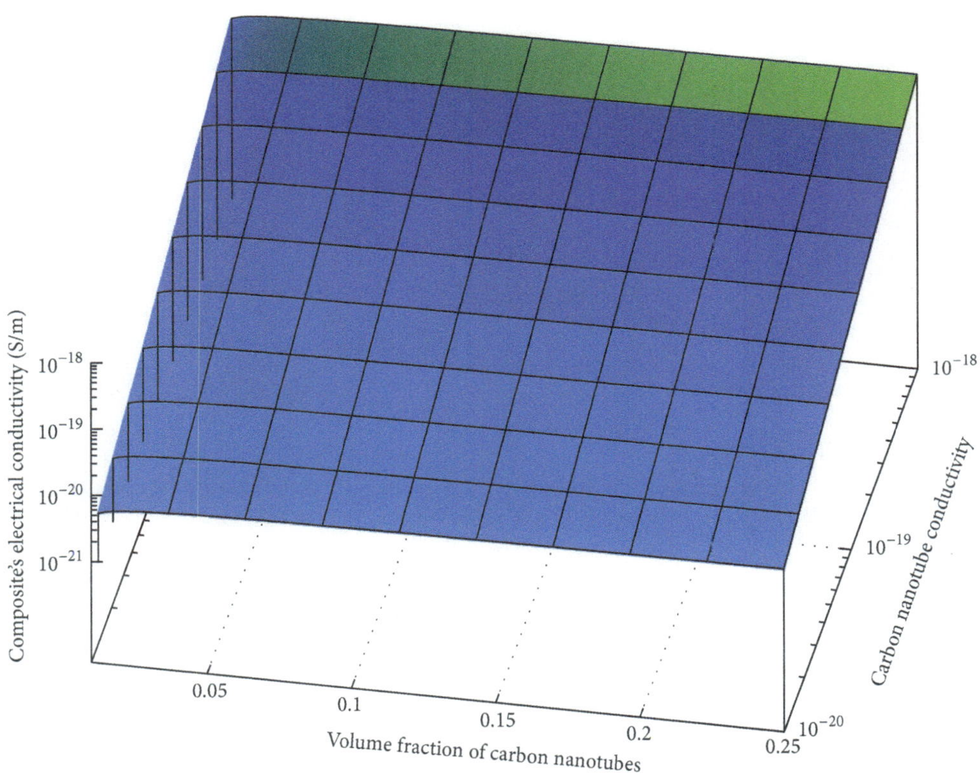

FIGURE 8: DC electrical conductivity of the nanocomposite when both the conductivity of the filler and the volume fraction of the filler are varied (Kirkpatrick's Model).

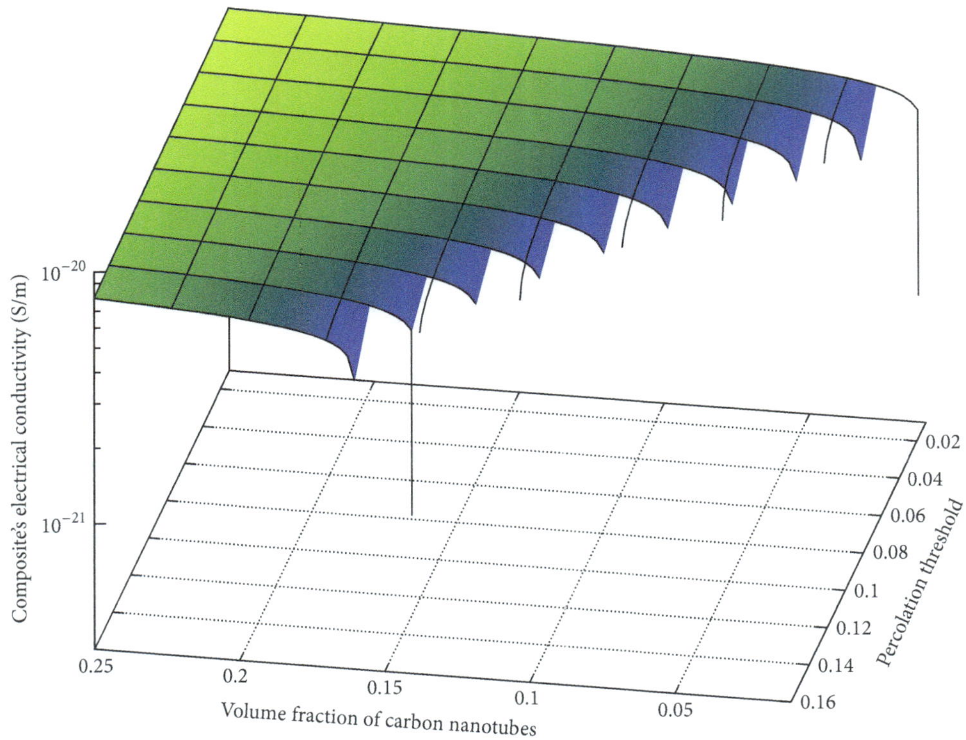

FIGURE 9: DC electrical conductivity of the nanocomposite at varying both the percolation threshold of the filler and the volume fraction of the filler (Kirkpatrick's model).

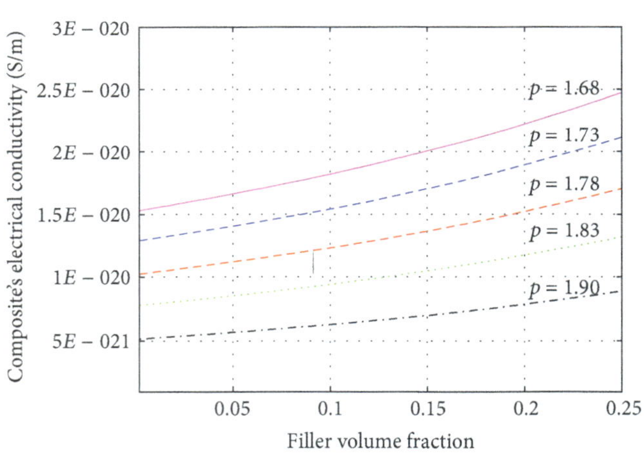

FIGURE 10: DC electrical conductivity of the nanocomposite when the critical exponent of the percolation is varied (McLachlan's model).

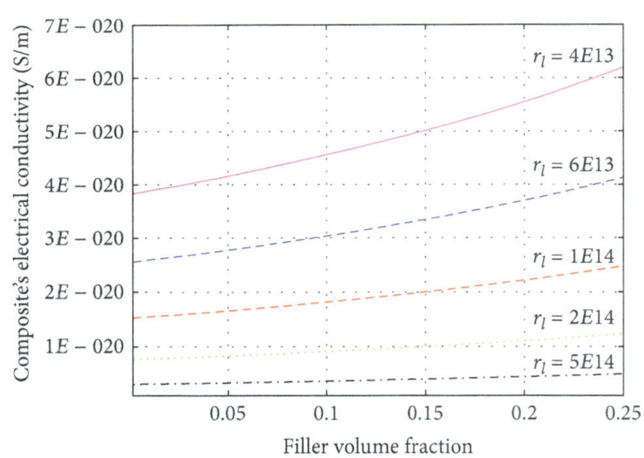

FIGURE 11: DC electrical conductivity of the nanocomposite at varying the conductivity of the matrix (McLachlan's model).

be repetitive to be exploited by electronic industry. Several models of DC electrical behavior have been proposed until now; unfortunately, a complete comparative study among these models that determines a strategic methodology of design does not exist.

In this paper, an explorative study of the DC electrical models of nanocomposites is realized with the aim of obtaining a complete list of design parameters to achieve electrical behavior well defined. In Section 2, the description of four DC electrical models for composites, which will be modeled, is realized. Next, these models are simulated with the aim of identifying main design parameters that increase electrical conductivity in Section 3. In Section 4, results of the simulation and their discussion are analyzed with details. Finally, conclusions highlighting the importance of the study realized are given in Section 5.

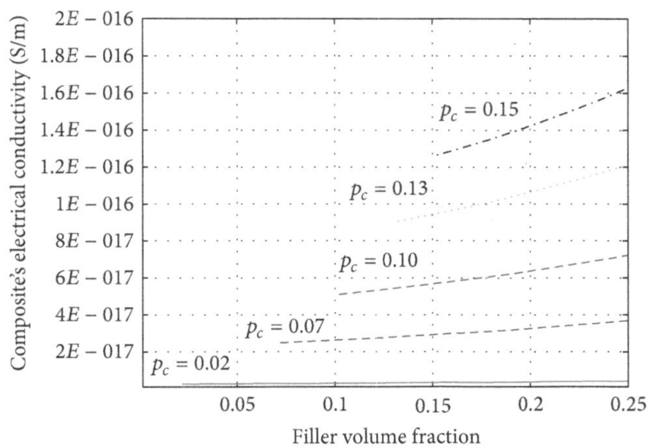

FIGURE 12: DC electrical conductivity of the nanocomposite when the percolation threshold of the filler is varied (McLachlan's model).

2. DC Electrical Models for Composites

An electrical model is a mathematical description that represents electrical behavior of an electrical device or system. Direct current (DC) is the unidirectional flow of electric charge. Direct current can flow through metals, semiconductors, insulators, and composites. The contribution to electrical conductivity of the polymer (or matrix) by part of the filler is not continuous and linear, rather it is discrete and nonlinear. There is a critical composition or also called percolation threshold, at which electrical conduction is increased by several orders of magnitude leading the composite from an insulating range to semiconductive and conductive ranges. In this paper, DC electrical models that will be analyzed are: Kirkpatrick's model, McLachlan's model, Mamunya's model, and Sigmoidal function model. The models are associated with an extended basic statistical percolation theory, and the last model makes use of a nonlinear regression.

2.1. Kirkpatrick's Model. This model predicts the DC electrical conductivity based on the likelihood of contact between particles of filler within the composite [3–6]. Its equation is expressed as a power law equation of the form

$$\sigma_m = A(\phi - V_{bc})^b, \quad (1)$$

where σ_m is the conductivity of the composite, A is the conductivity of the filler, ϕ is the volume fraction of filler, V_{bc} is the percolation threshold of filler, and b is the critical exponent which depends on the type of space dimension, and it is a characteristic value experimentally obtained. The percolation threshold is the minimum quantity of the filler required to form a continuous network of particles (as shown in Figure 1) to transport electrical charge between ends of the material being electrically polarized by a DC source. As it is illustrated in this figure, there are three main zones in which conductivity of nanocomposites can be found. At low filler concentrations, zone 1 in Figure 1, the conductivity of the composite is very close to the pure polymer matrix. In zone 2, the percolation threshold have been achieved and the electrical conductivity undergoes a drastic increase due to the start of the formation and consolidation of an electrically conductive network through the sample. Finally, in zone 3, the maximum electrical conductivity is obtained when a completely interconnected network is formed.

2.2. McLachlan's Model. Some physical models can describe the macroscopic properties of a medium in accordance with their properties and the relative fractions of its components under the name of effective medium approximations or effective medium theory. The properties under study, in particular, are usually the electrical conductivity or the dielectric constant of the medium. McLachlan et al. [7–11] proposed a statistical model of the DC electrical conductivity of the composite materials based on the effective medium theory by means of the equation

$$\frac{(1-\phi)\left(\rho_m^{1/p} - \rho_f^{1/p}\right)}{\rho_m^{1/p} + ((1-p_c)/p_c)\,\rho_f^{1/p}} + \frac{\phi\left(\rho_m^{1/p} - r_l^{1/p}\right)}{\rho_m^{1/p} + ((1-p_c)/p_c)\,r_l^{1/p}} = 0, \quad (2)$$

where ϕ is the volume fraction of filler, p_c is the percolation threshold of filler, ρ_f is the resistivity of the filler, r_l is the resistivity of the polymer or matrix, and p is the critical exponent which represents the volume fraction of the phases.

2.3. Mamunya's Model. The model suggested by Mamunya et al. [12–16] makes uses of the surface energy of the polymer and the filler, and it includes the aspect ratio of the filler. In addition, this theory relates the electrical conductivity of composites with the presence of clusters of connected particles. Therefore, there exists a relationship between the electrical conductivity of the composite and the volume fraction of the filler. Their equations are

$$\log(\sigma_m) = \log(\mathrm{SC}) + (\log(\mathrm{SF}) - \log(\sigma_c))\left(\frac{\phi - \phi_c}{\mathrm{F} - \phi_c}\right)^k, \quad (3)$$

where

$$k = \frac{K\phi_c}{(\phi - \phi_c)^{3/4}},$$
$$K = A - B\gamma_{pf},$$
$$\gamma_{pf} = \gamma_p + \gamma_f - 2(\gamma_p\gamma_f)^{1/2}, \quad (4)$$
$$\mathrm{F} = \frac{5}{75/(10 + \mathrm{AR}) + \mathrm{AR}}.$$

In this model, σ_m is the conductivity of the composite, SC is the conductivity at the percolation threshold, SF is the conductivity at the maximum packaging fraction (F), ϕ is the volume fraction, ϕ_c is the percolation threshold, γ_{pf} is the interfacial tension between polymer and filler, γ_p is the surface energy of the polymer, γ_f is the surface energy of the filler, A and B are constants, and AR is the aspect ratio. In Figure 2, the maximum packaging fraction (F) has been calculated to illustrate the effect of the aspect ratio of the filler.

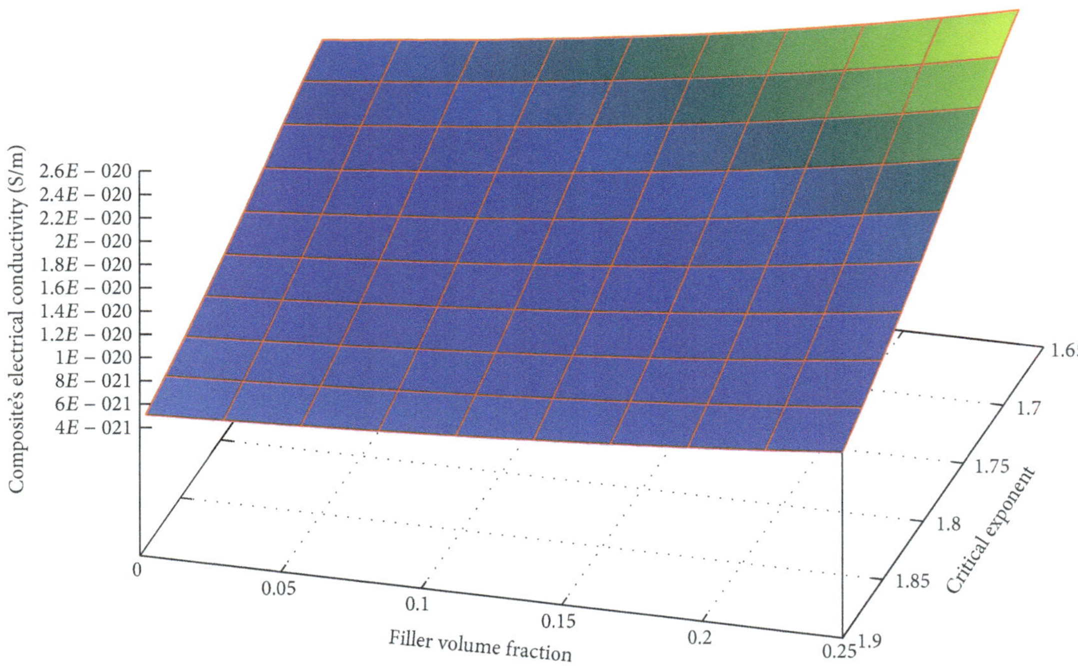

FIGURE 13: DC electrical conductivity of the nanocomposite at varying both the critical exponent of the percolation and the volume fraction of the filler (McLachlan's model).

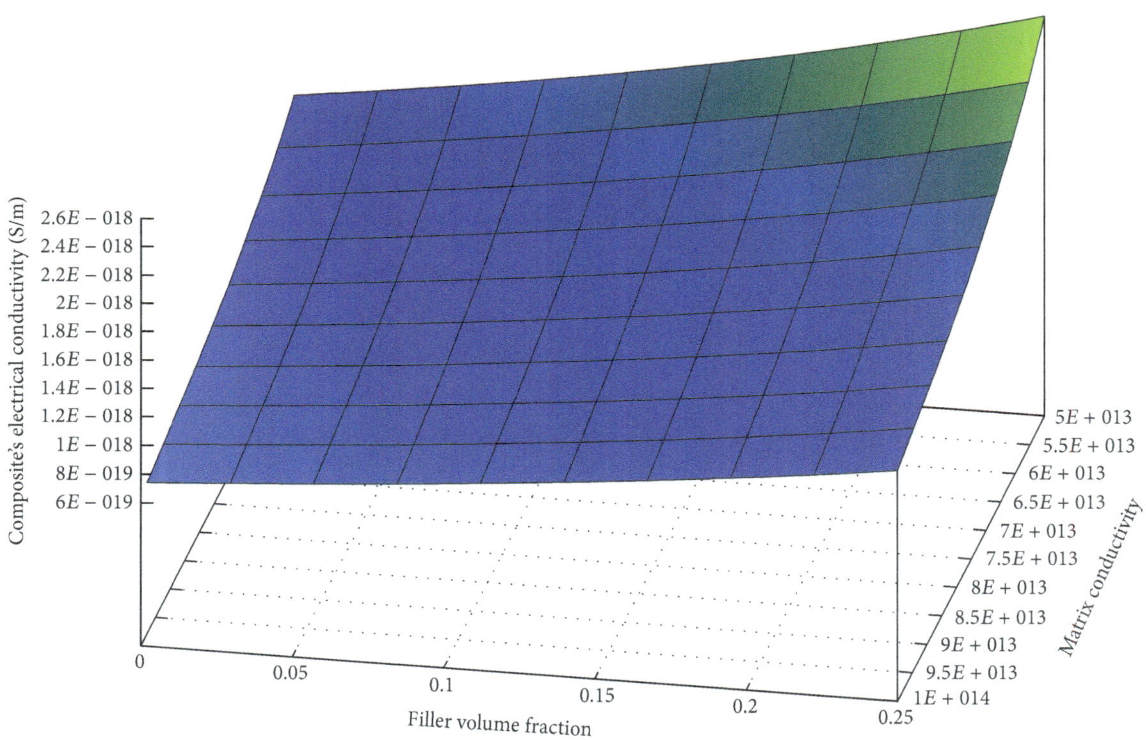

FIGURE 14: DC electrical conductivity of the nanocomposite when both conductivity of the matrix and the volume fraction of the filler are varied (McLachlan's model).

2.4. Sigmoidal Function Model. A sigmoidal function [17–21] is a mathematical logistic function with "S" shape and generically defined as

$$S(t) = \frac{a}{1 + e^{(-bt+c)}}. \qquad (5)$$

In particular, the model used in the simulations of DC electrical conductivity of nanocomposites was

$$\sigma_m = \sigma_p + \frac{\mathrm{SF} - \sigma_p}{1 + \exp\left[-(\phi - \mathrm{PCP})/w\right]}, \qquad (6)$$

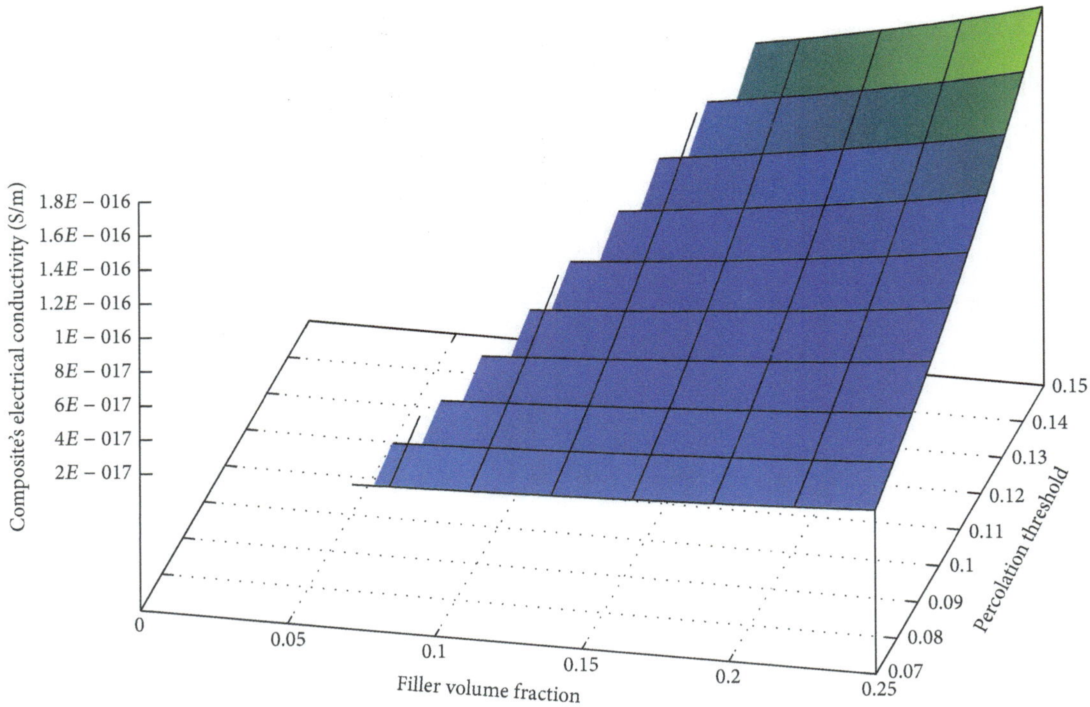

FIGURE 15: DC electrical conductivity of the nanocomposite at varying both the percolation threshold of the filler and the volume fraction of the filler (McLachlan's model).

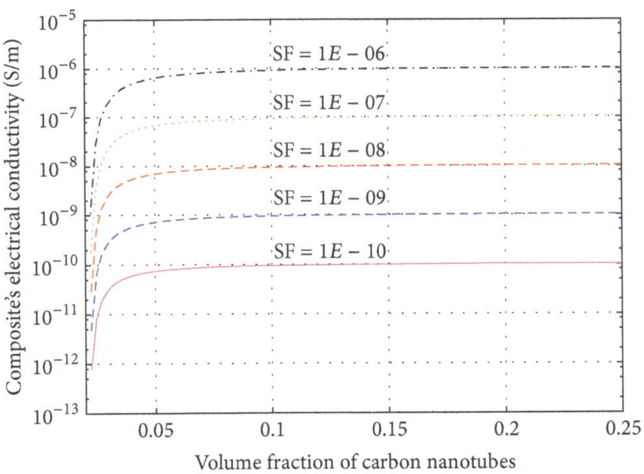

FIGURE 16: DC electrical conductivity of the nanocomposite when the conductivity at the maximum packaging fraction is varied (Mamunya's model).

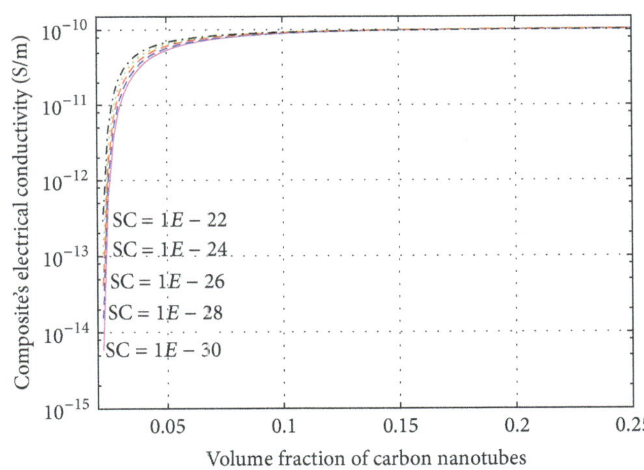

FIGURE 17: DC electrical conductivity of the nanocomposite at varying the conductivity at the percolation threshold (Mamunya's model).

where σ_m is the conductivity of the composite, SF is the conductivity of the filler, σ_p is the conductivity of the polymer, ϕ is the volume fraction of the filler, PCP is the volume fraction of the filler in the midpoint of the percolation, and w is the width of the percolation region. Therefore, σ represents the independent variable, and the constants can be identified as $a = \mathrm{SF} - \sigma_p, b = -1/w$, and $c = \mathrm{PCP}/w$. This equation is represented in Figure 3.

3. Simulation of the DC Conductivity Models for Nanocomposites

The first step taken in the analysis of DC electrical conductivity models was evaluating their behavior by means of the variation of their parameters, with the aim of determining which parameters modify in a great proportion the electrical conductivity of the nanocomposites. The models selected for this study include the models reported by Kirkpatrick et al. [3–6], McLachlan et al. [7–11], Mamunya et al. [12–16], and

Table 1

Mathematical models	Parameters			Electrical conductivity of the composite materials
	Conductivity of the Filler	Critical exponent	Percolation threshold	
Kirkpatrick's model	$5E-20$–$50E-20$			$1E-20$–$4.3E-19$
		0.1–0.8		$1.4E-21$–$4.3E-19$
			0.01–0.20	$2E-20$–$4.3E-19$
	Resistivity of the matrix	Critical exponent	Percolation threshold	
McLachlan's model	$4E13$–$5E14$			$0.3E-20$–$6.2E-20$
		1.68–1.90		$5E-21$–$2.5E-20$
			0.02–0.15	$0.2E-17$–$1.63E-16$
	Conductivity at the percolation threshold	Conductivity at the maximum packaging fraction	Packaging fraction	
Mamunya's model	$1E-30$–$1E-22$			$6E-15$–$1E-10$
		$1E-10$–$1E-06$		$9E-13$–$1E-06$
			0.055–0.64	$2E-13$–$1E-10$
	Midpoint of the percolation region's width	Conductivity of the filler	Width of the percolation region	
Sigmoidal function model	0.05–0.20			$2.8E-12$–$1E-10$
		$1.5E-15$–$1E-07$		$8E-15$–$1E-19$
			$5E-03$–$45E-03$	$1.7E-09$–$1E-10$

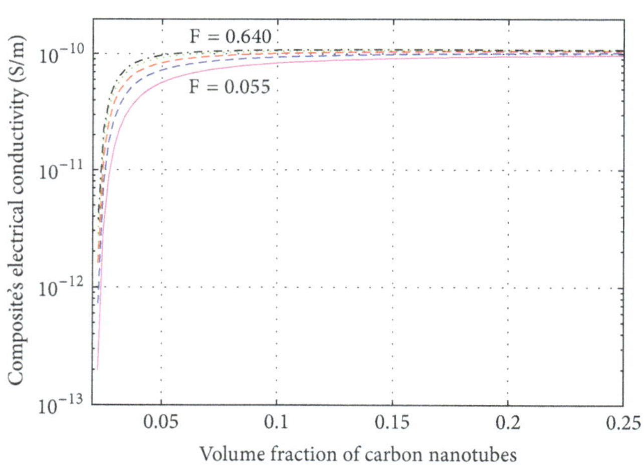

Figure 18: DC electrical conductivity of the nanocomposite when the packaging fraction of the filler is varied (Mamunya's model).

Sigmoidal function [17–21]. The functions were evaluated using Gnuplot 4.6.

3.1. Simulations of Kirkpatrick's Model. The evaluation of the Kirkpatrick's model was realized for three parameters: critical exponent, conductivity of the filler, and percolation threshold. The volume fraction was restricted to the range of 0.02 to 0.25. The critical exponent of the percolation was varied from 0.1 to 0.8. The conductivity of the filler was varied from $5E-20$ to $50E-20$ S/cm. The percolation threshold was varied from 0.01 to 0.20.

3.2. Simulations of McLachlan's Model. The evaluation of the McLachlan's model was realized for three parameters: critical exponent, conductivity of the matrix, and percolation threshold. The volume fraction was restricted to the range of 0.02 to 0.25. The critical exponent was varied from 1.68 to 1.90 S/cm. The resistivity of the matrix was varied from $4E13$ to $5E14$ S/cm. The percolation threshold was varied from 0.02 to 0.15.

3.3. Simulations of Mamunya's Model. The evaluation of the Mamunya's model was realized for three parameters: conductivity at the maximum packaging fraction, conductivity at the percolation threshold, and packaging fraction. The volume fraction was restricted to the range of 0.02 to 0.25. The conductivity at the maximum packaging fraction was varied from $1E-10$ to $1E-06$ S/cm. The resistivity at the percolation threshold was varied from $1E-30$ to $1E-22$ S/cm. The packaging fraction was varied from 0.055 to 0.64.

3.4. Simulations of Sigmoidal Function Model. The evaluation of the Sigmoidal function model was realized for three

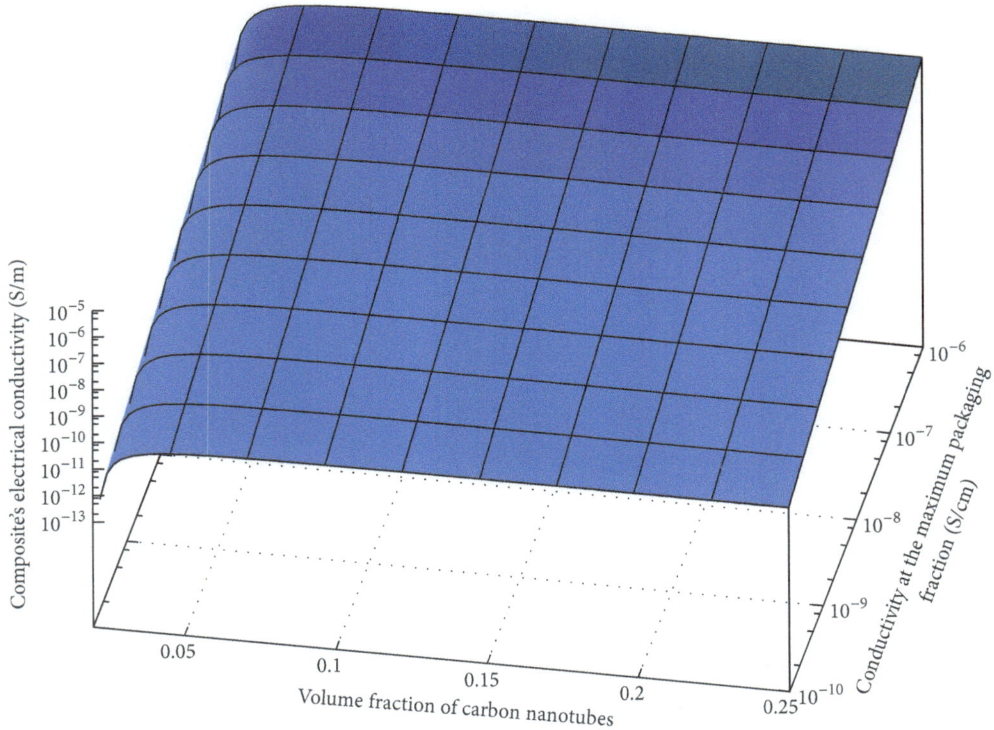

FIGURE 19: DC electrical conductivity of the nanocomposite at varying both the conductivity at the maximum packaging fraction of the filler and the volume fraction of the filler (Mamunya's model).

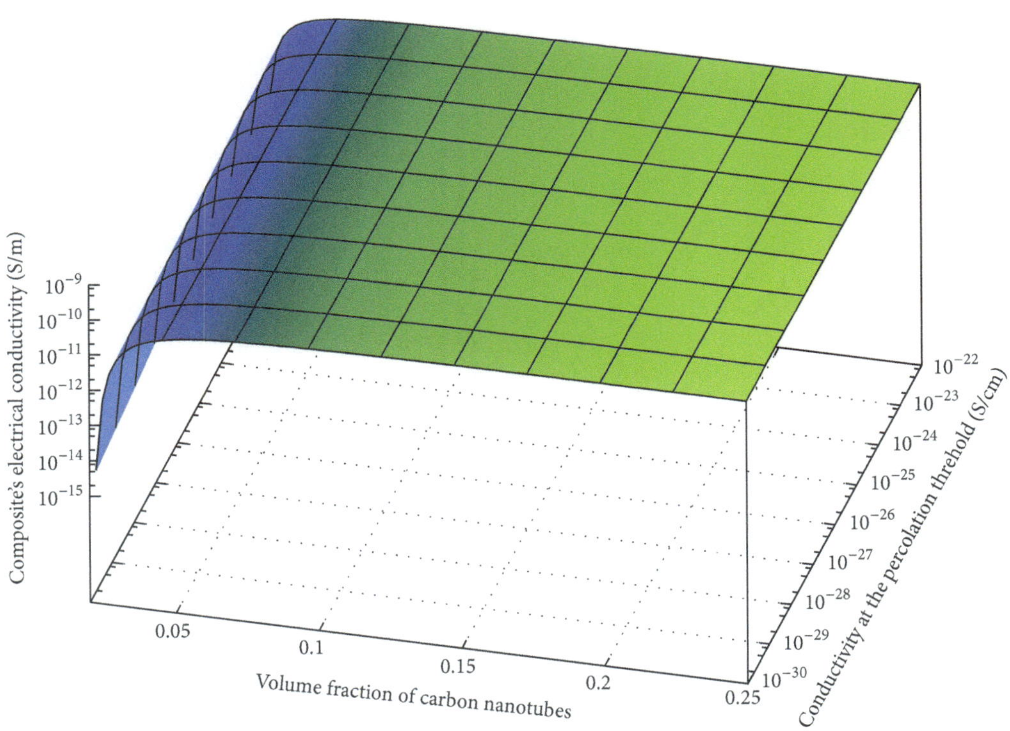

FIGURE 20: DC electrical conductivity of the nanocomposite when both the conductivity at the percolation threshold of the filler and the volume fraction of the filler are varied (Mamunya's model).

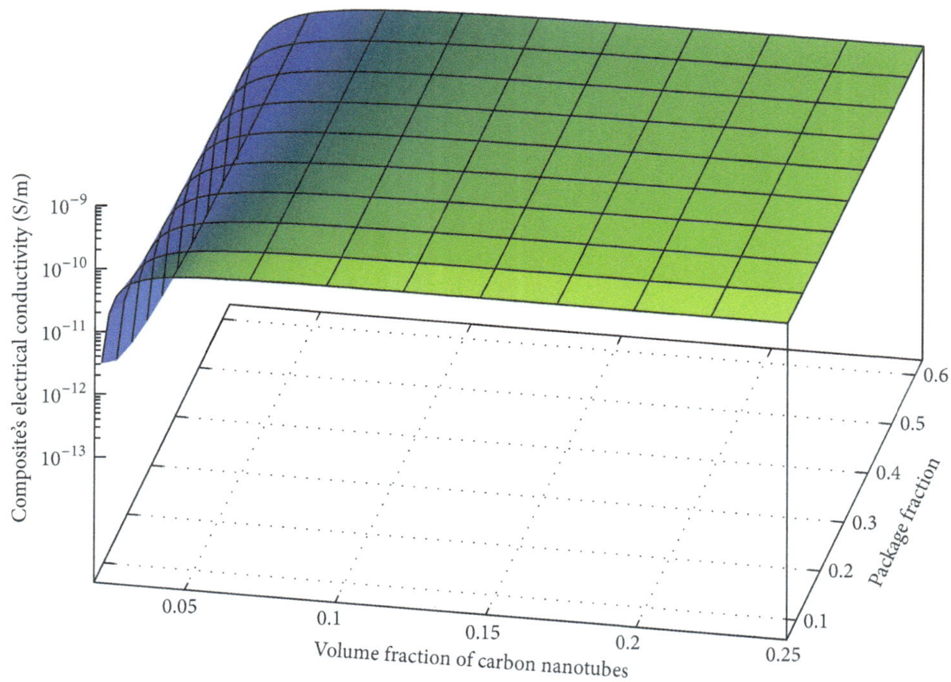

FIGURE 21: DC electrical conductivity of the nanocomposite at varying both the packaging fraction of the filler and the volume fraction of the filler (Mamunya's model).

parameters: midpoint of the percolation region's width, conductivity of the filler, and width of the percolation region. The volume fraction was restricted to the range of 0.02 to 0.25. The midpoint of the percolation region's width was varied from 0.05 to 0.20. The conductivity of the filler was varied from $1E-15$ to $1E-07$ S/cm. The width of the percolation region was varied from $5E-03$ to $45E-03$.

4. Results and Discussion

In this section, the results and discussion of these models is realized. Each model is analyzed separately with the aim of identifying their advantages and drawbacks in the modeling of the nanocomposites. In addition, this analysis allow us determine which are the main parameters that must be considered when a design process related with this type of materials is used to fabricate electronic devices for VLSI circuits and systems.

4.1. Analysis of the Kirkpatrick's Model. In Figure 4, the variation of the electrical conductivity of composites with respect to the value of the critical exponent of the percolation is illustrated. Higher electrical conductivities are obtained when the value of the critical exponent is low. Therefore, this parameter must be controlled to manipulate the value of the electrical conductivity of the resulting composite.

The value of the conductivity of the filler determines the maximum electrical conductivity obtained for the composite as shown in Figure 5. Thus, this parameter can be reinforced as part of a strategy to increase the electrical conductivity of the nanocomposite to be designed.

At varying the percolation threshold of the filler in the composite, as shown in Figure 6, the electrical conductivity is not changed, but rather the initial value, for which an electrical conductivity is achieved, is displaced in accordance with the value of the percolation threshold.

In Figure 7, it is depicted that the behavior is obtained when both the critical component of the percolation and the volume fraction of the filler are modified. It was found that when critical exponent is low and volume fraction is high, high electrical conductivities are established. In the case of the conductivity of the filler and the volume fraction of the filler, as shown in Figure 8, the electrical conductivity of the composite is at maximum, if at least the first parameter is great. Finally, the percolation threshold at minimum and the volume fraction at maximum can increase the electrical conductivity of the nanocomposite, as depicted in Figure 9.

4.2. Analysis of the McLachlan's Model. The critical value of the percolation increases the electrical conductivity of the resulting nanocomposite with better intensity for greater values of the volume fraction, as shown in Figure 10. Thus, it is feasible considering that this parameter must be favored in the design of this class of materials.

The conductivity of the polymeric matrix can be considered as a value to be reinforced during the design of nanocomposites, since it increases the electrical conductivity of them as shown in Figure 11. Therefore, a better electrical conductivity of the matrix will lead to better conductive nanocomposites.

In Figure 12, at varying the percolation threshold of the filler, it was found that the initial value of the conductivity

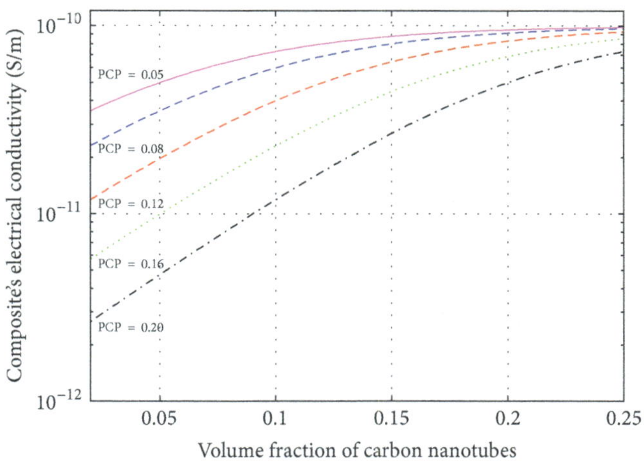

FIGURE 22: DC electrical conductivity of the nanocomposite when the midpoint of the percolation region's width is varied (Sigmoidal function model).

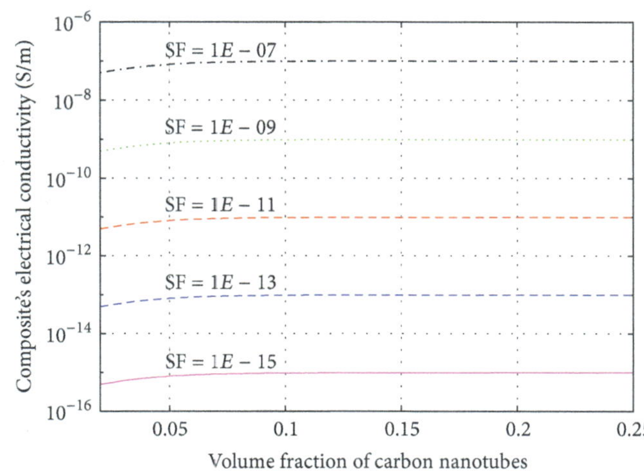

FIGURE 23: DC electrical conductivity of the nanocomposite at varying the conductivity of the filler (Sigmoidal function model).

is restricted for values of the volume fraction greater to itself established in the simulation. Therefore, this parameter holds the increasing trend but the electrical conductivity is not favored.

Tridimensional changes in this model for three different parameters can be visualized in Figures 13, 14, and 15. In Figure 13, DC electrical conductivity is increased when critical exponent of the percolation and volume fraction of filler are bigger. The conductivity of the polymeric matrix intensifies the electrical conductivity of the composite if both it and the volume fraction are larger. When percolation threshold of the filler and volume fraction are varied, then the electrical conductivity of the nanocomposite is favored if both values are greater.

4.3. Analysis of the Mamunya's Model. In Figure 16, it is depicted that the change of DC electrical conductivity when the conductivity at the maximum packaging fraction is done. Therefore, it can be concluded that this parameter considerably increases electrical conductivity.

The conductivity at the percolation threshold uniquely modifies the initial value of the conductivity in the percolation threshold, as shown in Figure 17. Thus, an insignificant change in the electrical conductivity for all values of this parameter can be established.

The packaging fraction of the filler only changes the edge of the curve of the conductivity, but does not affect the maximum value of electrical conductivity of the nanocomposite, as shown in Figure 18.

In Figures 19, 20, and 21, the tridimensional changes of these behaviors are illustrated with the aim of emphasizing that the maximum electrical conductivity is achieved when the volume fraction and the conductivity at the maximum packaging fraction are great. Insignificant changes have been found when the conductivity at the percolation threshold of the filler and the packaging fraction are increased, since the maximum conductivity is achieved for almost any value of them.

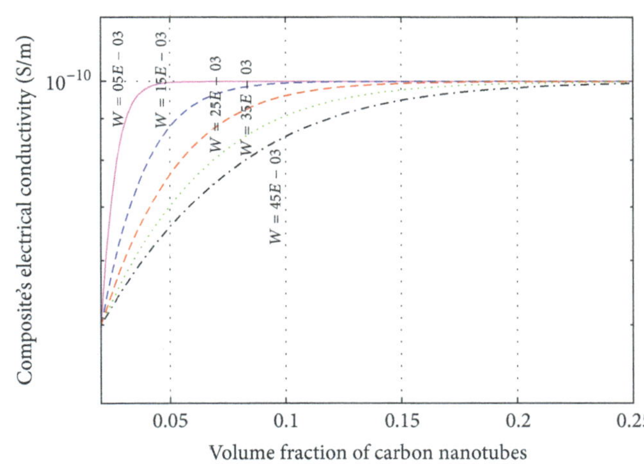

FIGURE 24: DC electrical conductivity of the nanocomposite when the width of the percolation region is varied (Sigmoidal function model).

4.4. Analysis of the Sigmoidal Function Model. In Figure 22, it can be observed that varying the midpoint of the percolation region's width modifies the initially electrical conductivity for small values of volume fraction, but the maximum value can be achieved for any value of the midpoint. Therefore, this value has an insignificant effect on the maximum value of the conductivity.

The conductivity of the filler has an important effect on the maximum value of the electrical conductivity of the nanocomposite, as shown in Figure 23. When the value of the conductivity of the filler is great, the conductivity also is increased. Thus, the filler's conductivity is a very important parameter in the design of these materials.

In the case of the width of the percolation region (Figure 24), it has a reduced effect on the range of values of the volume fraction of the filler since the slope of the curve is more abrupt or not, although the maximum value of the conductivity is not affected by this parameter. Therefore, this

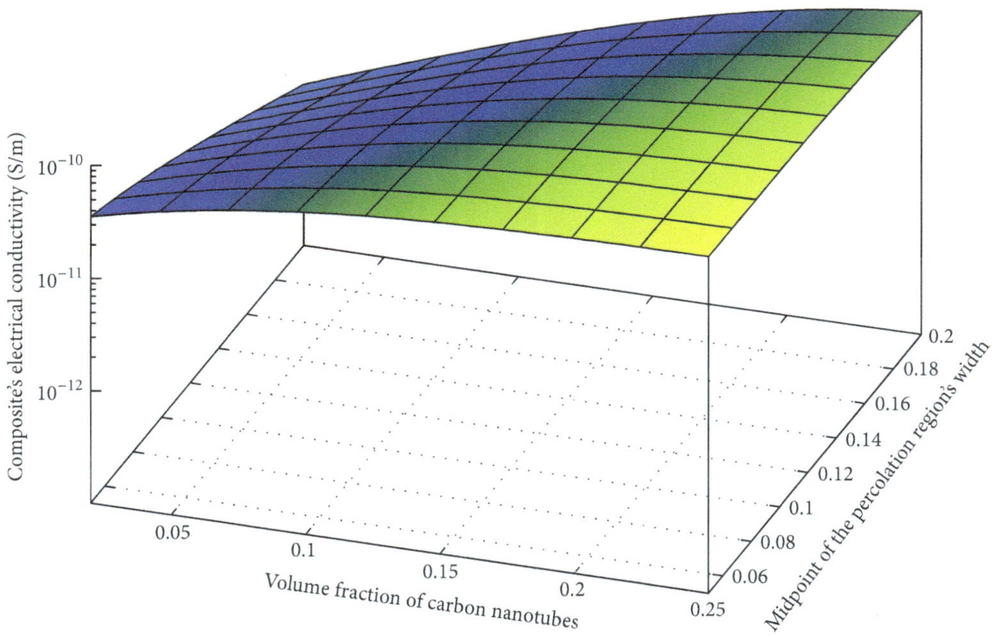

FIGURE 25: DC electrical conductivity of the nanocomposite at varying both the midpoint of the percolation region's width and the volume fraction of the filler (Sigmoidal function model).

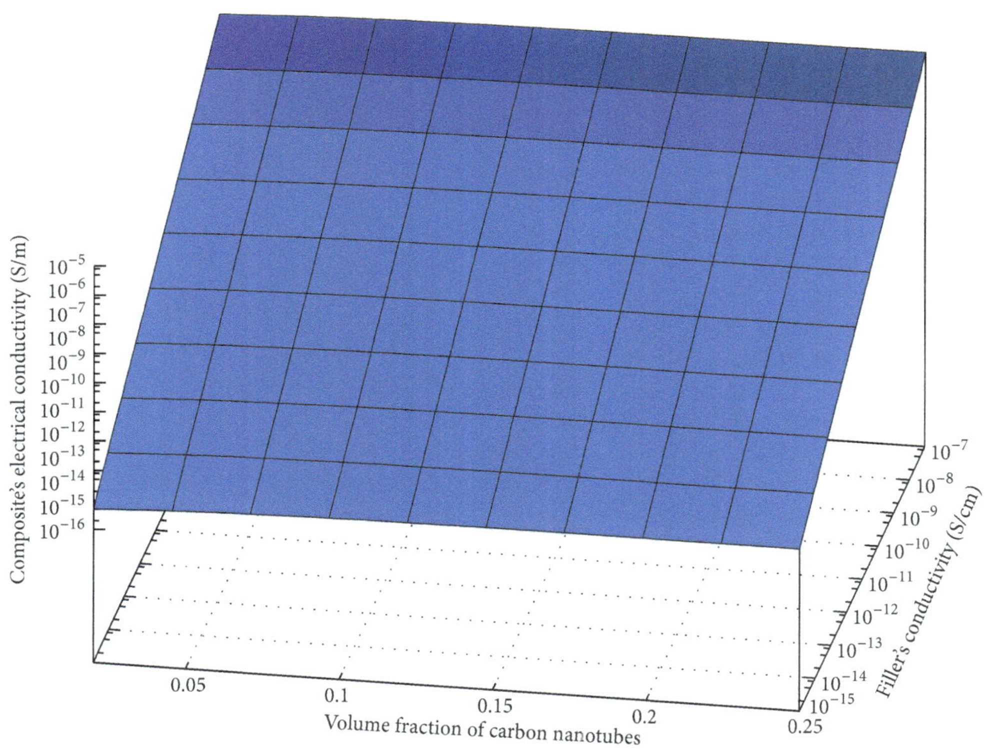

FIGURE 26: DC electrical conductivity of the nanocomposite when both the conductivity of the filler and the volume fraction of the filler are varied (Sigmoidal function model).

value has an insignificant effect on the maximum value of the conductivity of the nanocomposite.

In Figures 25, 26, and 27, a tridimensional representation of these changes allows the designer of these materials to take into account the regions where electrical conductivity will be maximum, middle, and minimum. In Figure 25, the maximum value of electrical conductivity is achieved when the volume fraction of the filler used was at maximum, and the midpoint of the percolations region's width is at maximum. The maximum value of conductivity can be

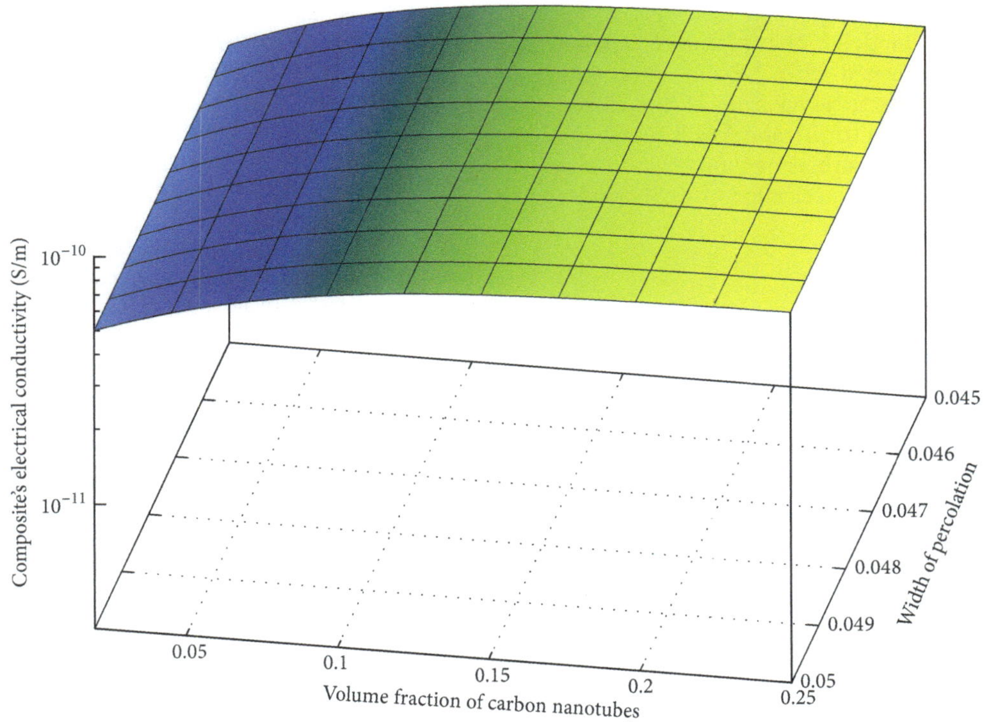

FIGURE 27: DC electrical conductivity of the nanocomposite at varying both the width of the percolation region and the volume fraction of the filler (Sigmoidal function model).

achieved when the conductivity of the filler is great, as shown in Figure 26. Finally, in Figure 27 it can be visualized that three different regions are found, but the maximum value is obtained when volume fraction is great.

4.5. Discussion of the Results. In Table 1, it can be found that a recompilation of the effect of each one of the parameters associated with mathematical models is used to determine electrical conductivity of the composite materials based on carbon nanotubes and polymers. It is necessary to remember that volume fraction of carbon nanotubes in the composite is restricted in the range of 0.02 to 0.25.

For Kirkpatrick's model, an increase in the electrical conductivity can be achieved. In the case of the McLachlan's model, a similar behavior can be found and specific values must be considered since two solutions can be obtained: (1) an increased behavior or (2) a decreased behavior. In addition, imaginary values in electrical conductivity can be achieved for determined ranges of parameters used in the mathematical model. These details have not been reported previously by different authors using this model. In particular, Mamunya's model has electrical conductivity restricted to the range where conductivity is at the maximum packaging fraction. Finally, Sigmoidal function model can be used for all values of concentration of the filler, and its behavior can be adjusted only when experimental values have been obtained since it belongs to the group of models based on nonlinear regression.

5. Conclusions

The DC electrical conductivity of nanocomposites based on polymers and carbon nanotubes has been studied with the aim of determining which are the main parameters of fabrication that must be considered for truly modifying the electrical behavior and how these change such behavior. This study was realized by computer simulation through four different models at varying three parameters in each one of them. Parameters such as volume fraction of the filler, percolation threshold of the filler, critical exponent of percolation, conductivity of the filler, conductivity of the matrix, conductivity at the maximum packaging fraction of the filler, conductivity at the percolation threshold, packaging fraction of the filler, midpoint of the percolation region's width, and width of the percolation region were analyzed. The main parameters involved in the increase of electrical conductivity of the nanocomposites are critical exponent of percolation, conductivity of the filler, conductivity of the matrix, conductivity at the maximum packaging fraction, conductivity at the percolation, and width of the percolation region. Unfortunately, a model that includes all these parameters does not exist until now. In McLachlan's model, specific values must be considered since two solutions can be possible: (1) an increased behavior or (2) a decreased behavior. In addition, imaginary values in electrical conductivity can be achieved for determined ranges of parameters used in the McLachlan's model. These details have not been reported previously by different authors using this model.

The results presented here suggest the development of new electrical conductivity models for nanocomposites, where all these parameters are related with the aim of providing a better approach to the electrical behavior of these materials. A more sophisticated electrical model will allow to lead nanocomposites based on carbon nanotubes and polymers to the electronic industry for fabrication of VLSI circuits and systems.

Acknowledgments

This work was supported by the National Council of Science and Technology of Mexico (CONACYT) under project no.152524 under their support to basic research. The authors desire to thank the collaboration of Rodolfo Angulo Olais, M.S. (CINVESTAV-IPN Zacatenco, Mexico), Alfredo Aguilar Elguezabal, Ph.D. and José Martín Herrera Ramírez, Ph.D. (CIMAV-Chihuahua, México), Ana Arizmendi Morquecho (CIMAV-Monterrey, México), Bárbara Bermúdez Reyes, Ph.D. (Universidad Autónoma de Nuevo León, México), and Carlos A. De la Cruz-Blas, Ph.D. (Universidad Pública de Navarra, Spain).

References

[1] B. P. Grady, *Carbon Nanotube-Polymer Composites: Manufacture, Properties, and Applications*, John Wiley & Sons, New York, NY, USA, 2011.

[2] N. Grossiord, J. Loos, L. Van Laake et al., "High-conductivity polymer nanocomposites obtained by tailoring the characteristics of carbon nanotube fillers," *Advanced Functional Materials*, vol. 18, no. 20, pp. 3226–3234, 2008.

[3] M. Sahimi, *Applications of Percolation Theory*, Taylor and Francis, Boca Raton, Fla, USA, 1994.

[4] S. Kirkpatrick, "Percolation and conduction," *Reviews of Modern Physics*, vol. 45, no. 4, pp. 574–588, 1973.

[5] R. Zallen, *The Physics of Amorphous Solids*, Wiley-Interscience, New York, NY, USA, 1998.

[6] D. Stauffer and A. Aharony, *Introduction To Percolation Theory*, Taylor & Francis, London, UK, 2003.

[7] D. S. McLachlan, M. Blaszkiewicz, and R. E. Newnham, "Electrical resistivity of composites," *Journal of the American Ceramic Society*, vol. 73, no. 8, pp. 2187–2203, 1990.

[8] D. S. McLachlan, "Analytical functions for the DC and AC conductivity of conductor-insulator composites," *Journal of Electroceramics*, vol. 5, no. 2, pp. 93–110, 2000.

[9] D. S. McLachlan, C. Chiteme, C. Park et al., "AC and DC percolative conductivity of single wall carbon nanotube polymer composites," *Journal of Polymer Science B*, vol. 43, no. 22, pp. 3273–3287, 2005.

[10] R. L. Barton, J. M. Keith, and J. A. King, "Development and modeling of electrically conductive carbon filled liquid crystal polymer composites for fuel cell bipolar plate applications," *Journal of New Materials for Electrochemical Systems*, vol. 10, no. 4, pp. 225–229, 2007.

[11] D. S. McLachlan and G. Sauti, "The AC and DC conductivity of nanocomposites," *Journal of Nanomaterials*, Article ID 30389, 9 pages, 2007.

[12] E. P. Mamunya, V. V. Davidenko, and E. V. Lebedev, "Effect of polymer-filler interface interactions on percolation conductivity of thermoplastics filled with carbon black," *Composite Interfaces*, vol. 4, no. 4, pp. 169–176, 1997.

[13] Y. P. Mamunya, V. V. Davydenko, P. Pissis, and E. V. Lebedev, "Electrical and thermal conductivity of polymers filled with metal powders," *European Polymer Journal*, vol. 38, no. 9, pp. 1887–1897, 2002.

[14] J. M. Keith, J. A. King, and B. A. Johnson, "Electrical conductivity modeling of carbon filled polypropylene based resins for fuel cell bipolar plate applications," *Journal of New Materials for Electrochemical Systems*, vol. 11, no. 4, pp. 253–257, 2008.

[15] K. Kalaitzidou, H. Fukushima, and L. T. Drzal, "A route for polymer nanocomposites with engineered electrical conductivity and percolation threshold," *Materials*, vol. 3, no. 2, pp. 1089–1103, 2010.

[16] A. Merzouki and N. Haddaoui, "Electrical conductivity modeling of polypropylene composites filled with carbon black and acetylene black," *ISRN Polymer Science*, Article ID 493065, 7 pages, 2012.

[17] E. Tjørve, "Shapes and functions of species-area curves: a review of possible models," *Journal of Biogeography*, vol. 30, no. 6, pp. 827–835, 2003.

[18] F. A. Graybill and H. K. Iyer, *Regression Analysis: Concepts and Applications*, Duxbury Press, Belmont, Calif, USA, 1994.

[19] J. O. Rawlings, S. G. Pantula, and D. A. Dickey, *Applied Regression Analysis: A Research Tool*, Springer, New York, NY, USA, 1998.

[20] G. A. F. Seber and C. J. Wild, *Nonlinear Regression*, Wiley-Interscience, Hoboken, NJ, USA, 2003.

[21] S. Chatterjee and A. S. Hadi, *Regression Analysis By Example*, Wiley-Interscience, Hoboken, NJ, USA, 2006.

13

A New Chaos-Based Image-Encryption and Compression Algorithm

Somaya Al-Maadeed,[1] Afnan Al-Ali,[2] and Turki Abdalla[2]

[1] *Department of Computer Science and Engineering, Qatar University, P.O. Box 2713, Doha, Qatar*
[2] *Department of Computer Engineering, College of Engineering, University of Basrah, B.P. 49, Basrah, Iraq*

Correspondence should be addressed to Somaya Al-Maadeed, s_alali@qu.edu.qa

Academic Editor: Ahmed Bouridane

We propose a new and efficient method to develop secure image-encryption techniques. The new algorithm combines two techniques: encryption and compression. In this technique, a wavelet transform was used to decompose the image and decorrelate its pixels into approximation and detail components. The more important component (the approximation component) is encrypted using a chaos-based encryption algorithm. This algorithm produces a cipher of the test image that has good diffusion and confusion properties. The remaining components (the detail components) are compressed using a wavelet transform. This proposed algorithm was verified to provide a high security level. A complete specification for the new algorithm is provided. Several test images are used to demonstrate the validity of the proposed algorithm. The results of several experiments show that the proposed algorithm for image cryptosystems provides an efficient and secure approach to real-time image encryption and transmission.

1. Introduction

An increasing amount of information is being transmitted over the Internet, including not only text but also audio, image, and other multimedia files. Images are widely used in daily life, and, as a result, the security of image data is an important requirement [1]. In addition, when either communication bandwidth or storage is limited, data are often compressed. In particular, when a wireless communication network is used, low-bit-rate compression algorithms are needed as a result of bandwidth limitations. Encryption is also performed when it is necessary to protect user privacy [2]. Image-encryption algorithms are used to provide this security; for our purposes, these algorithms can be divided into two groups with respect to the approach used to construct the encryption scheme: chaos-based methods and nonchaos-based methods. Image encryption can also be divided into full encryption and partial encryption (also called selective encryption) schemes according to the percentage of the data that is encrypted. Encryption schemes can also be classified as either combined-compression methods or noncompression methods.

Several reviews have been published on image and video encryption, including selective (or partial encryption) methods, providing a fairly complete overview of the techniques developed to date [3]. Kunkelmann [4] and Qiao and Nahrstedt [5] provide overviews, comparisons, and assessments of classical encryption schemes for visual data, with an emphasis on MPEG. Bhargava et al. [6] review four MPEG-encryption algorithms published by the authors themselves from 1997 to 1999. More recent MPEG encryption surveys are provided by But [7] (in which the suitability of available MPEG-1 ciphers for streaming video is assessed). Other data formats have also been discussed with respect to selective encryption. Coding schemes based on wavelets [8], quad trees [9, 10], iterated function systems (fractal coding) [11], and vector quantization [12] have been used to create selective encryption schemes.

In 1997, two kinds of schemes based on higher-dimensional chaotic maps were proposed in which a discretized chaotic map of the pixels in an image is permuted by several rounds of shuffling operations [2]. Between every two adjacent rounds of permutations, a diffusion process is performed, which can significantly change the distribution of

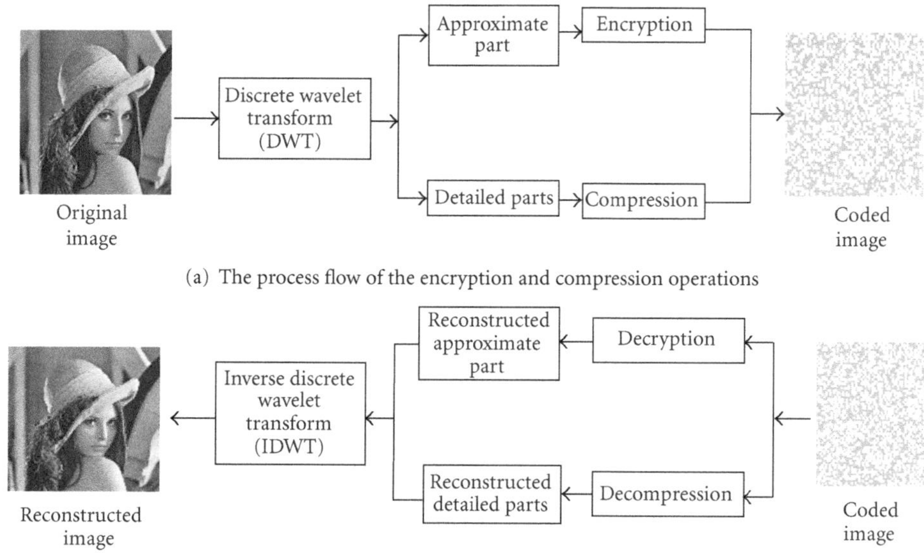

(a) The process flow of the encryption and compression operations

(b) The steps of decryption and decompression taken to reconstruct the image

Figure 1: The processes of the new algorithm.

the image histogram, thereby making a statistical decryption attack impossible. Empirical testing and cryptanalysis have both demonstrated that the chaotic Baker and Cat maps are good candidates for this kind of image encryption. Similar ideas have also appeared, including a 1998 paper [13], in which a rapid bulk-data encryption scheme was designed by combining chaotic. Kolmogorov flows with an adaptation of a very fast shift-register-based pseudorandom number generator [14]. Recently, a Feed-Back Chaotic Synchronization (FCS) for designing a real-time secure symmetric encryption scheme has been implemented in hardware [15–17].

The basic principle of encryption with chaos is based on the ability of some dynamic systems to produce sequence of numbers that are random in nature. This sequence is used to encrypt messages. For decryption, the sequence of random numbers is highly dependent on the initial condition used for generating this sequence. A very minute deviation in the initial condition will result in a totally different sequence. This sensitivity to initial condition makes chaotic systems ideal for encryption.

In this paper, we present an approach for encrypting images, our approach is based on chaotic maps, where we suggest to use the wavelet transform for image decomposition as we will discuss later and then we propose to use one, two, or three external encryption keys to encrypt the important part using one-dimensional chaotic map. We will provide a comparison in terms of the correlation between the original image and the encrypted image for all the three cases of encryption by the multiple keys when the number of external encryption keys increases in each time.

Next section is an overview of our proposed method of selective encryption of an image. The rest of this paper is organized as follows: Sections 3 and 4 give a brief discussion to the wavelet transform and chaos theory, respectively: Test, verification, and efficiency of the proposed new encryption algorithm are given in Section 5. Finally, Section 6 concludes this paper.

2. System Overview

In this paper, we propose an efficient, selective chaos-based image-encryption and compression algorithm. The block diagram of this algorithm is shown in Figure 1. The encryption and compression operations is shown in Figure 1(a). First, we have set the chaotic map parameters (the initial values) as well as to select the external encryption keys. The chaos-based encryption algorithm is as follows: after wavelet transformation, an amount of 25% (in case of one-level decomposition) or 6.25% (in case of two-level decomposition) of the original image, which corresponds to the important part, is transformed into one-dimensional vector with a dimension equal to the important part size, let us call it W. The chaotic map will be run for W iterations and generate a vector of ones and zeros of dimension equal to W too; it will be called a *threshold vector*, then the pixels of the image vector will be encrypted with one, two, or three external encryption keys according to that threshold vector.

From Figure 1(b), the secret image of size (128×128) pixels is decomposed using discrete wavelet transform (DWT). In practice, this process of decomposition usually repeated n times, and it is repeated just on the LL-sub band as mentioned for octave-band decomposition. Here in this work, the decomposition will be for one-level or two-level decomposition in order to compare different amounts of encrypted data and examine their effect on security. The image is decomposed into four subimages: the approximation component (LL) and three detail components (horizontal, vertical, and diagonal). The most important component, the LL component, is then encrypted using a chaos-based image-encryption algorithm, and the other three components are subsequently compressed using wavelet analysis. The implementation of the algorithm achieves high encryption rates, as discussed in Section 5. The steps of decryption and decompression taken to reconstruct the image are described at Figure 1(b).

3. Wavelet Transform

A wavelet is a small wave with its energy concentrated in time, providing a tool for the analysis of time-varying phenomena. A wavelet not only has an oscillating characteristic but also has the ability to allow for simultaneous time and frequency analyses with a flexible mathematical foundation. The first step of a general image-compression technique is the wavelet transform, which is designed to distinguish between the visually important information and unimportant information. The transform is also intended to reduce the statistical dependence between coefficients so that the source coding will be more efficient.

The important of wavelet as a multiresolution technique comes from its decomposition of the image into multilevel of the independent information with changing the scale-like geographical map, in which the image has nonredundant information due to the changing of the scale. In this way, every image will be transformed in each level of decomposition to a one low information image and three details image in horizontal, vertical, and diagonal axis image, also the low information image can be decomposed into another four images. These approaches of decomposition process provide us a number of unrealizable features in the original image, which appear in their levels after the application of transformation [18].

There are different types of wavelet transforms, including continuous wavelet transform (CWT) and the discrete Wavelet Transform (DWT). The CWT is used for signals that are continuous in time, and the DWT is used when a signal is being sampled, such as during digital signal processing or digital image processing [19]. In this work, discrete wavelet transform is used. In a discrete wavelet transform, an image can be analyzed by passing it through an analysis filter bank, which consists of a low-pass and high-pass filter at each decomposition stage. When a signal passes through these filters, it is split into two bands. The low-pass filter, which corresponds to averaging operation, extracts the coarse information of the signal. The high-pass filter, which corresponds to a differencing operation, extracts the detail information of the signal. The output of the filtering operations is decimated by 2. A two-dimensional transform can be accomplished by performing two separate one-dimensional transforms. First, the image is filtered along the x-dimension using low pass and high pass analysis filters and decimated by 2. Low pass filtered coefficients are stored on the left part of the matrix and high pass filtered on the right. Because of decimation, the total size of the transformed image is same as the original image. Then, it is followed by filtering the subimage along the y-dimension and decimated by 2. Finally, the image has been split into four bands denoted by LL, HL, LH, and HH,

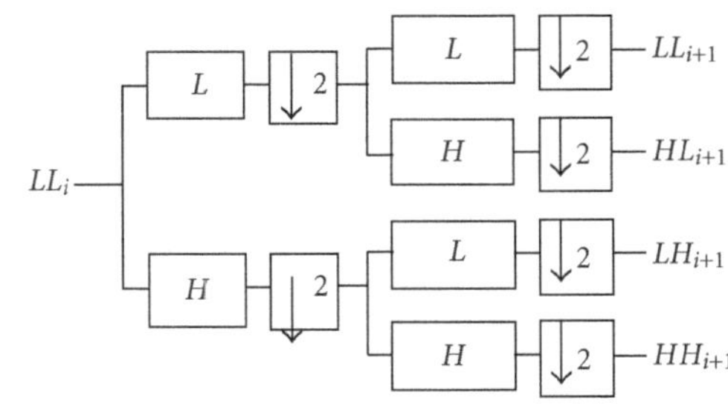

Figure 2: Splitting of subband into next higher Level four subbands.

after one-level decomposition. The LL band is again subject to the same procedure. The reconstruction of the image can be carried out by reversing the above procedure, and it is repeated until the image is fully reconstructed [18, 19]. This procedure of wavelet splitting is shown in Figure 2.

4. Image Encoding Using 1D-Discrete Chaotic Maps

In mathematics and physics, chaos theory describes the behavior of certain nonlinear dynamical systems that, under certain conditions, exhibit dynamics that are highly sensitive to initial conditions. As a result of this sensitivity, the behavior of chaotic systems appears to be random as a result of the exponential growth of errors in the initial conditions. This apparently chaotic behavior occurs even though these systems are deterministic in the sense that their future dynamics are well defined by their initial conditions, and there are no random elements involved. A chaotic dynamical system is a deterministic system that exhibits seemingly random behavior as a result of its sensitive dependence on its initial conditions and can never be specified with infinite precision. The behavior of a chaotic system is unpredictable; therefore, it resembles noise [18]. The close relationship between chaos and cryptography makes a chaos-based cryptographic algorithm a natural candidate for secure communication and cryptography [1]. Chaotic maps and cryptographic algorithms (or, more generally, maps defined on finite sets) have similar properties, such as sensitivity to changes in the initial conditions and parameters, pseudorandom behavior, and unstable periodic orbits with long periods. In a cryptographic algorithm, repeated encryption rounds lead to the desired diffusion and confusion properties of the algorithm. Iterations of a chaotic map spread the initial region over the entire phase space. The parameters of the chaotic map may represent the key to the encryption algorithm.

An important difference between chaos and cryptography is that encryption transformations are defined on finite sets, whereas chaos has meaning only for real numbers. Moreover, at present, the notion of cryptographic security and the performance of cryptographic algorithms have no counterpart in chaos theory [20]. In summary, we note that chaos-based encryption techniques are considered promising candidates for practical applications because these techniques provide an effective combination of speed, high

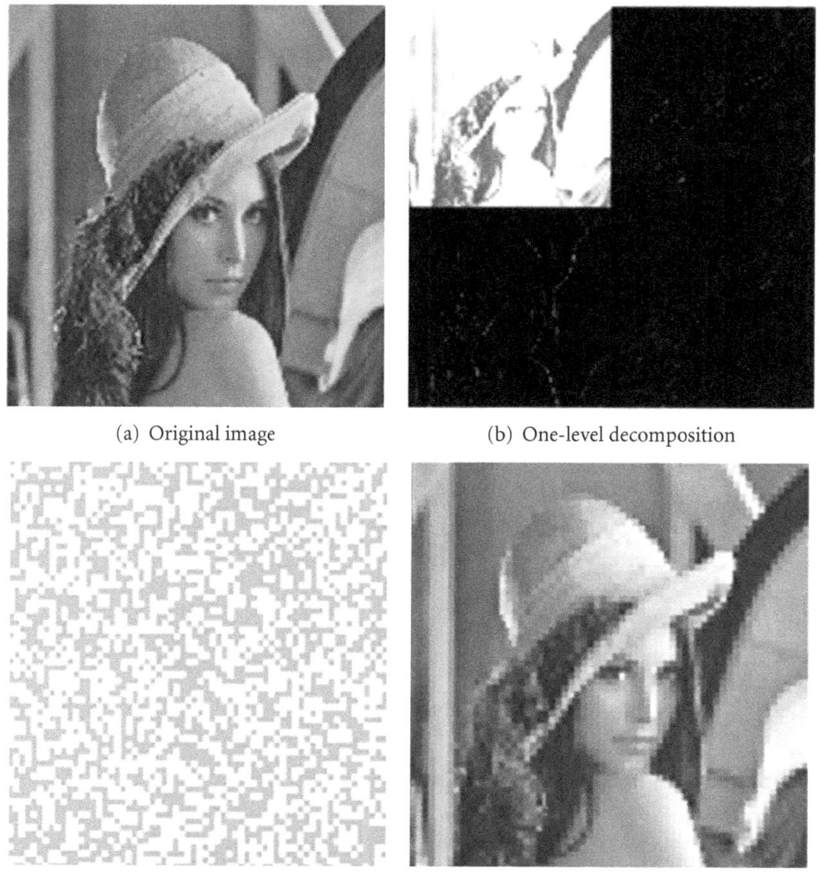

(a) Original image (b) One-level decomposition

(c) Coded image (d) Reconstructed image

Figure 3: Test image 1 (Lena.PGM) when $x(1) = 0.12345$, $\alpha = 4$, key1 = 1234567890223344, and key2 = 0987654321001122, Thr = 30, PSNR = 28.1240.

security, complexity, reasonable computational overhead and computational power [1]. Chaotic maps have attracted the attention of cryptographers as a result of the following fundamental properties.

(i) Chaotic maps are deterministic, meaning that their behavior is predetermined by mathematical equations.

(ii) Chaotic maps are unpredictable and nonlinear because they are sensitive to initial conditions. Even a very slight change in the starting point can lead to a significantly different outcome.

(iii) Chaotic maps appear to be random and disorderly but, in fact, they are not; beneath the random behavior is an order and pattern.

Chaotic phenomena are analogous to stochastic processes in that they appear in nonlinear dynamical systems; such processes are nonperiodic and nonconvergent and are extremely sensitive to initial conditions [20].

In cryptography, we focus on dynamic *discrete-time* system. A dynamic system is chaotic if all trajectories are bounded and nearby trajectories diverge exponentially at every point of the phase space. A chaotic system is given by an iterated function (chaotic map) f of a state space X. The iterated function transforms the current stage of the system into the next one, that is,

$$x_{n+1} = f(x_n), \quad (1)$$

where $x_n \in X$ denotes the system state at the discrete time. In chaos-based cryptography, the state space is typically a finite binary space

$$X = P = C = \{0,1\}^n, \quad n = 1, 2, \ldots, \quad (2)$$

where P = Plaintext and C = Ciphertext.

The initial condition is a vector $x_0 \in X$, and it is assigned to an internal state variable before the first iteration. The vector $c \in K = \{0,1\}^n$ contains the parameters of the dynamic system. The parameters are kept constant through all cycles (iterations) [21].

In this work, we are concerned with image encoding with the generation and use of 1D-chaotic maps. A one-dimensional dynamical system is a couple (I, ϕ), where I is real interval and ϕ is a transformation from I to I. Chaotic maps are nonlinear maps presenting the property of sensitivity to initial conditions.

FIGURE 4: Results of application of chaotic algorithm to SF.GIF. (a) and (d) original image. (b) and (e) Image resulting from encryption with L1 or L2, respectively, using only one external encryption key in the larger size. (c) and (f) Reconstructed image in each case.

FIGURE 5: Results of application of chaotic algorithm to SF.GIF. (a) and (d) original image. (b) Image resulting from encryption with L1 or L2, respectively, using two external encryption keys in the larger size. (c) and (f) Reconstructed image in each case.

FIGURE 6: Results of application of chaotic algorithm to SF.GIF. (a) and (d) original image. (b) image resulting from encryption with L1 or L2, respectively, using three external encryption keys in the larger size. (c) and (f) reconstructed image in each case.

The nonlinear transformation used here is defined as an iterative scalar map [22]

$$x_{n+1} = F(x_n, \alpha),$$
$$x_{n=0} = x_0, \tag{3}$$

where α is a set of real parameters. The use of (3) for image encoding means that the image is seen as a dynamical system; the main difference is that images are presented by integer values, while they are mapped into real values under system in (3).

One of the most known chaotic maps is the logistic map. It is defined by the following equation:

$$x_{n+1} = \alpha x_n (1 - x_n),$$
$$x_{n=0} = x_0. \tag{4}$$

The use of chaos for image encoding yields to three types of keys; these keys may be used together or separately, in order to enhance the privacy. They are the external or (control) parameter α, the initial state x_0, and the number of iterations [21]. The number of iterations is fixed during all the cycles and equal to the size of the important part of the image, which should be encrypted. In this paper, one two or three external encryption keys are used, then three types of approaches are performed as following: first using one external encryption key: in this type, the chaotic map will generate a threshold vector of two values: 0 and 1. The pixels of the image vector that are corresponding to the zeros of the threshold vector will be encrypted by one external encryption key and the other pixels will be normalized in such a way to hide the details of the image vector. Second, using two external encryption keys: in order to increase the security, the second type uses two external encryption keys such that the chaotic map will also generate a threshold vector of two values: 0 and 1, and the pixels of the image vector that are corresponding to the zeros of the threshold vector will be encrypted by the first key, and other pixels that are corresponding to the ones will be encrypted by the second key. Third, using three external encryption keys: in order to increase the security further, the third type uses three external encryption keys. In this case the chaotic map will generate a threshold vector of three different values: 0, 0.5, and 1. The pixels of the image vector that are corresponding to the zeros will be encrypted using the first key; the pixels that are corresponding to 0.5 values will be encrypted by the second key, and finally the pixels that are corresponding to the ones will be encrypted by the third key. The encryption process is done for one-level decomposition and two-level decomposition; also it is repeated for two different sizes of external keys. The previous schemes will be applied for chaotic maps given in (4). The decryption process is the reverse operation to each process in the encryption procedure as shown in Figure 1(b).

Table 1: Test images.

#	Images	Names
1		SF.GIF
2		ROAR512.JPG
3		GREEN ROSE.JPG
4		GOLDRUSH.JPG

5. Experimental Results

In this work, a 2D grayscale image is divided into subbands and subsampled using a discrete wavelet transform. The test images and their sizes can be shown in Table 1. Each coefficient represents a spatial area of approximately 2×2 pixels from the original image. The total number of components remaining after the vertical and horizontal decompositions is four. These components are referred to as subbands. As shown in Figure 3, the most important component of these subbands is the LL component because it is much more similar to the original image; LL is called the *approximate image*. To provide security for this component, the second step is to encrypt this component using a chaos-based image-encryption algorithm, as described above. The algorithm described previously will be implemented for one 2D-gray scale image and one 2D-color image, each of them has size of (128×128) pixels. The test images and their sizes can be shown in Table 1.

The correlation coefficients between the original image and the encrypted image for all the three cases of encryption which are explained previously, are illustrated in Tables 2 and 3. The results obtained for one-level and two-level decomposition. Note that the initial state for which the encrypted images is computed is $x_0 = 0.65440$ and the number of iterations is fixed during all cycles equal to the size of the important part of the image, which should be encrypted. The chaotic map parameters are the initial values which are also used to select the external encryption keys (see Section 4). In this paper, the parameters are set as follows: the parameter $\alpha = 4$, one, two, or three external encryption keys. Here, our aim is to compare the output images for different values of parameters (we have considered six cases). They are: small keys sizes (key1 = 4567, key2 = 3336, and key3 = 5892) and large keys sizes (key1 = 4567332, key2 = 3336877, and key3 = 5892117). These results are obtained by applying chaotic partial encryption to the image after the wavelet transform operation is performed. Table 2 shows the results for the two grayscale images, while Table 3 shows the results for the two color images for small key size and large key size. Figures 4, 5, and 6 show the results for SF.GIF grayscale image in one, two, and three external encryption keys, respectively.

A good encryption procedure should be robust against all kinds of cryptanalytic, statistical, and brute-force attacks. In this section, we discuss the security analysis of our method.

From the results, we can confirm that our cipher is excellent, and this is clear from the below analysis to the results of our experiments.

For a secure image cryptosystem, the encryption key plays a very important role against the brute force attack; so for high security, the key space should be large enough to make the brute force attack infeasible but may decrease encryption/decryption speed. In this algorithm, we used a key space of 94 bit combination, then we increased the key space up to 97 bit combination and compared the results (Tables 2 and 3). Out of this comparison, we found that the correlation between the cipher image and the original image decreased when we increased the key space size. Moreover using more than one external encryption keys which is a perfect way to conclude that increasing the number of keys means increasing the security because this makes the intrud-suffers when trying to attack the cipher image and find the secret keys. In addition to the external encryption keys, we should mention that our algorithm depends also on the parameters of logistic map, which is sensitive to the initial condition that can be connected with confusion and diffusion property in good cipher.

We have performed statistical analysis by calculating the histograms and correlations of cipher and reconstructed image with the original one. The correlation between the original image and the cipher image is nearly equal to zero expressing the original image and cipher image being almost independence. The correlation between the original image and the reconstructed image is nearly equal to one that means a perfect reconstruction at the receiver and

TABLE 2: Results of encryption of different amount for grayscale SF.GIF and ROAR512.JPG images using small and large keys sizes.

Image	Key size	Amount %	One key		Two keys		Three keys	
			Cipher-image correlation	Reconstructed-image correlation	Cipher-image correlation	Reconstructed-image correlation	Cipher-image correlation	Reconstructed image correlation
1	Small	L1 = 25	0.07462337	0.99999286	0.03005224	0.99999287	0.00513192	0.99999287
		L2 = 6.25	0.14151426	0.99999784	0.13672353	0.99999784	0.0433832	0.99999784
1	Large	L1 = 25	0.01197778	0.99999286	0.00645548	0.99999287	0.00495562	0.99999287
		L2 = 6.25	0.07438078	0.99999784	0.02457648	0.99999784	0.00654208	0.99999784
2	Small	L1 = 25	0.13040771	0.99999835	0.05808717	0.99999836	0.04030919	0.99999836
		L2 = 6.25	0.24379369	0.99999949	0.15640678	0.99999949	0.07057224	0.99999949
2	Large	L1 = 25	0.01142504	0.99999835	0.00668093	0.99999836	0.00121170	0.99999836
		L2 = 6.25	0.10455903	0.99999491	0.07915913	0.99999492	0.02147876	0.99999492

TABLE 3: Results of encryption of different amount for color GREEN ROSE.JPG and GOLDRUSH.JPG images using different keys.

Image	Key size	Amount %	One key		Two keys		Three keys	
			Cipher-image correlation	Reconstructed-image correlation	Cipher-image correlation	Reconstructed-image correlation	Cipher-image correlation	Reconstructed-image correlation
3	Small	L1 = 25	0.06636301	0.99999996	0.04991651	0.99999897	0.04777646	0.99999897
		L2 = 6.25	0.13200788	0.99999998	0.08473995	0.99999967	0.07023566	0.99999967
3	Large	L1 = 25	0.02938994	0.99999996	0.00519053	0.99999897	0.00418059	0.99999897
		L2 = 6.25	0.03215505	0.99999998	0.02981104	0.99999967	0.01996734	0.99999967
4	Small	L1 = 25	0.03293707	0.99999995	0.02655446	0.99999833	0.00998768	0.99999833
		L2 = 6.25	0.06094592	0.99999998	0.03180833	0.99999470	0.01329204	0.99999947
4	Large	L1 = 25	0.00525827	0.99999995	0.00494368	0.99999833	0.00218012	0.99999833
		L2 = 6.25	0.02447215	0.99999998	0.01329412	0.99999470	0.00979585	0.99999470

the encryption algorithm works in a very good manner to protect the image. Out of the results of our experiments, one can see that as the amount of the encrypted part is decreased (from L1 = 25% to L2 = 6.25%), the execution time is decreased too (about 0.218 sec for one external encryption key, 0.453 sec for two external encryption keys, and 0.5 sec for three external encryption keys), which indicates a high speed in performing encryption and decryption process and in getting the results, but also we should mention that decreasing the amount of encrypted part makes the correlation between the cipher image and the original image increase. All the proposed algorithms were programmed in MATLAB version 7.0. Performance was measured on a 2.0 GHz. Pentium IV with 1 GB of RAM running Windows XP Professional.

6. Conclusion

In this paper, we have proposed a method for the selective encryption of an image combined with a compression method. The technique used here results in a significant reduction in encryption and decryption time. This method consists of three steps; in each step, we used selected properties that served the goal of this paper, which were to obtain an effective cipher and high-quality image compression to achieve both security against unauthorized access during data transmission through an unsecured channel and high compression to allow for a low transmission rate. For the encryption, we used an efficient algorithm based on a chaotic map to encrypt the low subband of the image; this algorithm has very good diffusion and confusion properties, and we obtained a very good cipher of the subband, as evidenced by the low correlation between the original image and the coded image. For the compression, we used the wavelet transformation, and the results were highly satisfactory; this method allowed us to achieve a perfect reconstruction with a good PSNR. The basic idea of this work is to show the influence of using multiple keys in increasing security by increasing the number of external encryption keys in each time such that the important part of the image is encrypted with one, two, or three external encryption keys; the encryption is done by a new 1D chaotic map. We show from the results that the correlation coefficients between the original image and the encrypted image are decreased when the number of external encryption keys is increased, and this increases the security. Also, we show how the correlation coefficient changed exponentially when using different values of the control parameter in each time. In future work, we can use more than 128 bits for the external keys to increase overall security, and we can use another method for compression.

References

[1] H. Hossam El-din, H. M. Kalash, and O. S. Farag Allah, "An efficient chaos-based feedback stream cipher (ECBFSC) for image encryption and decryption," *Informatica*, vol. 31, no. 1, pp. 121–129, 2007.

[2] J. Fridrich, "Symmetric ciphers based on two-dimensional chaotic maps," *International Journal of Bifurcation and Chaos in Applied Sciences and Engineering*, vol. 8, no. 6, pp. 1259–1284, 1998.

[3] R. Pfarrhofer and A. Uh, "Selective image encryption using JBIG," in *Proceedings of the Communications and Multimedia Security (CMS '05)*, pp. 98–107, September 2005.

[4] T. Kunkelmann, "Applying encryption to video communication," in *Proceedings of the Multimedia and Security Workshop at (ACM Multimedia '98)*, pp. 41–47, England, UK, September 1998.

[5] L. Qiao and K. Nahrstedt, "Comparison of mpeg encryption algorithms," *Computers and Graphics*, vol. 22, no. 4, pp. 437–448, 1998.

[6] B. Bhargava, C. Shi, and S. Y. Wang, "MPEG video encryption algorithms," *Multimedia Tools and Applications*, vol. 24, no. 1, pp. 57–79, 2004.

[7] J. But, "Limitations of existing MPEG-1 ciphers for streaming video," Tech. Rep. CAIA 040429A, Swinburne University, Melbourne, Australia, April 2004.

[8] A. Pommer and A. Uhl, "Selective encryption of wavelet-packet encoded image data: efficiency and security," *Multimedia Systems*, vol. 9, no. 3, pp. 279–287, 2003.

[9] H. Cheng and X. Li, "Partial encryption of compressed images and videos," *IEEE Transactions on Signal Processing*, vol. 48, no. 8, pp. 2439–2451, 2000.

[10] X. Li, J. Knipe, and H. Cheng, "Image compression and encryption using tree structures," *Pattern Recognition Letters*, vol. 18, no. 11-13, pp. 1253–1259, 1997.

[11] S. Roche, J.-L. Dugelay, and R. Molva, "Multi-resolution access control algorithm based on fractal coding," in *Proceedings of the IEEE International Conference on Image processing (ICIP'96)*, pp. 235–238, IEEE Signal Processing Society, Lausanne, Switzerland, September 1996.

[12] T. S. Chen, C. C. Chang, and M. S. Hwang, "A virtual image cryptosystem based upon vector quantization," *IEEE Transactions on Image Processing*, vol. 7, no. 10, pp. 1485–1488, 1998.

[13] J. Scharinger, "Fast encryption of image data using chaotic Kolmogorov flows," *Journal of Electronic Imaging*, vol. 7, no. 2, pp. 318–325, 1998.

[14] Y. Mao and G. Chen, "Chaos based image encryption," in *Handbook of Computational Geometry for Pattern Recognition, Computer Vision, Neural Computing and Robotics*, E. B. Corrochano, Ed., Springer, Heidelberg, Germany, 2004.

[15] M. S. Azzaz, C. Tanougast, S. Adoudi, A. Bouridane, and A. Dandache, "An FPGA implementation of a feed-back chaotic synchronization for secure communications," in *Proceedings of the 7th International Symposium on Communication Systems, Networks and Digital Signal Processing (CSNDSP '10)*, pp. 239–243, July 2010.

[16] M. S. Azzaz, C. Tanougast, S. Sadoudi, A. Dandache, and F. Monteiro, "Real-time image encryption based chaotic synchronized embedded cryptosystems," in *Proceedings of the 8th IEEE International NEWCAS Conference (NEWCAS '10)*, pp. 61–64, June 2010.

[17] S. Sadoudi, C. Tanougast, M. S. Azzaz, A. Dandache, and A. Bouridane, "Embedded genesio-tesi chaotic generator for ciphering communications," in *Proceedings of the 7th International Symposium on Communication Systems, Networks and Digital Signal Processing (CSNDSP '10)*, pp. 234–238, July 2010.

[18] A. Cohen and J. Kovacevic, "Wavelets: the mathematical background," *Proceedings of the IEEE*, vol. 84, no. 4, pp. 514–522, 1996.

[19] G. Lekutai, *Adaptive self-tuning neuro wavelet network controllers*, Ph.D. thesis, Electrical Engineering Department, Virginia

Polytechnic Institute and State University Blackburg, Virginia, Va, USA, March 1997.

[20] C.-H. Hsu, *A study of chaotic image encryption algorithm*, M.S. thesis, Electrical Engineering Department, Chung Yuan Christian University, 2004.

[21] F. Belkhouche and U. Qidwai, "Binary image encoding using 1D-chaotic maps," in *Proceedings of the IEEE Region 5 Annual Technical Conference*, pp. 39–42, New Orleans, La, USA, April 2003.

[22] R. M. May, "Simple mathematical models with very complicated dynamics," *Nature*, vol. 261, no. 5560, pp. 459–467, 1976.

14

Feature Point Extraction from the Local Frequency Map of an Image

Jesmin Khan,[1] Sharif Bhuiyan,[1] and Reza Adhami[2]

[1] *Department of Electrical Engineering, Tuskegee University, Tuskegee, AL 36088, USA*
[2] *Department of Electrical and Computer Engineering, University of Alabama in Huntsville, Huntsville, AL 35899, USA*

Correspondence should be addressed to Jesmin Khan, khanj@mytu.tuskegee.edu

Academic Editor: Tamal Bose

We propose a novel technique for detecting rotation- and scale-invariant interest points from the local frequency representation of an image. Local or instantaneous frequency is the spatial derivative of the local phase, where the local phase of any signal can be found from its Hilbert transform. Local frequency estimation can detect edge, ridge, corner, and texture information at the same time, and it shows high values at those dominant features of an image. For each pixel, we select an appropriate width of the window for computing the derivative of the phase. In order to select the width of the window for any given pixel, we make use of the measure of the extent to which the phases, in the neighborhood of that pixel, are in the same direction. The local frequency map, thus obtained, is then thresholded by employing a global thresholding approach to detect the interest or feature points. Repeatability rate, a performance evaluation criterion for an interest point detector, is used to check the geometric stability of the proposed method under different transformations. We present simulation results of the detection of feature points from image utilizing the suggested technique and compare the proposed method with five existing approaches that yield good results. The results prove the efficacy of the proposed feature point detection algorithm. Moreover, in terms of repeatability rate; the results show that the performance of the proposed method with respect to different aspect is compatible with the existing methods.

1. Introduction

Interest point detection is a recent terminology in computer vision that refers to the detection of interest points for subsequent processing. An interest point is a point in an image which has a well-defined position and can be robustly detected. This means that an interest point can be a corner but it can also be, for example, an isolated point of local intensity maximum or minimum, line endings, or a point on a curve where the curvature is locally maximal. In practice, most of the so-called corner detection methods detect interest points in general rather than corners in particular. In the literature, "corner," "interest point," and "feature" are used somewhat interchangeably. An interest point is a point in the image which in general can be characterized as follows:

(i) it has a clear, preferably mathematically well-founded, definition;

(ii) it has a well-defined position in image space;

(iii) the local image structure around the interest point is rich in terms of local information contents, such that the use of interest points simplify further processing in the vision system;

(iv) it is stable under local and global perturbations in the image domain, including deformations as those arising from perspective transformations (sometimes reduced to affine transformations, scale changes, rotations, and/or translations) as well as illumination/brightness variations, such that the interest points can be reliably computed with high degree of reproducibility.

Historically, the notion of interest points goes back to the earlier notion of corner detection, where corner features were in early work detected with the primary goal of obtaining robust, stable, and well-defined image features for object tracking and recognition of objects from two-dimensional images. In practice, however, most corner detectors are sensitive not specifically to corners, but to local image regions which have a high degree of variation in all directions. Today, a main application of interest points is to signal points/regions in the image domain that are likely candidates to be useful for image matching, object recognition, motion detection, tracking, image mosaicing, panorama stitching, 3D modeling, and so forth.

There are a wide variety of methods reported in the literature for interest point and corner detection in grey-level images. However, many true corner points are missed and false interest points are detected in an image. Current detection methods can be categorized into three types: contour-based, parametric-model-based, and intensity-based methods. Contour-based methods first extract contours and then search for maximal curvature or inflexion points along the contour chains or carry out some polygonal approximation and then search for intersection points. Parametric model methods fit a parametric intensity model to the signal. They often provide subpixel accuracy, but are limited to specific types of interest points, for example, to L-corners. Intensity-based methods compute a measure that indicates the presence of an interest point directly from the grey values. This type of detector does not depend on edge detection or mathematical models.

Our paper presents a novel affine invariant interest point detector. The main contribution of this work is the extraction of interest points from local frequency map of an image. The proposed technique cannot be explicitly placed into any of the three categories. But it can be considered as a contour-based method, because local frequency estimation detects edges of an image. For the computation of local frequency, we use local phase which is a local neighborhood property. Therefore, in order for the value of that property to be useful, it must be computed in a neighborhood, which requires the determination of the appropriate size of the window that defines the local structure. This requirement is the same as the problem of scale selection. In this paper we introduce a method of scale selection which is based on a newly defined local image property called *phase divergence*. Local phase of any 1D signal can be calculated by utilizing a quadrature filter called the Hilbert transformer. In our work we have defined an equivalent 2D Hilbert transformer for the computation of local phase of an image. By definition, the spatial derivative of local phase is called the local or instantaneous frequency. Finally, a global thresholding technique is applied to the local frequency image to extract interest points.

The requirement for good feature point detectors is that they should be invariant to image transformations and correspond to the same regions of an image for different viewpoints. That means, with the change in the shape, size, and orientation, the detector automatically adapts based on the underlying image intensities. In particular, considering images from different geometric transformations, the interest points detected after the transformations should be the same so that the transformed versions of the points detected in the original image and points detected after the transformation of the image commute. They have often been called invariant interest points in the literature, though in principle they change covariantly with the transformation. The fact is that, even though the regions themselves are covariant, the normalized image pattern they cover and the feature descriptors derived from them are typically invariant. The proposed method is largely invariant to significant affine transformation including large rotations, shearing, and scale changes. Such transformations introduce significant changes in point locations as well as in the scale and the shape of the neighborhoods of interest points. Our approach addresses these problems simultaneously and offers invariance to geometric transformation.

In this paper we evaluate the proposed method utilizing the "repeatability" [1] criteria, which directly measures the quality of the detected feature points for tasks such as image matching, object recognition, and 3D reconstruction. It is complementary to localization accuracy, which is relevant for tasks such as camera calibration and 3D reconstruction of specific scene points. Repeatability and localization are conflicting criteria: smoothing improves repeatability but degrades localization [2]. Repeatability explicitly compares the geometrical stability of the detected interest points between different images of a given scene taken under varying viewing conditions. An interest point is "repeated" if the 3D scene point detected in the first image is also accurately detected in the second transformed image. The repeatability rate is the percentage of the total observed points that are detected in both images. Our detector satisfies the criteria well. The proposed detector is compared to five existing methods which have been shown to yield good results. Utilizing repeatability the proposed method is shown to yield comparable to improved performance.

The remainder of the paper is organized as follows. Section 2 presents a state-of-the-art on some existing interest point detectors. The motivation behind this work has been described in Section 3. The proposed interest point detection algorithm is given in Section 4. Experimental results along with the performance evaluation employing repeatability rate are demonstrated in Section 5. The comparison with five other existing methods is also given in this section. Concluding remarks and recommendations for future improvement are given in Section 6.

2. Related Work on Interest Point Detection

In this section we briefly present some existing interest points detection methods for each of the three categories.

2.1. Contour-Based Methods. Contour-based methods have existed for some long time. Bandera et al. [3] and Rattarangsi and Chin [4] utilize chain codes or curvature to detect boundary corners of an object. Rattarangsi and Chin use boundary curvature as an input function for further Gaussian scale-space analysis to detect corners. Bandera et al. estimate the curvature of a contour in an adaptive fashion so that corners can be detected even if they appear at different natural scales. Horaud et al. [5] extract line segments from image contours that are grouped, and intersections of grouped line segments are utilized as interest points. Shilat et al. [6] first detect ridges and troughs in the images. High curvature points along ridges or troughs, or intersection points, are utilized as interest points. Mokhtarian and Suomela [7] describe an interest point detector based on two sets of interest points. One set consists of T-junctions extracted from edge intersections, and the second set is obtained using a multiscale framework. The two sets are compared, and close interest points are merged. Pikaz and Dinstein [8] propose an algorithm based on a decomposition of noisy digital curves into a minimal number of convex and concave sections. They use the properties of pairs of sections that are determined in an adaptive manner to detect interest points.

Besides the above-mentioned contour-based interest points detection schemes, we specially want to give particular emphasis to interest point detection methods based on wavelet transform because wavelet transforms (WTs) may be considered as a form of time-frequency representation, and local or instantaneous frequency is the term or parameter that accounts for the time-varying nature of the spectral characteristics (in particular the frequency of the spectral peaks) of nonstationary signal.

In the past decade, interest point detection schemes using the WT became popular as a result of the fact that the WT is able to decompose an input signal into smooth and detailed parts by low-pass and high-pass filters at multiresolution levels [9]. In this manner, local deviations can be easily captured at various detailed decomposition levels. Several wavelet-based approaches [10–12] directly utilized the slope between two boundary points (e.g., tangent orientation) on an object contour as input and analyzed the wavelet modulus sum utilizing an interlevel ratio between multiresolution levels. These approaches detect corner candidates (including true corners and circular arcs) in a first step and eliminate circular arcs in a second by predefined thresholds. For instances, Quddus and Gabbouj [13] utilize optimal scale and pre-defined threshold to localize true corners precisely for a complicated shape (e.g., a shark). Hua and Liao [14] use x and y parametric coordinates to replace curvatures and tangent orientations as an input to WT analysis. This approach achieves a higher identification rate for polygon-shaped objects. Loupias et al. [15, 16] present a salient point detector that extracts points where variations occur in the image, whether they are corner like or not. The detector is based on WT for the detection of global and local variations. After reviewing wavelet-based researches, Yeh [17] recently concluded that 1D WT is a robust corner detection scheme due to its excellent local deviation capturing capability. In their work they transform the 2D boundaries of an object into a 1D $\theta - P$ representation (where θ is the tangent angle variations of the arc length, P along the object's boundary) and they detect the feature points utilizing the 1D wavelet coefficients at the finest (first) detailed decomposition level.

2.2. Intensity-Based Methods. Moravec [18] develops one of the first signal-based interest point detectors. His detector is based on the autocorrelation function of the signal. Beaudet's [19] detector uses the second derivatives of the signal to compute the measure $\mathrm{DET} = I_{xx}I_{yy} - I_{xy}^2$, where $I(x,y)$ is the intensity surface of the image. Points where this measure is maximal are defined as interest points. Kitchen and Rosenfeld [20] define an interest point detector that uses the curvature of planar curves.

Several interest point detectors [21–24] are based on a matrix related to the autocorrelation function. The utilized matrix, A, averages derivatives of the signal in a window W around a point (x, y):

$$A(x,y) = \begin{bmatrix} \sum\limits_{W} (I_x(x_k, y_k))^2 & \sum\limits_{W} I_x(x_k, y_k) I_y(x_k, y_k) \\ \sum\limits_{W} I_x(x_k, y_k) I_y(x_k, y_k) & \sum\limits_{W} \left(I_y(x_k, y_k)\right)^2 \end{bmatrix}, \tag{1}$$

where $I(x, y)$ is the image function and (x_k, y_k) are the points in the window W around (x, y). This matrix captures the structure of the neighborhood. If this matrix is of rank two, an interest point is detected. A matrix of rank one indicates an edge and a matrix of rank zero a homogeneous region.

Heitger et al. [25] develop an approach that extracts 1D directional characteristics by convolving the image with orientation-selective Gabor-like filters. The first and second derivatives of the 1D characteristics are computed in order to obtain 2D characteristics. Smith and Brady [26] compare the brightness of each pixel in a circular mask to the center pixel to define an area, and two-dimensional features are detected from the size, centroid, and second moment of this area. The approach proposed by Laganiere [27] is based on a variant of the morphological closing operator which uses two closing operators and four structuring elements. The first closing operator is sensitive to vertical/horizontal L-corners and the second to diagonal L-corners. Lowe [28] presents a method for extracting distinctive invariant features from images by taking a cascade filtering approach, which has four major stages: (1) identifying potential interest points that are invariant to scale and orientation by searching across all possible scales, (2) fitting a detailed model at each candidate location to select keypoints based on measures of their stability, (3) assigning one or more orientations to each keypoint based on local image gradient directions, and (4) measuring the local image gradients at the selected scale around each keypoint to allow for significant local shape distortion and change in illumination.

2.3. Parametric-Model-Based Methods. The parametric model used by Rohr [29] is an analytic junction model convolved with a Gaussian kernel. Rohr uses an interest point detector which maximizes DET(A) of (1) for A as well as the intersection of line segments to determine the initial values for the model parameters. Deriche and Blaszka [30] develop an acceleration of Rohr's method. They substitute an exponential for the Gaussian smoothing function. Baker et al. [31] propose an algorithm that represents a feature as a densely sampled parametric manifold in a low-dimensional subspace. A feature is detected if the projection of the surrounding intensity values in the subspace lies sufficiently close to the feature manifold. Parida et al. [32] use a deformable template to detect radial partitions. The minimum description length principle determines the optimal number of partitions that best describes the signal.

3. Motivation

Local frequency representation is a useful preprocessing operation for a number of image processing and image exploitation tasks. The local frequency extracted from a given image provides a very useful and robust feature for execution of tasks that involve comparison of image frames for classification purposes and/or combining distinct frames into a single better quality image [33]. In particular, image registration and image fusion are specific tasks that can be executed with greater precision by employing local frequency representations obtained from the component image frames to be registered. Recent research has shown that registration algorithms that implement interest point matching ideas can be tailored to deliver improved performance (more accurate estimation of registration parameters, registration of dissimilar imagery data captured by sensors operating in different modalities, etc.) when the interest points used for matching are obtained from local frequency representations [34].

The following properties of local frequency make it a valuable candidate for an invariant image representation: (1) local frequency estimate, unlike gradient, is relatively invariant to signal energy, in other words, also independent to the image contrast. This makes a notable advantage over ordinary gradient-based edge detector; for example, in some images there are some important features that reside in a region with a quiet small contrast against background and gradient methods will probably fail to detect them efficiently, while local frequency representation can do this work well; (2) local frequency estimation can detect the structure of the scene in the image (high frequency features, that is, edge, ridge, and texture information at the same time). This imposes an advantage over the methods, where the edge and ridge have to be extracted by different operators; (3) the third notable advantage that makes local frequency representations valuable is that it is relatively invariant to illumination changes and has a good localization to the original signal in the spatial domain (thus enjoying structure-preserving properties). Based on the above facts, in this paper local frequency map has been suggested to be employed as an efficient approach for extracting interest points.

Despite the usefulness of these representations, there do not exist computationally efficient procedures for extraction of local frequencies from digital imagery data, especially for large-format images. Given a digital image, computation of local frequency typically involves evaluating the gradient of the phase of the analytic form of the image [33], which in turn is obtained by processing the image with a bank of analytic filters, typically a set of 2D Gabor filters characterized by a set of parameters (spatial frequency, orientation angle, and the filter bandwidth), and each tuned to a specific frequency value or to a range of frequencies. Tuning this set of Gabor filters that can cover the entire frequency space for the given image is critical for obtaining an accurate representation of the local frequency; however, choosing the size of the filter bank is often done in an ad hoc manner and one usually errs on the conservative side by employing a larger than needed set of filters in order to cover the entire range of frequencies. In this paper, we propose a computationally efficient approach for developing the local frequency representation from a given image. In doing so, we formulate a synonymous 2D interpretation of the quadrature filter (Hilbert transformer). Moreover, this new algorithm provides an interesting solution for choosing the size of the quadrature filter adaptively for each pixel based on the image content and hereby ensures the generation of the precise value of the local frequency corresponding to the image intensity at each pixel of an image.

4. Proposed Algorithm

The outline of the interest point extraction algorithm presented in this paper is illustrated in Figure 1 that involves three stages. (1) The first step of the proposed approach for interest point extraction is to formulate a quadrature filter called the Hilbert transformer with customized size, $w \times w$ for each pixel. We consider 8 different sizes of the filter as $w \times w$, for $w = 2k - 1$ and $k = 1, 2, \ldots, 8$. In this process we need to compute local phase for each of the filter window size $w = 2k - 1$ at pixel (x, y), which is represented as $\phi_k(x, y)$. We make use of an image attribute named phase divergence, $\nabla \phi_k$ to select the appropriate size of the filter $\widehat{w}_k(x, y) \times \widehat{w}_k(x, y)$. (2) The next step is to compute the local phase $\widehat{\phi}_k(x, y)$ for the pixel (x, y) at the selected scale $\widehat{k}(x, y)$. Thereafter, the local frequency $\widehat{f}_k(x, y)$ for pixel (x, y) at the selected filter size is estimated by taking the spatial derivative of the local phase. (3) At the final step, a global thresholding is applied to the local frequency image to extract the interest points. Local frequency representation of an image efficiently detects all the important features, even those that reside in the regions with a quiet small contrast against background. Accordingly, at the true feature points, the local frequency layout of an image shows distinctly higher values than at the points which carry less, or no, information. Thus, points where the frequency measure of the image is greater than a pre-defined threshold are set as interest points. The following subsections discuss each step of the algorithm in detail.

4.1. Local Phase Measurement. In order to describe the local phase of a real signal, we will first introduce the concept of analytic signal in one-dimension (1D) which can be generalized to higher dimensions [33]. Given a signal $g(t)$ in 1D, its analytic signal is defined as $g_A(t) = g(t) - jg_H(t)$, where $g_H(t)$ is the Hilbert transform of $g(t)$. $g_H(t)$ is defined as $g_H(t) = (1/\pi t) ** g(t) = (1/\pi) \int_{\infty}^{\infty} (g(\tau)/(\tau - t)) d\tau$, where $**$ denotes 1D linear convolution. For a 1D discrete signal, the Hilbert transform is the convolution of the signal with a discrete filter, h_1, of size $1 \times w$, where

$$h_1(n) = \begin{cases} 1 & \text{if } n = 0 \\ \dfrac{1}{\pi n} & \text{if } n \neq 0,\ n \in \left\{ \dfrac{w-1}{2}, \dfrac{w+1}{2} \right\}. \end{cases} \quad (2)$$

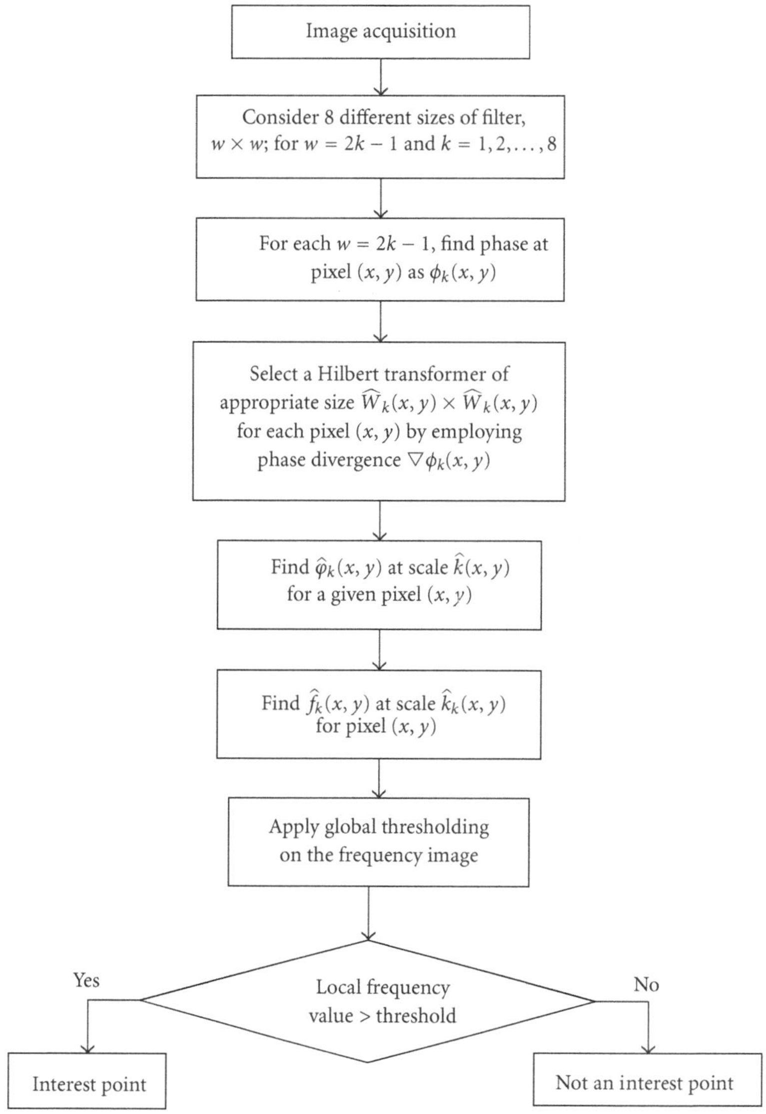

FIGURE 1: Block diagram of the proposed algorithm.

Although the above discussion is given for a one-dimensional signal for the sake of simplicity, it can be generalized to higher dimensions readily. For image-processing applications, evaluation of the local frequency for a given image can be performed using 2D functions.

In two-dimensional (2D), the Hilbert transform is not uniquely defined and the extension of this transform into 2D is not well known in the area of digital image processing [35]. Hence, we have devised an equivalent representation of the 2D version of the quadrature filter (Hilbert transformer) for the computation of local phase of an image. As we mentioned, the 2D discrete quadrature filter, H_2,of size $w \times w$ is defined as, $H_2 = h_1' * h_1$, where $'$ and $*$, respectively, denote matrix transpose and matrix multiplication.

To get the Hilbert transform of discretely sampled image data, we convolve the image $I(x, y)$ with this filter. For the image $I(x, y)$, the corresponding analytic image $I_A(x, y)$ can be defined as $I_A(x, y) = I(x, y) - jI_H(x, y)$, where $I_H(x, y)$ is the Hilbert transformation of $I(x, y)$. The Hilbert transform, $I_H(x, y)$, is defined by $I_H(x, y) = H_2 ** I(x, y)$, where $**$ denotes 2D linear convolution. Thus, the transformation of a given image $I(x, y)$ to the corresponding analytic image $I_A(x, y)$ can be regarded as the result of convolving the real image with a complex filter, such as the Hilbert transformer. The argument of $I_A(x, y)$ is referred to as the local phase of $I(x, y)$, which is defined in the spatial domain. For the window size $w = 2k - 1$, we represent the phase at pixel (x, y) by $\phi_k(x, y)$.

4.2. Scale Selection. We have proposed a new image property called *phase divergence* that is used for selecting the controlling size of the filter window, which is also called the integration scale [21, 36]. Phase divergence is a measure of the extent to which the phase angles in a certain window around a pixel are in the same direction with respect to the phase angle of that pixel. There are other methods that have already been utilized in the literature for scale selection; for example, in [37, 38] the image property called *polarity* for selecting scale is used. In that paper, the polarity is computed in the neighborhood of a given pixel with respect to the dominant orientation of that pixel. The scale selection method in [36] is based on localizing the peak across scale of the determinant or trace of the second moment matrix computed for each pixel within a window. The phase divergence at pixel (x, y) for a given window size $w = 2k - 1$, as we define it, is

$$\nabla\phi_k(x, y) = \frac{|\phi_{k+}(x, y) - \phi_{k-}(x, y)|}{\phi_{k+}(x, y) + \phi_{k-}(x, y)}, \tag{3}$$

where $\phi_{k+}(x, y)$ is the number of phase angles in the window w, that are on the positive side of the phase angle at pixel (x, y) and $\phi_{k-}(x, y)$ is the number of phase angles in the window, w, that are on the negative side of the phase angle at pixel (x, y).

The phase divergence, $\nabla\phi_k$, ranges from 0 to 1 and it varies as the size of the window, w, changes. By computing the phase divergence at all pixels of the image for $w = 2k - 1$, $k = 1, 2, \ldots, 8$, we produce a stack of phase divergence images across scale, k. Then, the phase divergence image at scale k is smoothed by convolving with a Gaussian filter of size, w. Finally the scale is selected based on the derivative of the phase divergence with respect to scale. For a given pixel (x, y), the scale $\hat{k}(x, y)$ is selected as the first value of $k(x, y)$ for which the difference between values of phase divergence at successive scales $(\nabla\phi_k(x, y) - \nabla\phi_{k-1}(x, y))$ is less than 1 percent. In the uniform regions of an image, the selected scale should be 1, because uniform region appears not to change across scale. A region can be declared to be uniform if the mean contrast (standard deviation of intensity) of that region across scale is less than 0.1 [37].

4.3. Frequency Selection. Local frequency can be estimated by using quadrature filters [33, 34]. Given a real signal, its corresponding analytic signal is complex with the real part being the original signal itself and the imaginary part being its Hilbert transform obtained by convolving the signal with a quadrature filter. As discussed before, the argument of the analytic signal is referred to as the local phase of the original signal. The spatial derivative of the local phase gives the local frequency of an image. For the scale k we denote the frequency at pixel (x, y) as $f_k(x, y)$, which is estimated by computing the gradient of the phase values in the window of size w.

The gradient image is simply the derivative of the local image values. Mathematically, the gradient of a two-variable function (here the image intensity function) is a 2D vector at each image point with the components given by the derivatives in the horizontal and vertical directions. At each image point, the gradient vector points in the direction of largest possible intensity increase, and the length of the gradient vector corresponds to the rate of change in that direction.

To estimate local frequency $f_k(x, y)$, we use the most common method of computing the gradient of the local phase image which is the simple local derivative

$$\begin{aligned} \text{grad}(\phi_k(x, y)) &= \left(\frac{\partial\phi_k(x, y)}{\partial x}, \frac{\partial\phi_k(x, y)}{\partial y}\right) \\ &= \left(\nabla\phi_{kx}(x, y), \nabla\phi_{ky}(x, y)\right) \\ f_k(x, y) &= \sqrt{\nabla\phi_{kx}^2(x, y) + \nabla\phi_{ky}^2(x, y)}, \end{aligned} \tag{4}$$

where $\nabla\phi_{kx}(x, y) = \partial\phi_k(x, y)/\partial x = \phi_k(x+1, y) - \phi_k(x, y)$ and $\nabla\phi_{ky}(x, y) = \partial\phi_k(x, y)/\partial y = \phi_k(x, y+1) - \phi_k(x, y)$ are the numerical gradients (differences) in the x and y directions, respectively.

The local frequency images are computed for $w = 2k - 1, k = 1, 2, \ldots, 8$, and thus there are a stack of local frequency images across scale, k. After selecting the appropriate scale $\hat{k}(x, y)$ for a given pixel (x, y), we compute frequency $f_{\hat{k}}(x, y)$ at that scale, and finally we have the accurate value of the local frequency at each pixel of the image.

4.4. Interest Point Determination. The local frequency image representation is found to be stable to describe and capture all the meaningful and salient features of an image. For an image, which is a kind of nonstationary signal, the local frequency is an important characteristic; it is a time-varying parameter which defines the location of the signals spectral peak as it varies with time. Conceptually, at the edges, ridges and corners (they are the dominant features of an image and associated with high-frequency components), the local frequency representation has high values.

As an example for a one-dimensional signal the local frequency estimation is demonstrated in Figure 2. Figure 2(a) is the original one-dimensional signal. It contains both edges and ridges with different contrasts. Figure 2(b) shows the final local frequency representation at the selected scale $\hat{k}(n)$ customized for each point n of that 1D signal.

To evaluate the performance resulting from the present approach for extracting local frequency representation from an image data and to establish its robustness to scene details, we conducted experiments on images, and the results are presented in Figure 3. For the reference image shown in Figure 3(a), Figure 3(b) shows the local frequency representation at the selected scale $\hat{k}(x, y)$ customized for each point (x, y) of that 2D signal in Figure 3(a). As we have seen for 1D signal, once again the local frequency representation given in Figure 3(b) clearly displays the edges, ridges, and corners which are extracted perfectly by the algorithm. These results serve to illustrate the versatility of the approach used here to process 1D or 2D signals and the robustness, accuracy, and utility of the algorithm to extract local frequency information.

Three noticeable characteristics are demonstrated in this experiment: (1) the local frequency representation yields outputs at the edges, ridges, and corners of a signal; (2) the local frequency, unlike gradient, is relatively independent of signal contrast; (3) the localization of the detector is quiet faithful to the original signal. The latter property is due

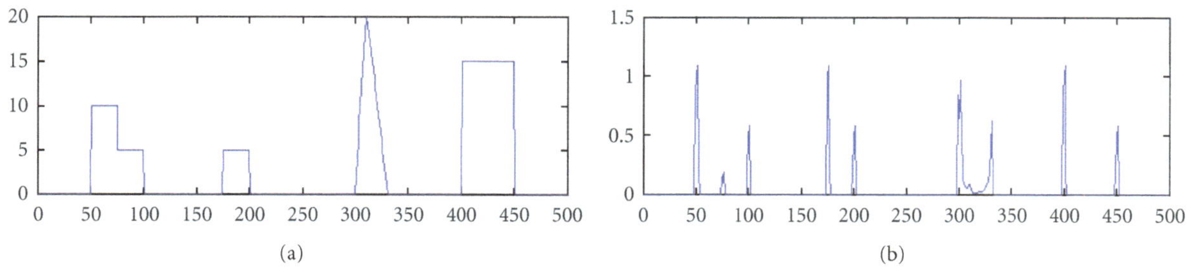

FIGURE 2: (a) A 1D signal and (b) local frequency map.

FIGURE 3: (a) A 2D image and (b) local frequency map.

to the stability of phase information [9]. The scale space properties of local frequency estimation, as shown in Figures 2 and 3, have long been exploited in the computer vision community both from a computational and localization of feature point of view. The scale-space, that is, the multiresolution properties of local frequency estimation, leads to computationally efficient algorithms and alleviates the problem of local minima entrapment when used in the context of interest point detection problem as described below.

For the selection of the interest or feature points, the local frequency image, $\hat{f}_k(x, y)$, is thresholded. A global non-histogram-based thresholding technique has been incorporated rather than local (adaptive) thresholding in this step [39]. The threshold level, γ, is determined by $\gamma = \sum(\hat{f}_k(x, y) \cdot s)/\sum s$, where "$\cdot$" denotes pixel wise multiplication and s is given by $s = \max(|g_1 ** \hat{f}_k(x, y)|, |g_2 ** \hat{f}_k(x, y)|)$, where $g_1 = [-101]$, $g_2 = [-101]^T$, and "$**$" denotes two-dimensional linear convolution. The binary image $I_{bf}(x, y)$ is then given by

$$I_{bf}(x, y) = \begin{cases} 1 & \text{if } \hat{f}_k(x, y) > \gamma, \\ 0 & \text{otherwise.} \end{cases} \tag{5}$$

A thinning algorithm is applied to this binary image yielding $I_f(x, y)$, which contains the extracted interest points. The algorithm for finding true interest points makes three passes through the thinned local frequency image. First, points are selected if the value of the frequency exceeds a threshold determined globally. Second, the selected points that are not locally maximum in the original intensity image in its 3×3 neighborhood are deleted. In the final pass, the subset of pixels are kept such that the minimum distance between any pair of points is larger than a given threshold.

Consider an image of size $M \times N = n_1$ pixels such that $F_1 = \{p_i = (x_i, y_i);\ i = 1, 2, 3 \ldots n_1\}$ contains all the pixel coordinates. Let F_2 be the first selection of the feature points from the local frequency image, which forms the set $F_2 \subset F_1$

$$\begin{aligned} F_2 &= \left\{p_i : I_f(x_i, y_i) = 1\right\} \\ &= \{p_i = (x_i, y_i);\ i = 1, 2, 3, \ldots, n_2 : n_2 < n_1\}, \end{aligned} \tag{6}$$

where n_1 is the number of all pixels and n_2 is the number of selected points after discarding unwanted points by thresholding.

At the next two passes, from the points in F_2 we retain those points that are locally maximum in their $W_m = 3 \times 3$ neighborhood with the restriction that the distance between any two feature points is larger than a given threshold (this is set to 5 pixels in our experiment). Thus, $F_3 \subset F_2$ is the set of feature points that are locally maximum in the frequency image, $\hat{f}_k(x, y)$, and $F_f \subset F_3$ is the final set of feature points after discarding those closely spaced points

$$\begin{aligned} F_3 &= \left\{p_i = (x_i, y_i) : \hat{f}_k(x_i, y_i) = \max_{\{j \in i - W_m/2,\, i + W_m/2\}} \hat{f}_k\left(x_j, y_j\right)\right\} \\ F_3 &= \{p_i = (x_i, y_i);\ i = 1, 2, 3, \ldots, n_3 : n_3 < n_2 < n_1\}, \end{aligned} \tag{7}$$

where n_3 is the number of selected points after discarding redundant points from F_2

$$F_f = \{p_i : |p_i - p_{i-1}| > 5\,\text{pix},\ |p_i - p_{i+1}| > 5\ \text{pix}\},$$
$$F_f = \left\{p_i = (x_i, y_i);\ i = 1, 2, 3, \ldots, n_f : n_f < n_3 < n_2 < n_1\right\}, \tag{8}$$

where n_f is the number of selected points after discarding redundant points from F_3.

Following the above procedure, we select the interest points from an image.

5. Experimental Results

In this section, we provide the results of experiments which are conducted to test the efficacy of the proposed interest point detection algorithm. In order to test the immunity of our algorithm to transformation, the original images are scaled, rotated, and sheared. We evaluate the proposed detector using the repeatability rate for image rotation and scale change. Besides these, the repeatability rates of five interest point detectors are compared with our method under different image rotation and scale changes.

For comparison we have chosen five detectors such that those detectors have already been reported to offer good performance. Among the five chosen detectors Harris', Lowe's, and Tomasi's methods are intensity-based methods. These are chosen because Harris' method has been reported to be better than any other detector, Lowe's algorithm also known as scale-invariant feature transform (SIFT) is the best scale invariant detector and Tomasi's detector is the best for tracking applications. The other two detectors are chosen because (1) those are contour-based methods like our method and (2) those methods use wavelet decomposition.

Simulation results of the five methods on real image are presented in Figure 4. For the reference image shown in Figure 3(a), feature points extracted by Harris' approach, Lowe's procedure, Tomasi' method, Loupias' technique, and Yeh's algorithm are shown in Figures 4(a), 4(b), 4(c), 4(d), and 4(e), respectively. The points detected by the presented method are given in Figure 4(f). From this figure it can be seen that points selected by the proposed method cover all the curvatures of object boundaries and yield the true corner points.

For the evaluation of rotation invariance of the detector, Figures 5 and 6 show the detection results for two rotated versions of the reference image. Figures 5(f) and 6(f) give the results of our method, where the rotation angle for the images in Figure 5 is 40° and for the images in Figure 6 is 110°. The performance of Harris', Lowe's, Tomasi's, Loupias', and Yeh's methods are given in Figures 5(a), 5(b), 5(c), 5(d), and 5(e), respectively, for the rotation 40° and in Figures 6(a), 6(b), 6(c), 6(d), and 6(e), respectively, for the rotation 110°. From the figure, it can be said that Harris's, Lowe's, and the proposed methods give the best result for both rotations. The performance of Tomasi's technique is better than Loupias' and Yeh's methods.

The effect of image scale change on detection result is tested and demonstrated in Figures 7 and 8. The detected points by the proposed technique are shown in Figures 7(f) and 8(f), where the scale change for Figure 7 is 1.5 and for Figure 8 is 3.4. Points detected by Harris', Lowe's, Tomasi', Loupias' and Yeh's methods are presented in Figures 7(a), 7(b), 7(c), 7(d), and 7(e), respectively, for the scale change 1.5 and in Figures 8(a), 8(b), 8(c), 8(d), and 8(e) respectively for the scale change 3.4. It can be seen from the figure that all the methods are scale invariant.

To evaluate the functioning of all of the five detectors as affine-invariant systems, nonuniform scaling is applied in some directions to have shearing in the reference image shown in Figure 3(a). Figure 9(f) displays the feature points detected in the sheared image by the proposed method. The results of detection by Harris', Lowe's, Tomasi's, Loupias', and Yeh's methods are presented in Figures 9(a), 9(b), 9(c), 9(d), and 9(e), respectively. By examining the figure it can be said that the proposed method performs satisfactorily for detecting interest points from the sheared image as well.

From the figures it can be concluded that the proposed method can be used as an affine- and scale-invariant detector. To complement the subjective evaluations, we present a qualitative performance evaluation of the proposed affine-invariant detector. The stability and accuracy of the detectors are evaluated using the repeatability criterion [1]. The repeatability score for a given pair of images is the average number of corresponding points detected in those images under different geometric and photometric transformations. We take into account only the points located in the part of the scene present in both images. The repeatability rate is measured for a localization error of 1.5 pixels; that is the detected point lies in a very close pixel neighborhood of the predicted point. Measuring the repeatability rate within 1.5 pixels or less, the probability that two points are accidentally within the error distance is negligible.

We first compare the detectors for image rotation followed by scale change. The repeatability rate as a function of the angle of image rotation is displayed in Figure 10(a). The rotation angles vary between 0° and 180°. Under repeatability, Lowe's methods give the best results for all rotations, where this algorithm obtains a repeatability rate of about 78%. From the observation of the plot of the repeatability rate with image rotation, the proposed technique does not outperform the SIFT algorithm, but it yields the same performance as Harris' method with a repeatability rate of about 75% for all rotations. Notably, it offers better performance than Tomasi's, Loupious', and Yeh's techniques, where these approaches offer a repeatability rate of about 60%, 45%, and 50%, respectively.

Figure 10(b) shows the repeatability rate as a function of scale changes. The scale varies from 1 to 4 for the plot of repeatability rate as a function of scale. The results show that all the detectors are scale sensitive except Lowe's method. As the name implies, the SIFT algorithm proposed by Lowe offers the best performance with scale change. This method is the least dependent on the change in scale. Harris', Tomasi's, and the proposed detectors give reasonable results, with the repeatability rate as a decreasing function of scale change. Loupious' and Yeh's methods are very sensitive to scale change, and the results of these methods are hardly usable.

(a) (b) (c)

(d) (e) (f)

Figure 4: Feature points detected by (a) Harris' method, (b) Lowe's method, (c) Tomasi's method, (d) Loupias' technique, (e) Yeh's algorithm, and (f) proposed technique.

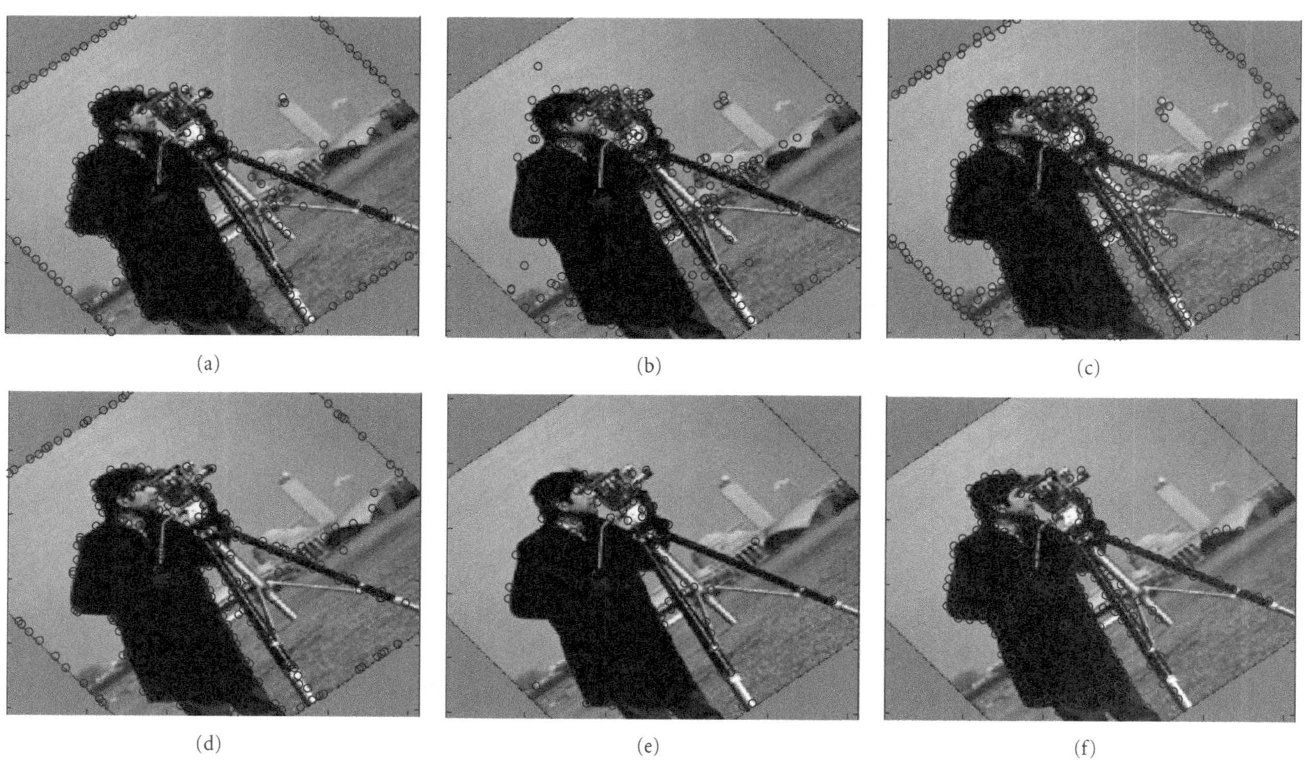

(a) (b) (c)

(d) (e) (f)

Figure 5: Feature points detected in 40° rotated image by (a) Harris' method, (b) Lowe's method, (c) Tomasi's method, (d) Loupias' technique, (e) Yeh's algorithm, and (f) proposed technique.

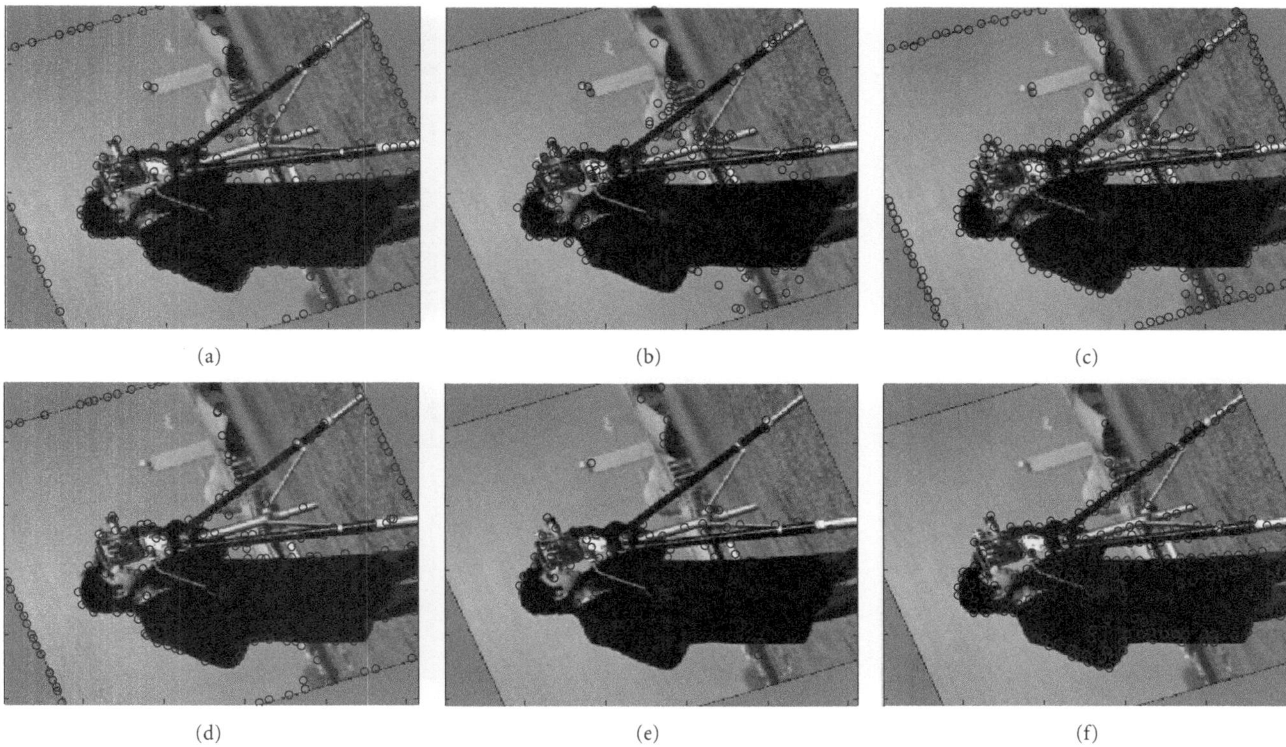

FIGURE 6: Feature points detected in 110° rotated image by (a) Harris' method, (b) Lowe's method, (c) Tomasi's method, (d) Loupias' technique, (e) Yeh's algorithm, and (f) proposed technique.

FIGURE 7: Feature points detected in 1.5 times scaled image by (a) Harris' method, (b) Lowe's method, (c) Tomasi's method, (d) Loupias' technique, (e) Yeh's algorithm, and (f) proposed technique.

Figure 8: Feature points detected in 3.4 times scaled image by (a) Harris' method, (b) Lowe's method, (c) Tomasi's method, (d) Loupias' technique, (e) Yeh's algorithm, and (f) proposed technique.

Figure 9: Feature points detected in sheared image by (a) Harris' method, (b) Lowe's method, (c) Tomasi's method, (d) Loupias' technique, (e) Yeh's algorithm, and (f) proposed technique.

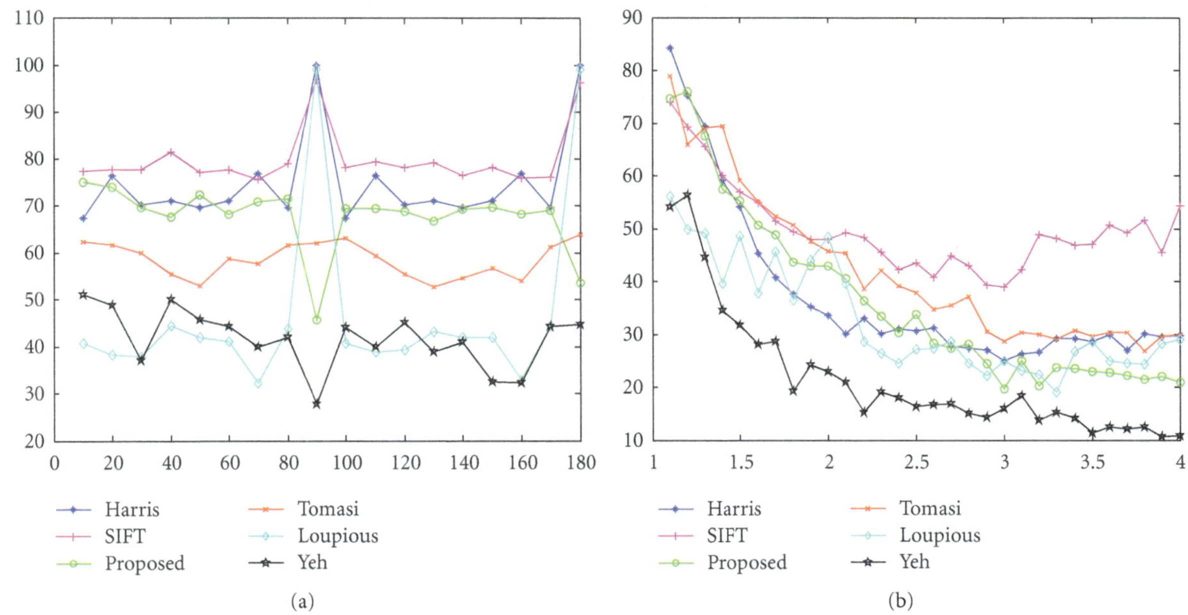

FIGURE 10: Plot for the repeatability rate as a function of (a) rotation angle and (b) change in scale.

FIGURE 11: Feature points detected in the second image by (a) Harris' method, (b) Lowe's method, (c) Tomasi's method, (d) Loupias' technique, (e) Yeh's algorithm, and (f) proposed technique.

Above a scale factor of about 2, the results from all the detectors, except Lowe's method, are mainly due to artifacts, because at larger scales many more points are found in the textured regions of the scene, so accidental correspondences are more likely.

For the evaluation of detection performance, the feature points extracted by the proposed method and the five other algorithms are presented for one more image in Figure 11, where the detected points are superimposed on the original image to evaluate the interest points location. From this

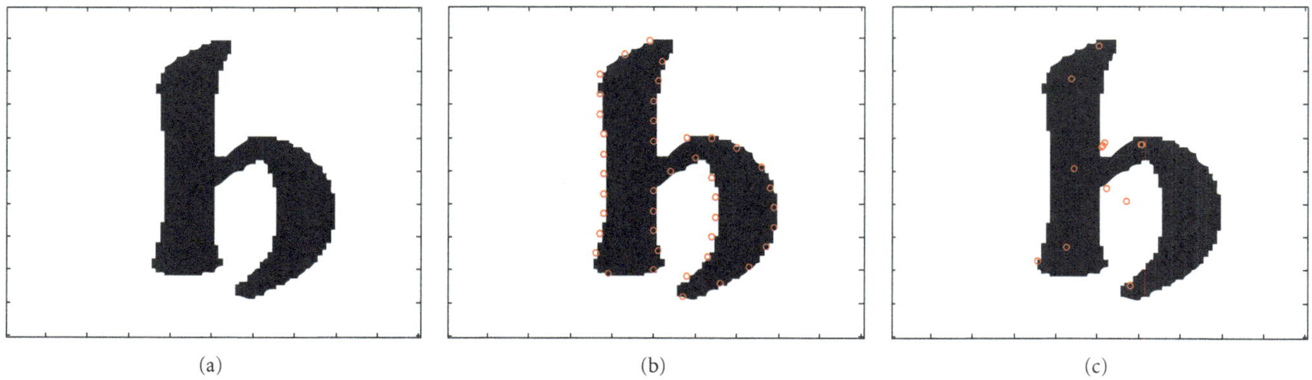

FIGURE 12: (a) Interest points detected from the binary image of letter "h" by (b) the proposed algorithm and (c) the SIFT algorithm.

figure it is observed that the proposed method extract points where variations occur in the image, that is, where the image information is supposed to be the most important. Additionally, the set of detected interest points are not cluttered in few regions but rather spread out at different parts of the image. Most importantly, the proposed method covers all the curvatures of object boundaries and yields at the true features i.e., the edges, ridges, and corners. Accordingly, the extracted points detect the structure of the scene and lead to a complete image representation.

From the parallel observation of the figures of detection results and the value of calculated repeatability rate, it can be claimed that the proposed method compares favorably against other well-known methods. Based on the plot for the repeatability rate, like Harris' and Lowe's algorithms, the proposed technique is minimally dependent on the rotations, which is a desirable and attractive characteristic for any feature point detector. Though the proposed technique does not outperform the SIFT algorithm, on average, it offers performance similar to Harris' method and it yields better results than Tomasi's, Loupious' and Yeh's techniques. Notably, it offers better performance than the other two contour-based methods: Loupious' and Yeh's approaches.

Therefore, as a contour-based technique, the presented approach can be expected to perform well for the applications where true corner points are needed to be detected from the edges, ridges, and corners for further processing. As an example, for a binary image, interest points should not be found in the uniform region of constant intensity, that is, either in the background or foreground. They would rather lie only on the edges. As shown in Figure 12, for the binary image of the letter "h," the SIFT algorithm detects feature points in the uniform region. But the proposed method, as a contour-based technique, extracts interest points only from the edges, which signifies the performance variation of different algorithms depending on the types of image and/or applications. From the simulation results, it can be concluded that the proposed method can be utilized as a dependable affine-invariant interest point detector.

In addition to the accuracy, the speed, in other words, the computational cost of the proposed algorithm is measured. The interest point detection process takes 25 seconds for an image of size 100 × 100 pixels. The processing time is recorded based on the simulation of MATLAB code on a 2.59 GHz Pentium IV machine. For the same image, the computation time for Harris', Lowe's, Tomasi's, Loupias' and Yeh's methods is 0.8 seconds, 1.8 seconds, 2.2 seconds, 3.6 seconds, and 1.4 seconds, respectively. We are working on making the proposed interest point detection scheme faster by parallel processing of an image with the employment of digital signal processor and C++ programming language.

6. Conclusion

This research presents a robust, rotation-invariant, and scale-invariant corner detection scheme on image based on the local or instantaneous frequency representation of an image. The image is convolved with the Hilbert transformer customized for each pixel to calculate the local phase image. We have formulated a scale selection technique based on the image property phase divergence, which is used to select the controlling size of the filter window. The phase image is differentiated to obtain the local frequency representation of the image. By the employment of a global thresholding to the frequency image, the proposed technique localizes all the true corner points on edges, ridges, and corners. Under appropriate image resolution and region of support, the proposed approach precisely captures the true corner points and is free from the false alarms on circular arcs for both simple and complicated objects in varying rotation and scale conditions.

The proposed technique detects feature points globally from the whole image at a time based on the value of the frequency at a given pixel, and the detected interest points are largely independent of the imaging conditions; therefore, points are geometrically stable. This approach has strong robustness of detecting good feature points, as it detects feature points from the regions with even a quiet small contrast against background. Experimental results also suggest that the proposed local-frequency-based corner detection approach is stable and efficient. The proposed method is a generic concept and can find its application in many matching and recognition problems.

In future work, we want to tackle a number of challenges for further improvement in the detection of interest points. In this respect, we are planning to address the following issues: (1) to improve the repeatability rate with both rotation and scaling, (2) to demonstrate the application of the proposed interest point detector for object recognition, and (3) to reduce the computation cost by parallel processing and implementing the algorithm in C++.

References

[1] C. Schmid, R. Mohr, and C. Bauckhage, "Evaluation of interest point detectors," *International Journal of Computer Vision*, vol. 37, no. 2, pp. 151–172, 2000.

[2] J. Canny, "A computational approach to edge detection," *IEEE Transactions on Pattern Analysis and Machine Intelligence*, vol. 8, no. 6, pp. 679–698, 1986.

[3] A. Bandera, C. Urdiales, F. Arrebola, and F. Sandoval, "Corner detection by means of adaptively estimated curvature function," *Electronics Letters*, vol. 36, no. 2, pp. 124–126, 2000.

[4] A. Rattarangsi and R. T. Chin, "Scale-based detection of corners of planar curves," *IEEE Transactions on Pattern Analysis and Machine Intelligence*, vol. 14, no. 4, pp. 430–449, 1992.

[5] R. Horaud, T. Skordas, and F. Veillon, "Finding geometric and relational structures in an image," in *Proceedings of the 1st European Conference on Computer Vision*, pp. 374–384, Antibes, France, 1990.

[6] E. Shilat, M. Werman, and Y. Gdalyahu, "Ridge's corner detection and correspondence," in *Proceedings of the IEEE Computer Society Conference on Computer Vision and Pattern Recognition*, pp. 976–981, Puerto Rico, Puerto Rico, USA, June 1997.

[7] F. Mokhtarian and R. Suomela, "Robust image corner detection through curvature scale space," *IEEE Transactions on Pattern Analysis and Machine Intelligence*, vol. 20, no. 12, pp. 1376–1381, 1998.

[8] A. Pikaz and I. H. Dinstein, "Using simple decomposition for smoothing and feature point detection of noisy digital curves," *IEEE Transactions on Pattern Analysis and Machine Intelligence*, vol. 16, no. 8, pp. 808–813, 1994.

[9] A. Bruce and H. Y. Gao, *Applied Wavelet Analysis with SPLUS*, Springer, New York, NY, USA, 1996.

[10] J. S. Lee, Y. N. Sun, and C. H. Chen, "Multiscale corner detection by using wavelet transform," *IEEE Transactions on Image Processing*, vol. 4, no. 1, pp. 100–104, 1995.

[11] C. H. Chen, J. S. Lee, and Y. N. Sun, "Wavelet transformation for gray-level corner detection," *Pattern Recognition*, vol. 28, no. 6, pp. 853–861, 1995.

[12] A. Quddus and M. M. Fahmy, "Fast wavelet-based corner detection technique," *Electronics Letters*, vol. 35, no. 4, pp. 287–288, 1999.

[13] A. Quddus and M. Gabbouj, "Wavelet-based corner detection technique using optimal scale," *Pattern Recognition Letters*, vol. 23, no. 1–3, pp. 215–220, 2002.

[14] J. P. Hua and Q. M. Liao, "Wavelet-based multi scale corner detection," in *Proceedings of the 5th International Conference on Signal Processing*, pp. 341–344, 2000.

[15] E. Loupias, N. Sebe, S. Bres, and J. M. Jolion, "Wavelet-based salient points for image retrieval," in *Proceedings of the IEEE Conference on Image Processing (ICIP '00)*, pp. 518–521, September 2000.

[16] Q. Tian, N. Sebe, M. S. Lew, E. Loupias, and T. S. Huang, "Image retrieval using wavelet-based salient points," *Journal of Electronic Imaging*, vol. 10, no. 4, pp. 835–849, 2001.

[17] C. H. Yeh, "Wavelet-based corner detection using eigenvectors of covariance matrices," *Pattern Recognition Letters*, vol. 24, no. 15, pp. 2797–2806, 2003.

[18] H. P. Moravec, "Towards automatic visual obstacle avoidance," in *Proceedings of the 5th International Joint Conference on Artificial Intelligence*, p. 584, Cambridge, Mass, USA, 1977.

[19] P. R. Beaudet, "Rotationally invariant image operators," in *Proceedings of the 4th International Joint Conference on Pattern Recognition*, pp. 579–583, Tokyo, Japan, 1978.

[20] L. Kitchen and A. Rosenfeld, "Gray-level corner detection," *Pattern Recognition Letters*, vol. 1, no. 2, pp. 95–102, 1982.

[21] W. Forstner, "A framework for low level feature extraction," in *Proceedings of the 3rd European Conference on Computer Vision*, pp. 383–394, Stockholm, Sweden, 1994.

[22] C. Harris and M. Stephens, "A combined corner and edge detector," in *Proceedings of the Alvey Vision Conference*, pp. 147–151, 1988.

[23] C. Tomasi and T. Kanade, "Detection and tracking of point features," Tech. Rep. CMU-CS-91-132, Carnegie Mellon University, 1991.

[24] J. Shi and C. Tomasi, "Good features to track," in *Proceedings of the IEEE Computer Society Conference on Computer Vision and Pattern Recognition*, pp. 593–600, Seattle, Wash, USA, June 1994.

[25] F. Heitger, L. Rosenthaler, R. von der Heydt, E. Peterhans, and O. Kubler, "Simulation of neutral contour mechanisms: from simple to end-stopped cells," *Vision Research*, vol. 32, no. 5, pp. 963–981, 1992.

[26] S. M. Smith and J. M. Brady, "SUSAN—a new approach to low level image processing," *International Journal of Computer Vision*, vol. 23, no. 1, pp. 45–78, 1997.

[27] R. Laganiere, "Morphological corner detection," in *Proceedings of the 6th IEEE International Conference on Computer Vision*, pp. 280–285, Bombay, India, January 1998.

[28] D. G. Lowe, "Distinctive image features from scale-invariant keypoints," *International Journal of Computer Vision*, vol. 60, no. 2, pp. 91–110, 2004.

[29] K. Rohr, "Recognizing corners by fitting parametric models," *International Journal of Computer Vision*, vol. 9, no. 3, pp. 213–230, 1992.

[30] R. Deriche and T. Blaszka, "Recovering and characterizing image features using an efficient model based approach," in *Proceedings of the IEEE Computer Society Conference on Computer Vision and Pattern Recognition*, pp. 530–535, New York, NY, USA, June 1993.

[31] S. Baker, S. K. Nayar, and H. Murase, "Parametric feature detection," *International Journal of Computer Vision*, vol. 27, no. 1, pp. 27–50, 1998.

[32] L. Parida, D. Geiger, and R. Hummel, "Junctions: detection, classification, and reconstruction," *IEEE Transactions on Pattern Analysis and Machine Intelligence*, vol. 20, no. 7, pp. 687–698, 1998.

[33] G. Granlund and H. Knutsson, *Signal Processing for Computer Vision*, Kluwer Academic, Dodrecht, The Netherlands, 1995.

[34] J. Liu, "Regularized quadrature filters for local frequency estimation: application to multimodal volume image registration," in *Proceedings of the Conference on Vision Modeling and Visualization*, pp. 507–514, November 2001.

[35] M. R. Arnison, C. J. Cogswell, N. I. Smith, P. W. Fekete, and K. G. Larkin, "Using the Hilbert transform for 3D

visualization of interference contrast microscope images," *Journal of Microscopy*, vol. 199, no. 1, pp. 79–84, 2000.

[36] J. Gårding and T. Lindeberg, "Direct computation of shape cues using scale-adapted spatial derivative operators," *International Journal of Computer Vision*, vol. 17, no. 2, pp. 163–191, 1996.

[37] C. Carson, S. Belongie, H. Greenspan, and J. Malik, "Blobworld: image segmentation using expectation-maximization and its application to image querying," *IEEE Transactions on Pattern Analysis and Machine Intelligence*, vol. 24, no. 8, pp. 1026–1038, 2002.

[38] T. Leung and J. Malik, "Detecting, localizing and grouping repeated scene elements from an image," in *Proceedings of the European Conference on Computer Vision*, pp. 546–555, 1996.

[39] S. U. Lee, S. Yoon Chung, and R. H. Park, "A comparative performance study of several global thresholding techniques for segmentation," *Computer Vision, Graphics and Image Processing*, vol. 52, no. 2, pp. 171–190, 1990.

15

An Efficient Renewable Energy Management and Sharing System for Sustainable Embedded Devices

Ting Zhu[1] and Chang Zhou[2]

[1] *Department of Computer Science, State University of New York, Binghamton, NY 13905, USA*
[2] *College of Information Engineering, China Jiliang University, Hangzhou, Zhejiang 310018, China*

Correspondence should be addressed to Ting Zhu, tzhu@binghamton.edu

Academic Editor: Qinru Qiu

We present the design and evaluation of a capacitor-driven efficient renewable energy management and sharing system, called REMS system, that is designed to replace battery-based energy storage units. We first identify potential applications and requirements of the REMS system. Then we describe the design of REMS system which aims at minimizing the energy leaked away and providing accurate and fine-grained energy sharing between embedded devices. We extensively evaluate our system under different real-world settings. Results indicate that our charging and discharging control can effectively minimize the energy leaked away. Moreover, the energy sharing design can efficiently share renewable energy among multiple devices and significantly extend the lifetime of a system.

1. Introduction

With the increasing need for ubiquitous computing, embedded devices are heavily involved in a lot of applications such as data collection [1, 2], weather condition monitoring [3], road service monitoring [4], online game [5], VoIP service [6], mobile payment [7], and social activities [8]. Due to the mobility requirement, embedded devices are normally equipped with limited and rechargeable energy storage units, such as rechargeable batteries.

However, the electro-chemical conversion inside the rechargeable batteries during the charging stage significantly reduces the energy efficiency to as low as 6% [9]. Moreover, rechargeable batteries have limited recharge cycles due to cyclic memory and crystalline formation (e.g., a Li-ion battery has 500 cycles and NiMH 300 cycles), which translates to a lifetime of 2 to 3 years. Given the huge amount of the embedded devices on the market, the relative short lifetime of rechargeable batteries results in a huge amount of disposed batteries each year. The chemical materials (e.g., nickel, mercury, etc.) inside these batteries can cause very serious environmental problem. Moreover, the innovation in lithium-ion battery capacity has essentially stalled for more than 4 years, with capacity improvement increasing by less than 5% per year [10].

Compared with rechargeable batteries, ultracapacitors possess a set of advantages: they (i) have high charge efficiency (i.e., more than 90%); (ii) have more than 1 million recharge cycles which translates to a lifetime of more than 10 years; (iii) can be charged very fast; (iv) are made from carbon nanotubes which does not have toxic-waste-disposal problems when the capacitors reach the end of their lifetime. Furthermore, recent advances in ultracapacitor technology make it possible to use ultracapacitors as the *only* energy storage device. For instances, the research groups at MIT [11] and the University of Maryland [12] have announced nanotube-based ultracapacitors, which can provide energy storage densities comparable to those of batteries. In 2006, a US patent [13] was issued for an electrical energy storage unit using an ultracapacitor that has an energy/weight value of about 342 W·h/kg, twice that of Li-ion batteries. The largest capacitance available on the market increased from 3,000 F [14] to 5,000 F [15] in 2009. Powered by a 5,000 F capacitor, a mobile phone can work for more than 11 days under standby mode with a single initial charge of the capacitor. It is highly possible that in the near future we will witness a paradigm shift from a battery-based to an ultracapacitor-based design for all kinds of embedded devices. Therefore, it is essential to explore this frontier in advance.

To utilize the aforementioned advantages of ultracapacitors, we designed a capacitor-driven renewable energy management and sharing system, called *REMS* system. The system can be charged within seconds and have high charge efficiency. Moreover, the systems can conduct fine-grained and efficient energy sharing among themselves. Given limited energy inside each individual embedded device, we believe that energy sharing between embedded devices to efficiently utilize the energy is the future trend. On October 22, 2009, the International Telecommunication Union announced a standard for a universal charger for mobile handsets [16], which provides a general hardware interface for applying our energy sharing approach.

In summary, our contributions are as follows.

(i) To our knowledge, energy sharing is a new concept that can effectively utilize the limited energy and improve the performance of multiple embedded devices as a whole. With fine-grained and efficient energy sharing, energy can quantitatively flow back and forth from one device to another device.

(ii) Our hardware and control layer design is simple and general. It can easily be extended to support n capacitors. Furthermore, our design affords simultaneous energy sharing among multiple embedded devices.

(iii) We evaluate our system extensively in multiple real-world testbeds and simulation. The results indicate that our system can effectively minimize the energy leaked away and provide accurate and efficient energy sharing among multiple systems.

The paper is organized as follows. Section 2 examines why energy sharing is necessary for embedded devices and discusses the need and the issue for capacitor-driven energy storage system. Section 3 gives an overview of the system architecture. Hardware design, control and network layers are presented in Sections 4, 5 and 6, respectively. System implementation and evaluation are detailed in Sections 7 and 8. Related work is discussed in Section 9. Finally, Section 10 concludes the paper.

2. Motivation

Although ultracapacitors have some advantages over rechargeable batteries, they also have a major limitation (i.e., leakage) which introduce a design challenge for capacitor-driven energy storage system. In this section, we (i) investigate the leakage property of ultracapacitors, (ii) show the benefit of the capacitor-driven design by studying two real-world scenarios, and (iii) discuss the requirements of enabling such benefit.

2.1. The Challenge for Capacitor-Driven Design. Compared with rechargeable batteries, ultracapacitors have higher leakage current which is only crudely specified in the data sheets and introduce a challenge to the system design. To further investigate the leakage property of the ultracapacitor, we conducted experiments over a period of 6 weeks using ten types of ultracapacitors, ranging from 22 F to 3000 F. After an ultracapacitor was fully charged, we isolated it and continuously monitored its remaining voltage over 1000 hours.

Figure 1(a) shows that leakage is more severe for larger ultracapacitors than for smaller ultracapacitors. For example, after 1000 hours, the energy leaked away inside a 3,000 F capacitor is about 83 times of the energy leaked away inside a 22 F. Figure 1(a) also reveals that the ultracapacitors have very large amount of energy (e.g., 22.9% of total energy for 3000 F) leaked away during the first 10 hours, but little energy (e.g., 21.7% for 3000 F) leaked away during the rest 990 hours. These results indicate that ultracapacitors have leakage performance that is comparable to that of batteries (e.g., 15% per month for NiMH battery [17]) when voltage is controlled under an appropriate level.

However, ultracapacitors suffer severe energy leakage when it is charged to the limit. We investigated the correlation between the voltage and leakage of ultracapacitors. Figures 1(b) and 1(c) show that when the voltage of ultracapacitors approaches its limit, the value of leakage power increases significantly. For example, when the voltage of the 3000F ultracapacitor increases to 2.7 V (shown in Figure 1(c)), its leakage power is 145.2 mW, which is 5 times of the power needed for powering a Cingular 2125 smartphone to work under GSM Idle scenario [6].

Our empirical study concludes that (i) ultracapacitor is a good substitute for battery and (ii) energy leakage in ultracapacitors should not be overlooked.

2.2. The Benefit for Capacitor-Driven Design. Except the leakage issue, capacitor-driven energy storage system enables a new style of energy management: *energy sharing* which allows energy flow back and forth among multiple capacitors which belong to different energy storage systems. Energy sharing will be of benefit to a lot of applications, such as wearable computing [18, 19], portable weather monitoring system [3], and robot-sensor networks [1, 20, 21]. Due to the space constraint, we only study two real-world scenarios in this section.

Scenario 1. In the wearable computing application, Dr. Abowd's team [22] *has empirically studied the expected harvesting power from six body locations. Their results indicate that different devices have significantly different harvesting power* (e.g., 115 ± 106 μW, 1.01 ± 0.46 mW, *and* 4.9 ± 3.63 mW *for the wristwatch, the cell phone, and the shoe, resp.). Moreover, the harvesting power varies over time.*

In this scenario, we see the temporal and spatial diversity of the harvesting power. However, the energy harvested by these devices may not meet their energy consumption requirement. For example, the cell phone may have the highest energy consumption, while the shoe may have the lowest energy consumption. If we enable energy sharing among these devices, the energy generated from the shoe can be utilized to power the cell phone. Moreover, the

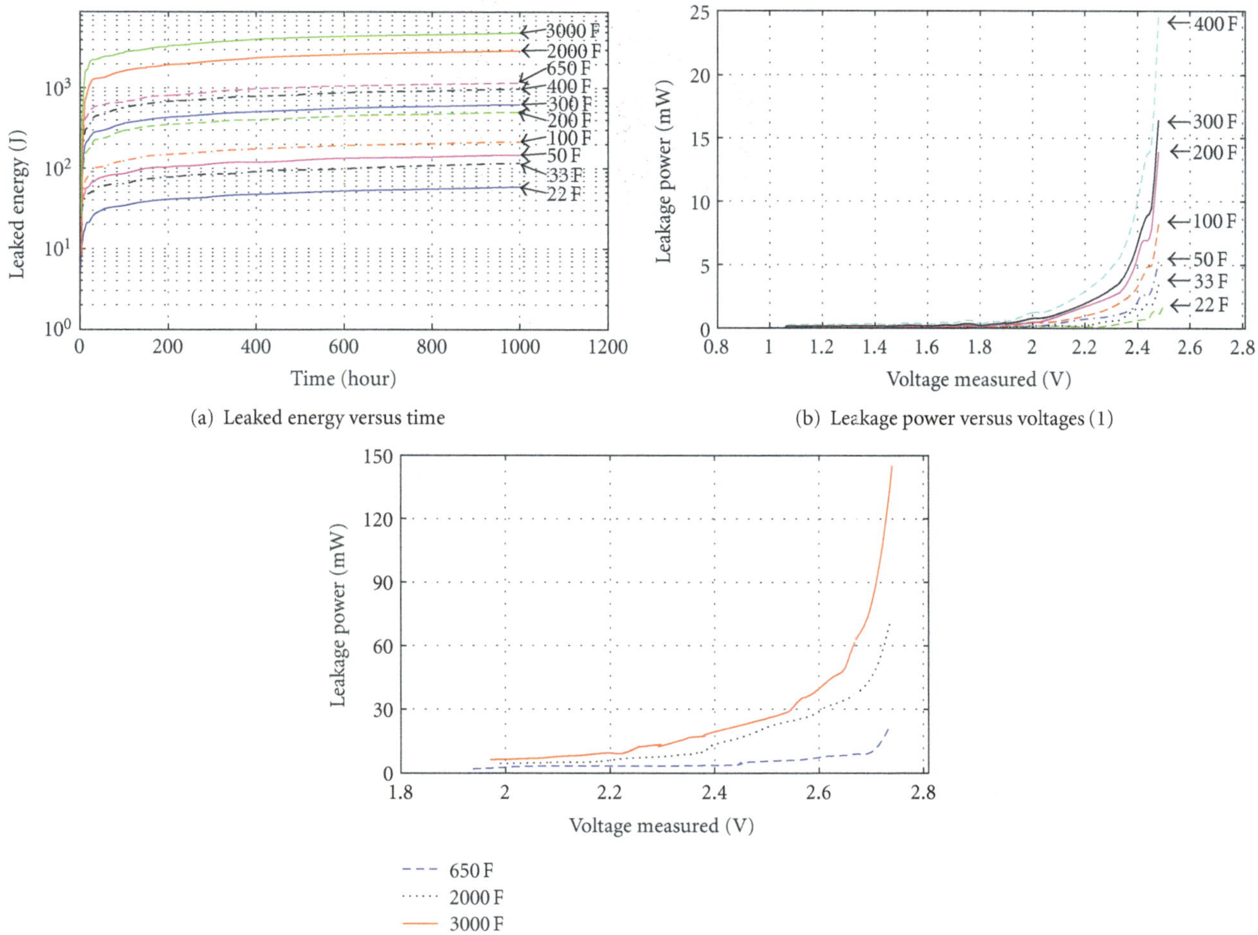

(c) Leakage power versus voltages (2)

FIGURE 1: Leakage property of ultracapacitor.

temporal diversity of the harvesting power also requires the wristwatch to share energy with the cell phone during the period when only the wristwatch can harvest energy. Given the low harvesting power, efficient and fine-grained energy sharing is needed to maintain the aliveness of all the devices.

Scenario 2. The Plant Care project [23] is an investigation of practical ubiquitous systems. In this system, a robot is employed to water the plants and to calibrate and recharge deployed sensors. To achieve sustainability, a power-line is applied to charge the robot periodically.

In this scenario, if the sensors and the robot are equipped with energy harvesting circuits (e.g., solar panels or wind generators), the system will keep on working even when a winter storm knocked out the power-line. Due to the location constraint, different sensors may have different harvesting power. If the energy can be shared among the sensors and the robot, the performance of the whole system can be balanced. Since the robot needs to interact with a lot of sensors during the dry season, the time duration of the energy sharing between the robot and sensors should be short.

The two scenarios discussed above illustrate the benefit of energy sharing feature and also indicate the requirements of the energy sharing: (i) highly efficient, (ii) fast, and (iii) fine-grained. Theoretically, the energy sharing concept can be applied to the battery-based systems, but the performance of these systems is significantly affected by the limitations of batteries: (i) low charge efficiency (as low as 6% [9]), (ii) limited charge current, and (iii) inaccurate remaining energy prediction which results in coarse granularity of energy sharing. In contrast, capacitors do not have all of the above limitations of the batteries, thus they meet the requirements of the energy sharing.

3. System Overview

The empirical study in Section 2 indicates that energy leakage is a critical hurdle that hinders designers exploiting the benefit of ultracapacitor. To address this hurdle, we propose a three-layer design as shown in Figure 2. The design goal of our system is to minimize the energy leaked away and efficiently utilize and share the energy among multiple embedded devices which form a network.

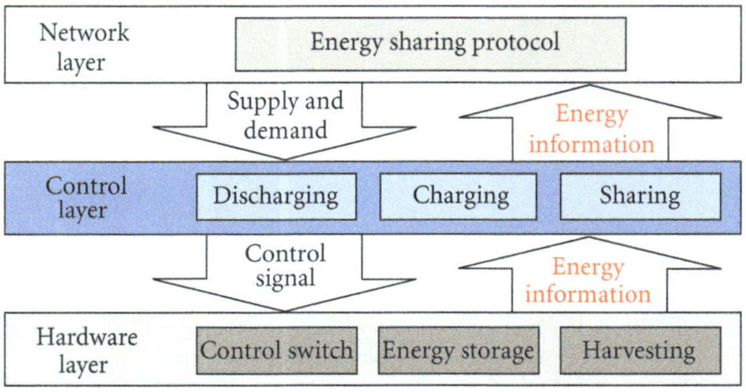

FIGURE 2: Overview of system architecture.

(1) The hardware layer uses (i) multiple ultracapacitors to form a capacitor array as the *only* energy storage unit to power the system, and (ii) a series of control switches to control different types of connections between the ultracapacitors (i.e., in parallel, serial, or hybrid). To achieve fast and efficient energy sharing, we use N-MOSFETs as switches which support large current, low energy consumption, and extremely short switching time (in the level of *ns*). This layer also provides an energy harvesting circuit (e.g., solar panel or wind generator) to harvest the environmental energy. Moreover, this layer (i) monitors the remaining energy inside each ultracapacitor and (ii) samples the harvesting power from the energy harvesting circuit. These two pieces of information are then fed into the control layer.

(2) The control layer serves as a bridge between the hardware layer and network layer. It plays two important roles: (i) it calculates the energy leakage rate online based on the remaining energy inside each ultracapacitor and then forwards the leakage information (together with the remaining and harvesting energy) to the network layer; (ii) based on the leakage model and the energy supply and demand request from the network layer, it controls the working state (i.e., discharging, charging, or energy sharing) of the energy storage system by sending the control signal to the hardware layer to control the on/off state of the control switches. Since the control layer has all of the energy information (including leakage information), it can conduct fine-grained control (including energy sharing).

(3) In the network layer, a running energy sharing protocol needs to effectively utilize the energy among multiple devices. In each embedded device, the energy sharing protocol calculates the difference between the actually available energy and the expected required energy by using the energy information forwarded from the control layer. Based on the energy difference, the energy sharing protocol decides the amount of energy needed to be shared among embedded devices and then determines the working state of the energy storage system by sending the quantitative energy supply and demand request to the control layer.

4. Hardware Layer

The design goal of the hardware layer is to provide a general, simple, and stable energy storage system with minimum leakage. To achieve this design goal, we designed the capacitor array circuit and built the hardware platform (i.e., Energy Mule).

4.1. Capacitor Array Circuit Design. A straightforward choice for energy storage is to use a single large ultracapacitor. However, this suffers from a slow boot-up time (to reach operational voltage), high remaining energy (detailed in Section 5.3), and inflexibility in fine-grained control (due to the limited granularity of the A/D converter inside sensor devices). For example, MICAz uses a 10-bit A/D converter. When being charged or discharged with the same amount of energy, a single large capacitor's voltage value does not change as significantly as a small capacitor, which causes inaccurate charging and discharging control (detailed in Sections 5.2 and 5.3). In our design, we build the Energy Mule using an array of ultracapacitors that have different levels of capacitances. This capacitor array is designed to satisfy the following three requirements.

(i) Generality. The circuit should be easily extended to n capacitors. Moreover, the capacitors can be connected in different types of connections (e.g., series, parallel, or hybrid) by using this circuit.

(ii) Simplicity. It should provide minimum number of control interface to the embedded device.

(Iii) Stability. It needs to have stable and simultaneous charging and discharging capability. In other words, during the charging state the circuit still needs to output energy to power the embedded device. Moreover, it is important to stably and smoothly change the connection type (e.g., from parallel to series) of the capacitors during the charging/discharging state.

The circuit we designed successfully achieves the above three design goals. Figure 3(a) shows the circuit for two capacitors which can be easily extended to support n capacitors (shown in Figure 3(b)). Moreover, this circuit can achieve different combinations of connections (shown in Table 1). Here we use 0/1 to indicate the on/off state of the switches. All of these switches can be controlled by a embedded device through the serial-to-parallel control signal converter (shown in Figure 4) which introduces minimum number of control interface to the embedded device. The output of the circuit is connected to a step-up DC/DC converter. Since a DC/DC converter normally has a very wide input voltage range (e.g., 0.7 V $\sim$ 5 V for MAX1676), our circuit can still provide a constant voltage from the output of the DC/DC converter when the capacitors are changing their connections (e.g., from parallel to series).

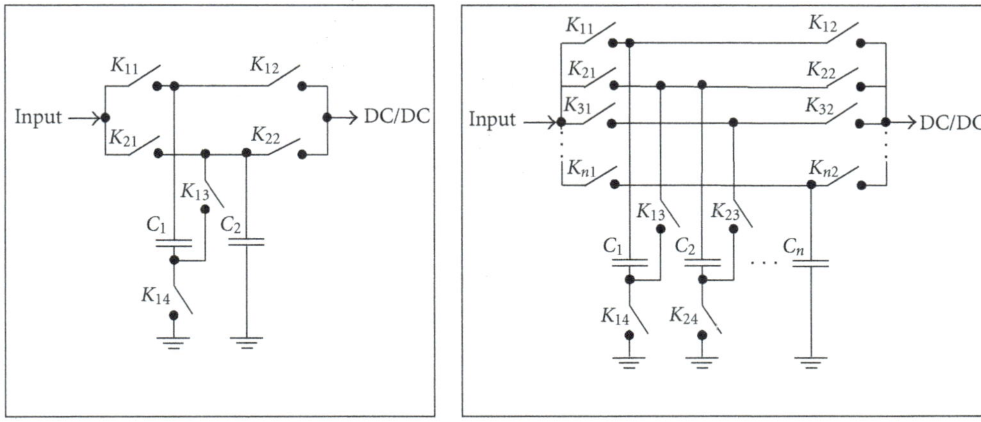

(a) Circuit for 2 capacitors (b) Circuit for n capacitors

FIGURE 3: Capacitor array circuit.

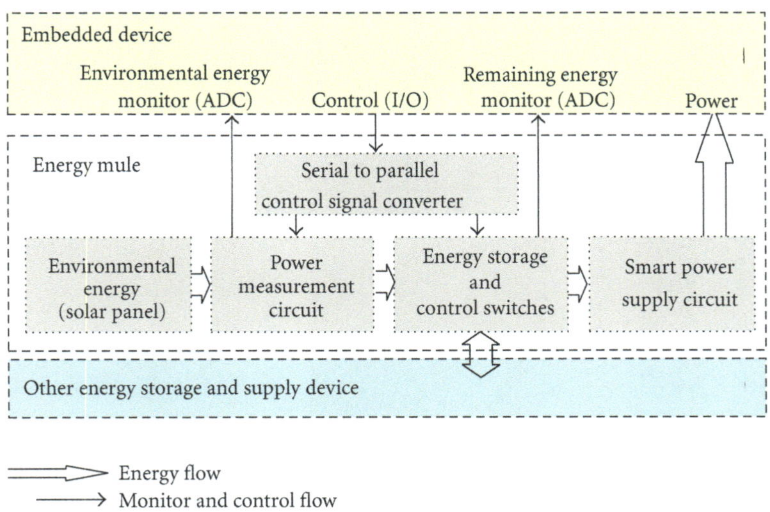

FIGURE 4: Energy mule hardware architecture.

4.2. Hardware Platform. Energy Mule is an add-on power board that supports multiple ultracapacitors to form an array as the only energy storage unit to overcome the intrinsic limitations of batteries (e.g., energy uncertainty, limited recharge cycles, low conversion efficiency, and environment unfriendliness). The architecture of Energy Mule is shown in Figure 4. It consists of (i) the solar panels and peripheral circuit for energy harvesting; (ii) the harvesting power measurement circuit; (iii) the ultracapacitor-based energy storage and the control switches which can directly share energy with the other energy storage and supply device; (iv) the serial to parallel control signal converter; (v) the smart power supply circuit with a DC/DC converter for powering the embedded device which is attached to the Energy Mule board.

5. Control Layer

By interacting with the hardware layer, the control layer achieves the following two design goals: (i) effectively controls the energy utilization of a single embedded device to minimize the energy leaked away inside the energy storage system of this device; (ii) efficiently shares energy between the embedded device itself and its neighboring devices while maintaining its own aliveness. In summary, the control layer contains three types of control: (i) charging, (ii) discharging, and (iii) energy sharing. To achieve effective control, leakage model is very important.

5.1. Energy Leakage Model. Based on our empirical study in Section 2.1, we formulate a leakage model to characterize the relationship between the voltage in the ultracapacitor and the leakage power. The embedded device builds the leakage model by storing the value of the leakage power P_L and the corresponding capacitor's voltage value V_c in its memory. Figure 5(a) shows the empirical leakage data of the leakage model for diverse capacitors. To express the model mathematically and reduce the storage space for the model, we use a piecewise linear approximation of the leakage curve. We define *turning points* as the points in the curve with

Table 1: Different types of connections.

Connection types	K_{11}	K_{12}	K_{13}	K_{14}	K_{21}	K_{22}
Discharge C_1, Disconnect C_2	0	1	0	1	0	0
Discharge C_1, Charge C_2	0	1	0	1	1	0
Charge and Discharge C_1 simultaneously	1	1	0	1	0	0
Discharge C_2, Disconnect C_1	0	0	0	1	0	1
Discharge C_2, Charge C_1	1	0	0	1	0	1
Charge and Discharge C_2 simultaneously	0	0	0	1	1	1
Discharges C_1, C_2 in parallel	0	1	0	1	0	1
Charge and Discharge C_1, C_2 in parallel	1	1	0	1	1	1
Discharges C_1, C_2 in serial	0	1	1	0	0	0
Charge and Discharge C_1, C_2 in serial	1	1	1	0	1	0

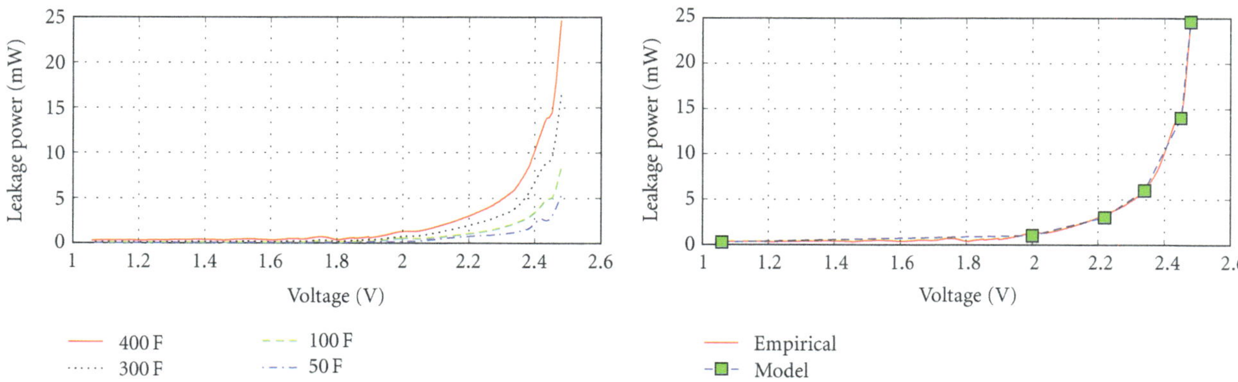

(a) Empirical leakage pattern

(b) Leakage model fitting for 400 F

Figure 5: Leakage model.

considerable slope change, which are used to decide the start and end points of line segments. Therefore, the whole leakage model can be represented by a set of linear functions, as follows:

$$P_L(V_c) = \begin{cases} a_1 \cdot V_c + b_1; & V_1 \le V_c < V_2 \\ a_2 \cdot V_c + b_2; & V_2 \le V_c < V_3 \\ \vdots & \vdots \\ a_n \cdot V_c + b_n; & V_n \le V_c \le V_{n+1}. \end{cases} \quad (1)$$

Here $V_1, V_2, \ldots, V_{n+1}$ are the remaining energy values corresponding to the turning points of the segments. $a_1, a_2, \ldots, a_n$ and $b_1, b_2, \ldots, b_n$ are the coefficients for each line segment. Figure 5(b) shows an example for the line-segment-based modeling of the leakage power for a 400F capacitor. The square dots represent the turning points. The model we built accurately matches the empirical results.

5.2. Charging. The design goal of the charging control is to efficiently charge the capacitor array based on the leakage model of each capacitor so that the amount of energy leaked away is minimized. As an example, Figure 6 shows the charging control of two capacitors (C_1 and C_2). Initially, switches K_{11} and K_{21} are turned off. When the charging control is conducted, the control layer measures the voltage values (V_{C_1} and V_{C_2}) of these two capacitors and calculates the corresponding leakage powers (P_{L_1} and P_{L_2}) based on the leakage model. If P_{L_1} is less than or equal to P_{L_2} (shown in Figure 6(a)), the control layer decides to charge C_1 by turning on K_{11}. In the case that P_{L_1} is greater than P_{L_2} (shown in Figure 6(b)), C_2 is charged. The charging control repeats with time interval T until there is no external energy available or all the capacitors are fully charged.

As an all-purpose energy storage system, it is important that the system can be charged by different energy sources (e.g., environmental energy or DC power supply). However, these diverse energy sources introduce a challenge to the design of charging control, because their charge current varies over four decades (ranging from less than 1 mA for environmental energy to more than 10 A for DC power supply). If the charging control is conducted with a fixed small time interval (in favor of the large charge current), unnecessary voltage measurements and comparisons are conducted when the capacitor is charged with a small current and its voltage increases slowly. Whereas, if the time interval is large (in favor of the small charge current), more energy is leaked away when the charge current is large.

To minimize the energy leaked away and reduce the control overhead, we propose the adaptive charging control

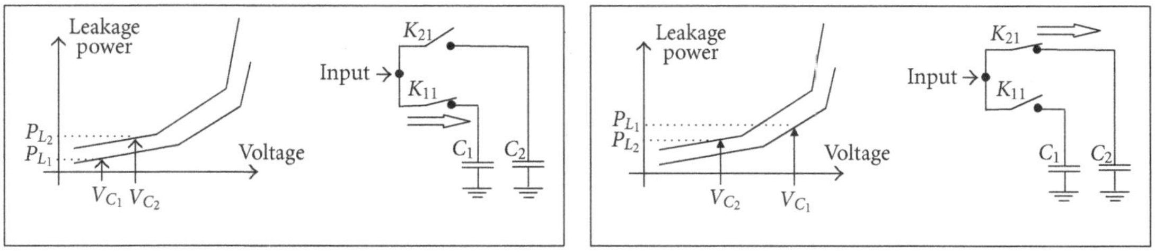

(a) Case 1 ($P_{L_1} \le P_{L_2}$)

(b) Case 2 ($P_{L_1} > P_{L_2}$)

FIGURE 6: Charging control example.

```
(1) for i ← 1 to n do
(2)     K_i1 ← 0;
(3)     Calculate P_Li based on measured V_ci;
(4) end
(5) m ← GetMinCapId(), K_m1 ← 1;
(6) Set timer with the smallest time interval T;
(7) while timer fired do
(8)     for i ← 1 to n do
(9)         Measure V_ci and update P_Li;
(10)    end
(11)    if m = GetMinCapId() then
(12)        T ← 2T,
(13)    else
(14)        K_m1 ← 0, m ← GetMinCapId();
(15)        K_m1 ← 1;
(16)    end
(17)    if Capacitors Full/No External Energy then
(18)        break;
(19)    else
(20)        Set timer with interval T;
(21)    end
(22) end
```

ALGORITHM 1: Adaptive charging control.

which conducts the charging control operation with an adaptive time interval based on the charge current. Algorithm 1 illustrates the control algorithm for n capacitors. Initially, the controller turns off all the switches and calculates the leakage power of all the capacitors based on the measured voltage values of these capacitors (Lines 1 to 4). Then, the controller turns on the switch which connects to the capacitor with the minimum leakage power and sets the timer with the smallest time interval T (Lines 5 and 6). Since we allow simultaneous charging and discharging, when the timer fired, the controller needs to recalculate the leakage power of all the capacitors based on the newly measured voltage values of these capacitors (Lines 8 to 10). If the capacitor with the minimum leakage power is the same as the one in the previous cycle (indicating the current time interval T is too short), the time interval is doubled (Lines 11 and 12). Otherwise, the controller turns off the switch that connects to the previous capacitor and turns on the switch that connects to the capacitor which currently has the minimum leakage power (Lines 13 to 16). The above control procedure repeats every time interval T, until all of the capacitors are fully charged or there is no external energy available (Lines 17 to 22).

5.2.1. Complexity and Feasibility Analysis. In the worst case, the timer is fired for k times and the control algorithm measures capacitors' voltage every time the timer is fired. Therefore, the complexity of the control algorithm is $O(kn)$, where n is the number of capacitors. In the algorithm, the smallest time interval T is 100 ms. It will exponentially increase if the capacitor with the minimum leakage power is the same as the one in the previous cycle. The value of time interval T can reach 1 hour within 16 iterations. Therefore, it is feasible for both indoor charging with DC power supply and outdoor charging with energy harvesting devices such as solar panels.

5.3. Discharging. This section discusses how to smartly discharge the capacitors to achieve the following two design goals: (i) the energy leaked away is minimized during the discharge stage, (ii) the remaining energy inside the capacitor

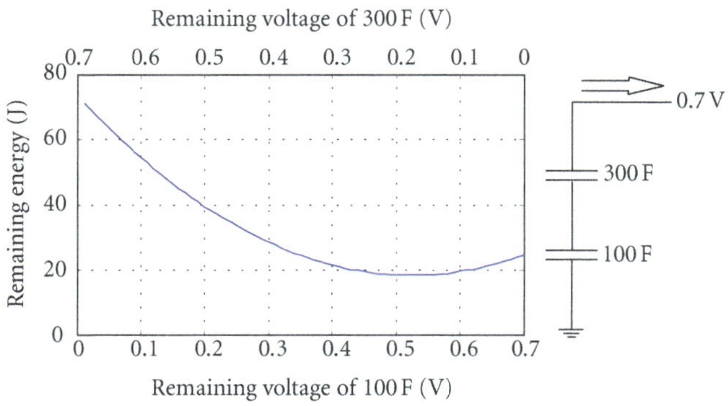

FIGURE 7: Different voltage combinations.

array is minimized at the end of the discharging process so that the lifetime of the system is maximized before the next charging process.

In order to minimize the energy leaked away, the controller uses adaptive discharging control algorithm which is very similar to the adaptive charging control algorithm, except that the capacitor with the highest leakage power needs to be discharged first.

Since the output of the capacitor array serves as the input of the DC/DC converter which has the minimum input voltage (e.g., 0.7 V for MAX1676), to minimize the remaining energy inside the capacitor array, the capacitors are connected in series so that the sum of their voltages is greater than the minimum input voltage of the DC/DC converter. However, for a fixed summation, different combinations of the capacitors' voltage value result in different remaining energy inside the capacitor array. As an example, Figure 6 shows that 100 F and 300 F capacitors are connected in series to provide the same sum of 0.7 V. However, the remaining energy is dramatically different for different combinations of voltage values of 100 F and 300 F capacitors. To minimize the remaining energy inside the series of capacitors, we need to determine the voltage value of each capacitor. The problem can be formulated as:

$$\begin{aligned} &\text{minimize} \quad V^T C V \\ &\text{subject to} \quad V \succcurlyeq 0 \\ &\qquad \sum_{i=1}^{n} V_{c_i} = V_{\min}. \end{aligned} \tag{2}$$

Here, we define $C = \frac{1}{2}\operatorname{diag}(c_1, c_2, \ldots, c_n)$, $V^T = [V_{c_1}, V_{c_2}, \ldots, V_{c_n}]$. $V_{\min}$ is the minimum input voltage of DC/DC converter. This is a quadratic programming problem, and the analytic solution is as follows.

$$\begin{aligned} V_{c_1} &= V_{\min} \cdot c_{\text{total}} \cdot \frac{1}{c_1}; \\ V_{c_2} &= V_{\min} \cdot c_{\text{total}} \cdot \frac{1}{c_2}; \\ &\vdots \\ V_{c_n} &= V_{\min} \cdot c_{\text{total}} \cdot \frac{1}{c_n}. \end{aligned} \tag{3}$$

Here c_{total} is the total capacitance of all the capacitors in series and $c_{\text{total}} = 1/(1/c_1 + 1/c_2 + \cdots 1/c_n)$.

From the above, we get the voltage values of a series of capacitors with the minimum remaining energy. However, initially all of the capacitors have voltage values larger than the minimum input voltage of the DC/DC converter. In order to minimize the energy leaked away, only the capacitor with the highest leakage power is selected as the input of the DC/DC converter, and all the other capacitors are originally disconnected from each other. To achieve the minimum remaining energy, we need to decide when to connect all of these capacitors in series. If we connect them in series too early (or late), their voltage values may be different which causes the remaining energy to be not minimum. By investigating (3) carefully, one can find that the voltage values of each capacitor share a common factor $V_{\min} \cdot c_{\text{total}}$. Moreover, these voltage values are reverse proportional to their capacitance. Our further experiments revealed that if the capacitors are discharged in series with their initial voltage values reverse proportionally to their capacitance, their voltage value will maintain this relation until the end of the discharge. Furthermore, the capacitor with the largest capacitance has the smallest voltage value. For example, in Figure 7, if the initial voltage values of the 100 F and 300 F capacitors are 2.1 V and 0.7V, respectively, after they are discharged in series, their voltage value will maintain the relation in which the voltage value of 100 F is 3 times more than the voltage value of 300 F.

Guided by the above observation, we can calculate the minimum voltage values of these capacitors before they are connected in series:

$$\begin{aligned} V_{1_{\min}} &= V_{\min} \cdot \frac{c_i}{c_1}; \\ V_{2_{\min}} &= V_{\min} \cdot \frac{c_i}{c_2}; \\ &\vdots \\ V_{i_{\min}} &= V_{\min}; \\ &\vdots \\ V_{n_{\min}} &= V_{\min} \cdot \frac{c_i}{c_n}. \end{aligned} \tag{4}$$

Here we assume that capacitor c_i has the largest capacitance which results in the smallest voltage value. Since all of the parameters in (4) are determined before conducting the discharging control, (4) can be solved off line by using personal computer. In summary, the discharging control algorithm is straightforward and can be described as follows.

Initially, all of the capacitors are adaptively discharged based on which capacitor has the highest leakage power. If the voltage value of the capacitor reaches the minimum voltage value calculated from (4), this capacitor is excluded from the adaptive discharging. When all of the capacitors have the voltage values less than or equal to the voltage in (4), these capacitors are connected in series and conduct discharge until their total voltage value is less than the minimum input voltage of the DC/DC converter.

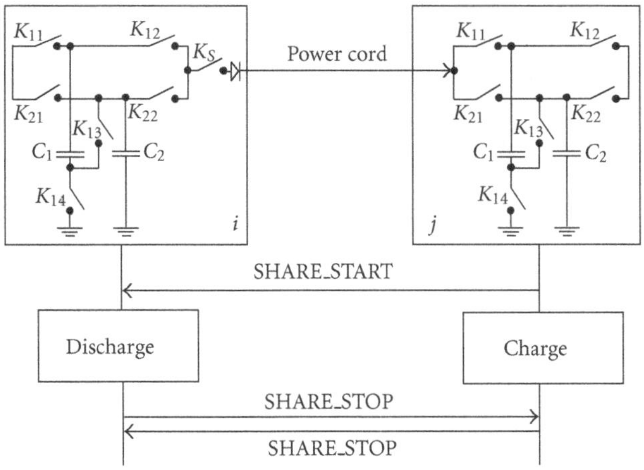

Figure 8: Energy sharing protocol.

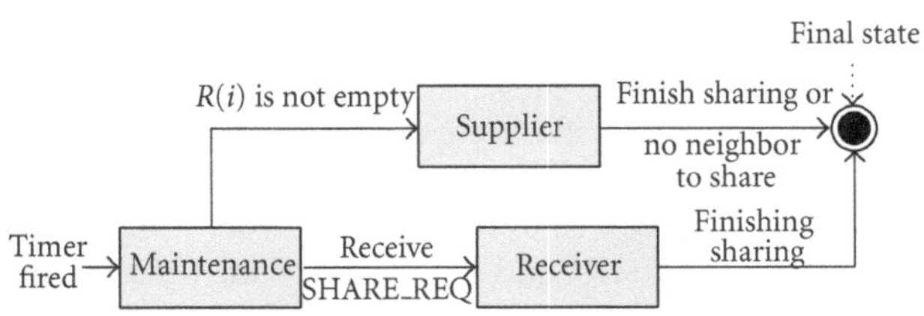

Figure 9: State machine diagram of network layer.

We note that there is a trade-off between minimizing the energy leaked away and minimizing the remaining energy. Here minimizing the remaining energy has higher priority than minimizing the energy leaked away. This is because when the capacitors start to be connected in series, their leakage power is very low.

5.4. Energy Sharing Protocol. Without losing generality, we assume that supplier i wants to share some energy with receiver j (shown in Figure 8). To achieve the design goal of being fast and highly efficient, instead of going through the DC/DC converter, the energy is directly shared between the capacitor arrays inside supplier i and receiver j. This is because the DC/DC converter consumes large amounts of energy, especially when its input voltage is low. Moreover, its output current is limited, which significantly increases the time duration of the energy sharing. As shown in Figure 8, the output of i is directly connected to the input of j through power cords. For high-efficiency purposes, we use the power cords instead of using the wireless charging interface.

The energy sharing protocol is a receiver-initiated protocol. After receiving the energy sharing control from the network layer, the receiver j sends out the *SHARE_START* packet to the supplier i and prepares for charging. Upon receiving the *SHARE_START* packet, i turns on the switch K_S and conducts the discharging control. During the discharging control, the supplier i periodically monitors the amount of energy transferred by sampling the voltage of its capacitors and calculating its own remaining energy until the expected amount of energy is transferred. Then the supplier turns off the switch K_S and sends out the *SHARE_STOP* packet to receiver j telling j the end of energy sharing.

On the receiver j side, the charging control is conducted. j also periodically monitors the energy received by sampling the voltage of its capacitors and calculating its remaining energy. When the expected energy is received, j turns off the charging switches (i.e., K_{11} and K_{21} in Figure 8) and sends out the *SHARE_STOP* packet to i telling i that the energy sharing is terminated. Since j can accurately monitor the increased remaining energy inside its capacitor, it can terminate the energy sharing process with high accuracy. In our energy sharing protocol, both the supplier and the receiver can decide when to terminate the energy sharing process. Therefore, our control is robust to the failure of supplier and receiver.

6. Network Layer

As discussed in Section 5, the control layer is designed to minimize the energy leaked away inside a single device and efficiently share energy between two devices. In this section, we introduce the network layer design which tries to efficiently utilize the energy inside the whole network and prolong the network lifetime. Here the network lifetime is defined as the period of time during which all the devices inside the network have energy to execute tasks.

In order to maintain the aliveness of all the devices and minimize the energy leaked away, it is very important for the network layer to balance the remaining energy among these devices. Given the dynamic change of the environmental energy and different working patterns of the devices, the challenge here is to design a general and light-weight network layer that works without knowing the information of environmental energy and the energy consumption of the other devices. To overcome this challenge, the network layer is designed to be a simple finite state machine. As shown in Figure 9, every device inside the network is in one of four states at any time: (i) maintenance, (ii) supplier, (iii) receiver, and (iv) final. Transitions between the states are triggered by events.

Each node's timer fires periodically with time interval T. After the timer is fired, the node enters the maintenance state, in which all of its neighbors' information is periodically maintained. Here, two nodes are considered as neighbors if they are directly connected by using power cords. Whenever the node detects that the energy waste can be reduced by sharing energy with its neighbor, the node enters the supplier state and sends out the *SHARE_REQ* message to the neighbor, which causes the least amount of energy waste. When a node receives the *SHARE_REQ* message from its neighbor, it enters the receiver state in which the node negotiates with the supplier node about the energy sharing. After energy sharing negotiation, the nodes in the supplier state and receiver state enter the final state and wait for the timer fires again. In the rest of this section, we explain the operation in each state in detail.

6.1. Maintenance State. Since the network layer is designed to work under unpredictable environmental energy and different working patterns of the nodes, maintenance is needed to keep track of the remaining energy inside each node and its neighbors. Each node periodically measures its remaining energy with time interval T. Without losing generality, suppose node i is one of the nodes inside the network and its neighboring nodes are $j, k, \ldots,$ and n (shown in Figure 10). Assume node i's remaining energy is E_{R_i} at time kT. Based on E_{R_i}, node i calculates the leakage power (P_{L_i}) of the ultracapacitors by using the leakage model which is discussed in Section 5.1. Similarly, node i's neighboring nodes can also calculate their own remaining energy and leakage power. Then these nodes share their remaining energy and leakage power information among themselves by embedding this information into their local broadcast packets.

After node i receives the remaining energy and leakage power information from its neighbor (assume node j), node i decides whether it needs to share energy with node j by comparing the energy waste in two scenarios: (i) no energy sharing with node j and (ii) energy sharing. In the first scenario, the energy leaked away inside nodes i and j during the time interval T is the energy waste which can be calculated by using the following equation:

$$W_{i,j} = \left(P_{L_i} + P_{L_j}\right) \cdot T. \quad (5)$$

In the second scenario, after sharing the energy with node j, node i's leakage power is decreased from P_{L_i} to P'_{L_i}. However, node j's leakage power is increased from P_{L_j} to P'_{L_j}. Therefore, the energy waste is the summation of the energy loss during the sharing procedure and the energy leaked away inside nodes i and j during the time interval T after the energy sharing. The energy waste can be calculated by using the following equation:

$$W'_{i,j} = \Delta E[i \Longrightarrow j] \cdot \left(1 - e_{i,j}\right) + \left(P'_{L_i} + P'_{L_j}\right) \cdot T. \quad (6)$$

Here $\Delta E[i \Rightarrow j]$ is the total amount of energy transferred from node i to node j, and $e_{i,j}$ is the energy sharing efficiency between i and j. Then $\Delta E[i \Rightarrow j] \cdot (1 - e_{i,j})$ is the energy loss during the sharing procedure. For example, if $\Delta E[i \Rightarrow j] = 100J$ and $e_{i,j} = 90\%$, the energy loss during the sharing procedure is $100 * (1 - 0.9) = 10J$.

If energy sharing wastes less amount of energy than no energy sharing (i.e., $W'_{i,j} < W_{i,j}$), node i will share energy with node j. Otherwise, energy sharing does not happen between i and j. The total energy gain during the energy sharing procedure can be calculated as follows:

$$G_{i,j} = W_{i,j} - W'_{i,j}. \quad (7)$$

Since node i may have multiple neighbors, i will send the *SHARE_REQ* message to its neighbor which results in the maximum energy gain. After sending out the *SHARE_REQ*, i enters the supplier state.

6.2. Supplier State. In the supplier state, a node needs to negotiate with its neighbors about the energy sharing. This is because the node's initially selected neighbor may receive multiple *SHARE_REQ* messages from other nodes. For example, node i selects node k to share the energy which can achieve the maximum energy gain between node i and node k from node i's perspective. However, node k may receive multiple *SHARE_REQ* messages from k's neighbors. From node k's perspective, sharing energy with node i may not achieve the maximum energy gain. Therefore, node k may deny node i's energy sharing request. In this case, node i needs to exclude node k and send the *SHARE_REQ* message to its neighbor that can achieve the maximum energy gain. This procedure is repeated until node i either can not find its neighbor to share the energy or receive the *SHARE_ACCEPT* message from one of its neighbors.

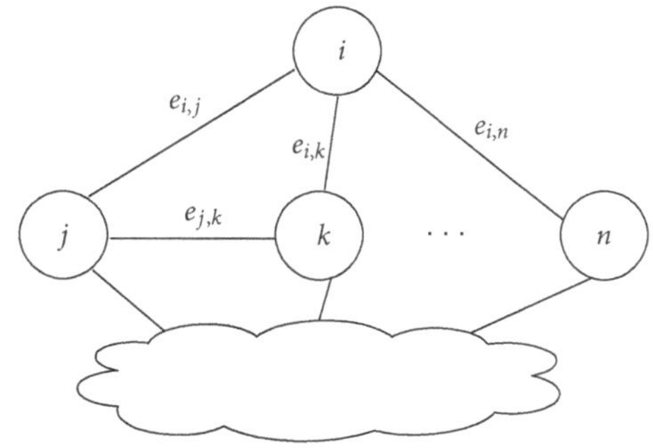

Figure 10: Energy sharing networks.

If node i can not find a neighbor to share energy, i terminates the energy sharing process and enters the final state. In the case that node i receives the *SHARE_ACCEPT* message from one of its neighbors, i indicates its control layer to conduct the energy sharing control and enters the final state.

6.3. Receiver State. After a node receives the *SHARE_REQ* message from its neighbor, this node enters the receiver state. If this node receives multiple *SHARE_REQ* messages from its neighbors, the node will (i) send *SHARE_ACCEPT* message to the neighbor which can reduce the largest amount of energy waste and (ii) send *SHARE_DENY* messages to the other neighbors. After sending out these messages, the node indicates its control layer to conduct energy sharing control and enters the final state.

In our design, the node in the receiver state can only accept one energy sharing request. This has two advantages.

(i) Accuracy. Since the node's remaining energy and leakage power will change after energy sharing, it is not accurate to conduct the second energy sharing.

(ii) Loop-Free. The node can either receive or transfer energy for a single time, thus no loop will be created. Therefore, energy can be efficiently utilized.

6.4. The Detailed Protocol. Combining all the design components, the network layer energy sharing protocol can be specified by the pseudocode shown in Algorithm 2. The

(1) **switch** *Event* **do**
(2) **case** *Timer fired*
(3) $R(i) \leftarrow \phi$, $S(i) \leftarrow \phi$, $\forall j \in N(i)$, $G_{i,j} \leftarrow 0$;
(4) Measure E_{R_i} and calculates P_{L_i};
(5) Share E_{R_i} and P_{L_i} information with neighbors;
(6) **for** $j \in N(i)$ **do**
(7) Update $G_{i,j}$ via Formula 5, 6, and 7;
(8) if $G_{i,j} > 0$ **then** $R(i) \leftarrow R(i) + \{j\}$;
(9) Call Receptor Selection method;
(10) **case** *Receive SHARE_DENY from node j*
(11) $R(i) \leftarrow R(i) - \{j\}$;
(12) Call Receptor Selection Method;
(13) **case***Receive SHARE_ACCEPT from node j*
(14) Conduct energy sharing control;
(15) **case***Receive SHARE_REQ from node k*
(16) $S(i) \leftarrow S(i) + \{k\}$;
(17) Wait until receives all the *SHARE_REQ* from neighbors;
(18) From $S(i)$, select k which has the maximum $G_{k,i}$;
(19) Send *SHARE_ACCEPT* to k;
(20) Conduct energy sharing control;
(21) $S(i) \leftarrow S(i) - \{k\}$;
(22) Send *SHARE_DENY* to all the nodes in $S(i)$;
(23) **Receptor Selection method:**
(24) **if** $R(i) \neq \phi$ **then**
(25) From $R(i)$, select node j which has the maximum $G_{i,j}$;
(26) Send *SHARE_REQ* to j;

ALGORITHM 2: Network layer energy sharing protocol.

design is simple and triggered by the events of either timer fired (Line 2) or receiving energy sharing messages (Lines 10, 13, and 15).

Lines 2 to 9 deal with the event of the timer fired and the node enters the maintenance state. Node i, which runs the protocol, maintains its receptor node set $R(i)$, supplier node set $S(i)$, and neighboring node set $N(i)$. After the timer fired, $R(i)$ and $S(i)$ are initially set to be empty, and the energy sharing gains between i and its neighbors are set to be 0 (Line 3). After sharing the remaining energy (E_{R_i}) and the leakage power (P_{L_i}) information with neighbors, i calculates the energy sharing gain from itself to all its neighbors by using Formula 5, 6, and 7 (Lines 5 to 7). If the neighbors result in positive energy sharing gains, i adds its neighbors into its receptor node set $R(i)$ (Line 8). After adding all the eligible neighbors into $R(i)$, i calls the *Receptor Selection* method (Lines 9). The *Receptor Selection* method tests whether the receptor node set $R(i)$ is empty. If $R(i)$ is not empty, the node with the maximum energy sharing gain will be selected and *SHARE_REQ* is sent to this node (Lines 23 to 26).

Lines 10 to 14 deal with the events when the node is in the supplier state. Lines 10 to 12 handle the event of receiving the *SHARE_DENY* message from a neighbor j. In this case, j is removed from the receptor node set $R(i)$, and a new receptor is selected by calling the *Receptor Selection* method. In the case that i receives the *SHARE_ACCEPT* message from a neighbor j, i indicates its control layer to conduct energy sharing control (Lines 13 and 14).

Lines 15 to 22 handle the event of receiving the *SHARE_REQ* message from a neighbor k when node i is in the receiver state. i adds node k into its supplier node set $S(i)$ (Line 16). After receiving all the *SHARE_REQ* messages from its neighbors, i sends the *SHARE_ACCEPT* message to the node which results in the maximum energy gain and sends the *SHARE_DENY* messages to all the other nodes in its supplier node set (Lines 17 to 22).

6.4.1. Complexity and Feasibility Analysis. The energy sharing protocol is triggered by the events of either timer fired or receiving energy sharing messages. In each case, there is a single loop. Therefore, the complexity of the algorithm is $O(N)$,where N is the number of neighbors a node has. In sensor networks, each node normally has limited number of neighbors. Moreover, the timer fires with long time interval (i.e., 5 minutes in our implementation). Thus, the overhead of the protocol is small. The impact of maintenance interval is evaluated in Section 8.4.

7. Implementation and Evaluation

In this section, we evaluated our system performance under different real-world settings.

7.1. Evaluation Baselines and Metrics. In our experiment, we set the total capacitance to be 400 F. To reduce the boot-up time and leakage, we use a combination of a 100 F and a 300 F capacitor, which are controlled by our efficient charging and discharging control algorithms. We call this approach the *efficient control* (*EC*), which is compared with the following baseline.

(i) No Efficient Control (NEC).This approach uses the same array of capacitors but provides no efficient control by simply connecting the capacitors in parallel to power the system.

The Key Advantage of Our System Design is That it Efficiently Minimizes the Energy Leaked Away Inside the Energy Storage System. Since it is Hard to Measure How Much Energy is Leaked Away, The Metric Used to Evaluate the System is the *Remaining Energy*, The Energy Inside the Capacitors That Can be Calculated Based on the Measured Voltage Values of the Capacitors. with the Same Amount of Energy Charged Both Into the System and Consumed, The Larger the Value of the Remaining Energy, The Less the Amount of Energy Leaked Away.

7.2. Implementation. Our hardware contains a custom circuit board to harvest (or share) energy to power the embedded device. We used a typical sustainable embedded device (i.e., MICAz node) as a controller which is powered by our energy storage system. We designed and implemented the control layer using TinyOS and NesC. The MICAz node normally works in low-duty cycle mode. It wakes up to conduct the charging, discharging, and energy sharing control. Besides the control operations, it also measures the remaining energy inside the capacitor-array, turns on the radio during the active time interval, and logs the data into the flash for the purpose of off-line analysis of the system performance.

7.3. Outdoor Energy Harvesting Experiment. The outdoor energy harvesting experiment supports such potential applications as data collection [2, 24] and habitat monitoring [25]. In this experiment, we deployed our systems inside a weatherproof enclosure that was set facing south and at a 40° angle to the ground (shown in Figure 11(a)). The total time duration of the experiment was 96 hours. Figure 12 shows the energy harvested by these systems. We placed the solar panels of these systems close to each other to harvest similar amounts of energy from the environment. The MICAz nodes of these systems ran at a 10% duty cycle to consume the same amount of energy all the time. As shown in Figure 11(b), the EC system can efficiently maintain the remaining energy at a higher level than the NEC systems. At the end of the experiment, the remaining energy inside the EC system is 872.8 J, which is 14.4% more than the remaining energy inside the NEC system. Figure 11(c) shows that our capacitor array circuit can smoothly conduct the charging and discharging controls.

7.4. Energy Sharing Experiments. The energy sharing experiment supports such potential applications as wearable computing [18, 19] and robot-sensor networks [1, 20, 21]. In this section, we evaluate the energy sharing from one device to another device.

As shown in Figure 13(a), the output of the supplier S_1 is connected to the input of the receptor R_1. The capacitors in S_1 are connected in parallel and their initial voltages are 2.5 V each. While the capacitors in R_1 are connected in series and their total voltage value is 1.6 V (i.e., 0.4 V for 300 F and 1.2 V for 100 F). The equivalent circuit is shown in Figure 13(b). Figure 13(c) shows the voltage values of the capacitors during the energy sharing process. The energy sharing process starts at time 1 second and ends at time 3.1 second with the total time duration of 2.1 seconds. At the end of the sharing process, the voltage value of the capacitors in S_1 is 2.37 V each, and the voltage values of the 300 F and 100 F capacitors in R_1 are 0.64 V and 1.71 V, respectively. From the voltage values, we can calculate the total energy lost in S_1 is 126.62 J and the total energy gained in R_1 is 113 J which can be used to power the Motorola V557 cell phone to work in standby mode for around 2 hours.

To investigate the impact of the length of interconnect wires, we connected two devices using an 8-inch and a 31-foot power cord. Then we measured these two devices' energy sharing efficiency which is 92.07% for an 8-inch power cord and 83.76% for a 31-foot power cord.

8. Simulation

System evaluation has demonstrated the effectiveness of our design in Section 7. However, due to physical constraints, we could not run the experiments for the applications such as wearable networks. Thus to understand the behavior of our design, extensive simulation is necessary. In this section, we choose wearable sensor networks as an example and evaluate its performance under different settings. Since this work is the first one investigating energy sharing applications, the state-of-art approaches (e.g., energy conservation) are complementary, which provide no appropriate baseline for comparison. Therefore, we compare the system that enables energy sharing (ES) with the same system that has no energy sharing (NES). In other words, the sensor devices can not share energy with each other in the no energy sharing wearable networks. The metrics used to evaluate the performance of our system is *Network Lifetime*: the time duration before the first device inside the network runs out of power.

8.1. Simulation Setup. In the simulation, the number of the sensor devices is 25. All of these devices are identical in terms of the energy storage capacity. We use the 3-day real energy traces (shown in Figure 14) collected from a solar panel which is mounted on a graduate student, as a baseline, and generate an energy pattern in which the harvested power varied by a random amount that is −10% to 10% of the baseline for each device. Each device works under a randomly generated working pattern with the mean value of the duty cycle varying from 0.5% to 5%. The same working pattern is applied to the systems with energy sharing and with no energy sharing. Each data point on a

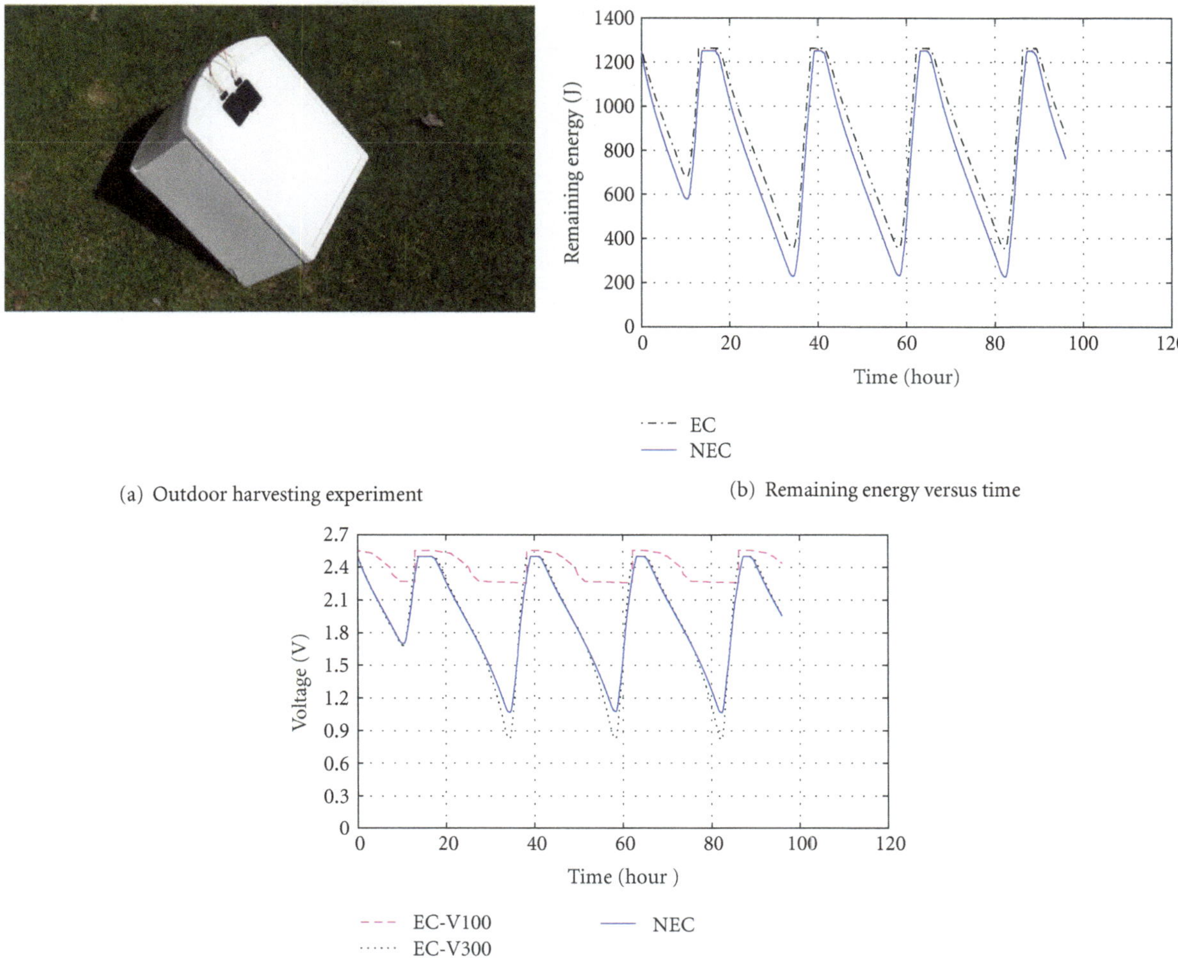

(a) Outdoor harvesting experiment

(b) Remaining energy versus time

(c) Capacitor voltage versus time

FIGURE 11: Outdoor energy harvesting.

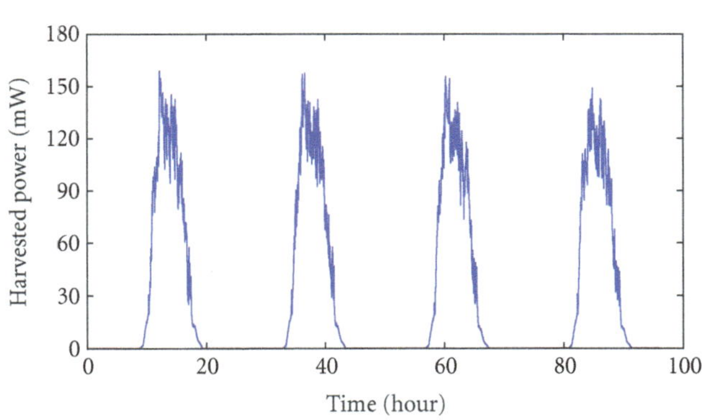

FIGURE 12: Harvested power over 4 days.

graph represents the averaged value of 10 runs, and the 95% confidence intervals of the data are within 3 ~ 8% of the mean shown.

8.2. Impact of Energy Sharing Efficiency. In this section, we analyzed the effect of energy sharing efficiency by varying the value of energy sharing efficiency from 0% to 100%. Figure 15 shows the performance of the systems with no energy sharing (NES) and with energy sharing (ES). When the energy sharing efficiency equals 0, the system with energy sharing has the same lifetime as the system with no energy sharing. As the efficiency increases, the network lifetime of the system with energy sharing also increases. When the efficiency equals 100%, all of the devices in the system with energy sharing is alive all the time (i.e., 72 hours). However, the system with no energy sharing only has a network lifetime equals to 11.3 hours. Since the charge efficiency of rechargeable batteries is as low as 6% [9], from Figure 15, one can tell that the battery-based energy storage system can not take the advantage of the energy sharing feature.

8.3. Impact of Energy Sharing Error. The granularity of the A/D converter may introduce the error during the energy sharing. In this experiment, we evaluate the effect of energy sharing error which varies from −25% to 25%. Here we use negative value to indicate that the actual energy gained is less than the energy expected. For example, if a device A wants to receive 100 J energy from a device B, −25% means that

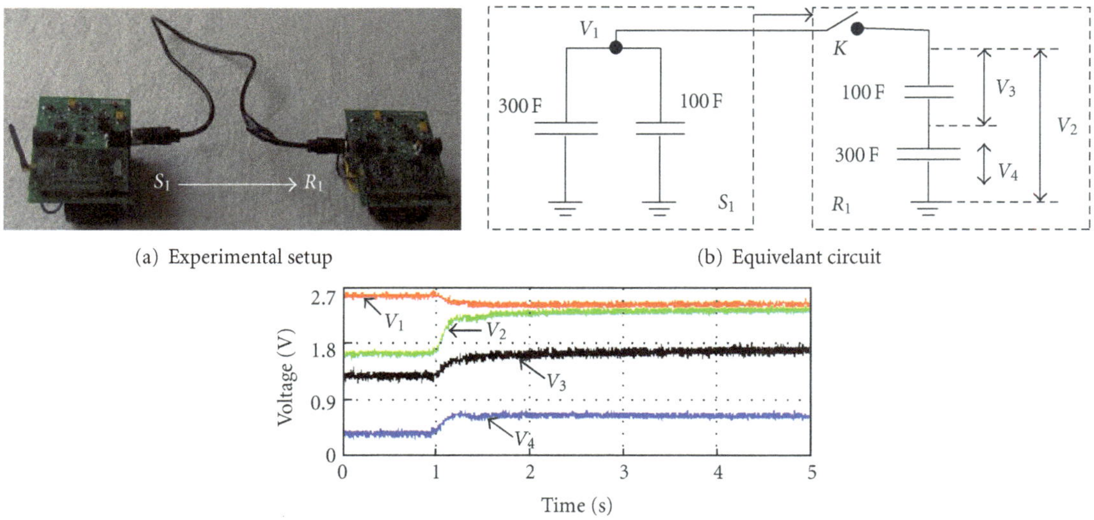

(a) Experimental setup

(b) Equivelant circuit

(c) Voltage versus time

FIGURE 13: Energy sharing between two devices.

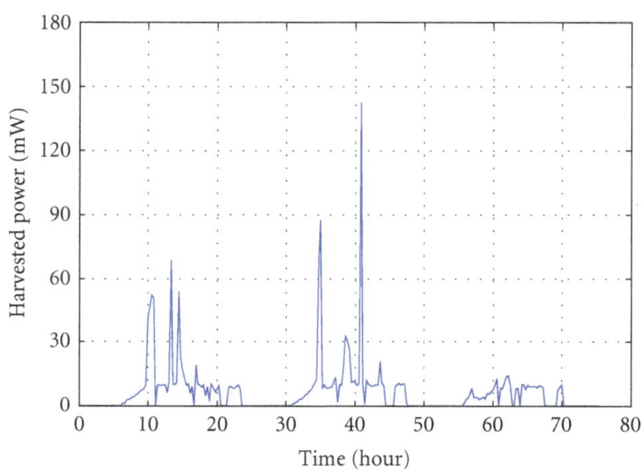

FIGURE 14: Harvested power versus time.

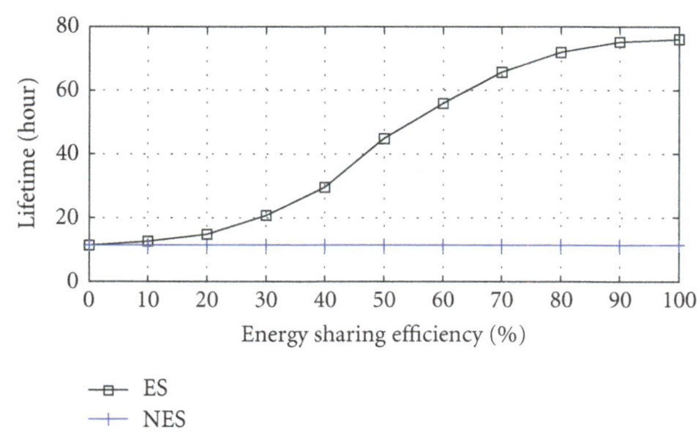

FIGURE 15: Impact of energy sharing efficiency.

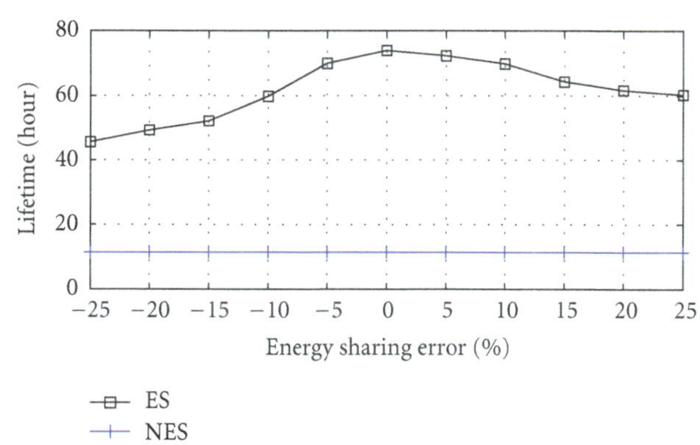

FIGURE 16: Impact of energy sharing error.

device A actually only receives 75 J. As shown in Figure 16, the energy sharing error could reduce the network lifetime. Moreover, the negative error has more impact on the system performance than the positive error. This is because less energy received by an energy deficient device could reduce the lifetime of the device.

8.4. Impact of Maintenance Time Interval. Maintenance time interval also affects the system performance. A short time interval may introduce unnecessary maintenance overhead. However, a long time interval may affect the system's lifetime. In this section, we evaluate the impact of maintenance time interval by varying the time interval value from 2 minutes to 16 minutes. As shown in Figure 17, the system's lifetime decreases when the maintenance time interval value increases. This is because some energy deficient nodes cannot timely receive energy from their neighboring nodes if the maintenance time interval is too long. Therefore, these nodes will run out of energy which causes the decrease of the system's lifetime. From the simulation result, setting the maintenance time interval value to be 6 minutes can not only reduce the overhead but also achieve a long lifetime.

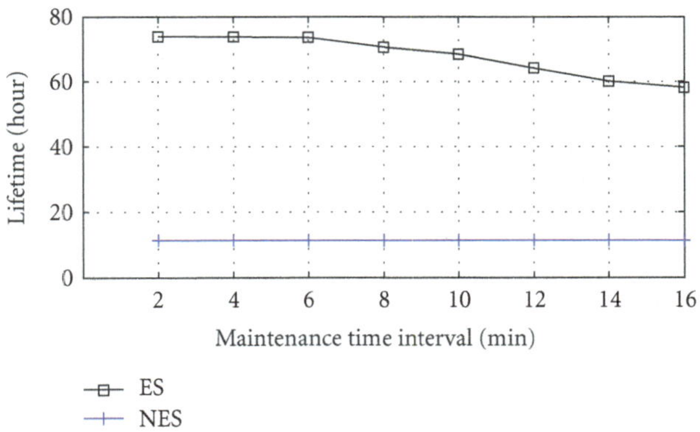

Figure 17: Impact of maintenance time interval.

9. Related Work

Energy harvesting is the conversion of ambient energy into usable electrical energy. Several technologies have been developed for extracting energy from the environment, including solar [26], wind [27], kinetic, and vibrational [28] energy. With these energy-harvesting technologies, researchers have designed various types of platforms to collect ambient energy from human activity or environments [23, 26, 27, 29]. Notable ones designed specially for sensor networks are Heliomote [23, 30], Prometheus [26], Trio [29], AmbiMax [27], and PUMA [31]. According to the type of energy storage used, these platforms can be separated into three categories: (i) rechargeable battery-based platforms [30], (ii) designs combining ultracapacitors and rechargeable batteries [26, 29], and (iii) capacitor-based designs.

(i) In rechargeable battery designs, such as Heliomote [23, 30], the energy harvesting panel is directly connected to its battery. As the primary energy storage device, the rechargeable battery is charged and discharged frequently, leading to low system lifetimes due to the physical limitation on the number of recharge cycles.

(ii) In designs that combine ultracapacitors and rechargeable batteries such as Prometheus [26], the solar energy is first stored in the primary energy buffer, which is one or multiple ultracapacitors. The rechargeable batteries are then used as the secondary energy buffer. This design inherits both the advantages and limitations of batteries and capacitors. For examples, it is difficult to predict remaining energy because of the inclusion of batteries, and the lifetime of energy storage subsystem is decided by the shelf time of batteries (in the order of a few years). Within this category, several similar systems have been built with a few enhanced features. For examples, AmbiMax [27] harvests energy from multiple ambient power sources (e.g., solar and wind generators), and PUMA [31] uses a power routing switch to route multiple power sources to multiple subsystems. The higher utilization of ambient power is achieved through a combination of MPPT and power defragmentation.

(iii) To our best knowledge, previous works have intentionally avoided capacitor-only design, citing the leakage issue [26]. To alleviate leakage, small capacitors are normally used, which makes secondary energy storage (batteries) necessary. However, the development of battery capacities is very slow and still leakage-prone. In addition, charging efficiency for batteries is comparatively low [9]. For example, according to the Natural Resources Defense Council [9], common battery chargers provide an efficiency between 6% and 40%. Different from previous works, this work investigates the frontiers of capacitor-only design and studies not only hardware designs but also related software control techniques to reduce the impact of energy leakage.

Energy conservation is an intensively studied area. Many solutions have been proposed at different layers, including high-efficiency hardware design [32], link layer design [33], topology management [34], node placement [35], network routing [36], sensing coverage [37], data aggregation [38], data placement [39, 40], operating system [41], and application-level energy-aware designs [42]. However, only a few works have focused on energy measurement [43] and energy adaptation [44–46]. Odyssey [44] and ECOsystem [45] demonstrate that application-level adaptations can successfully meet user-specified lifetimes for PDAs and laptops running Linux or other embedded OSes. The differences between these works and ours are that (i) they focus on battery modeling or application adaptations, while our work focuses on battery-free design and control and (ii) they currently do not consider energy harvesting issues.

However, only a few works have focused on energy adaptation [47, 48] and energy storage system [49]. The differences between these works and ours are that (i) they focus on battery modeling or application adaptations, while ours focuses on capacitor driven storage system design and control and (ii) they currently do not take the benefit of energy sharing among embedded devices.

The most closely related work is [50] which focuses on synchronizing the energy supply with demand in a small system powered by a single ultracapacitor. This paper focuses on the energy storage system design and proposes a generic solution for a large system which supports n capacitors and enables energy sharing.

10. Conclusion

The rapid advances in ultracapacitor design and the fast and highly efficient charging properties of ultracapacitor have motivated us to investigate the possibility of using capacitors as the sole energy storage and sharing system for embedded devices. In this paper, we build Energy Mule, an add-on power board that harvests energy from environment and employs an capacitor array as the only energy storage unit to power the embedded device and efficiently share energy among the neighboring embedded devices.

To minimize the energy leaked away inside a single embedded device, we propose adaptive charging and discharging control algorithms. Efficient, fast, and accurate energy sharing is conducted to prolong the service time of the system with multiple devices.

We invested significant amount of effort to evaluate our design in different real-world settings. The results indicate the effectiveness of our design. In future work, we shall focus more on the design of applications so that the system can get higher performance through energy sharing.

Acknowledgment

This paper was supported by NSF under Grant no. 1019343 to CRA for the CI Fellows Project.

References

[1] A. Kansal, A. A. Somasundara, D. D. Jea, M. B. Srivastava, and D. Estrin, "Intelligent fluid infrastructure for embedded networks," *Proceedings of the 2nd International Conference on Mobile Systems, Applications and Services (MobiSys '04)*, pp. 111–124, 2004.

[2] S. Jain, R. C. Shah, W. Brunette, G. Borriello, and S. Roy, "Exploiting mobility for energy efficient data collection in wireless sensor networks," *Mobile Networks and Applications*, vol. 11, no. 3, pp. 327–339, 2006.

[3] C. Hartung, R. Han, C. Seielstad, and S. Holbrook, "FireWxNet: a multi-tiered portable wireless system for monitoring weather conditions in wildland fire environments," in *Proceedings of the 4th International Conference on Mobile Systems, Applications and Services (MobiSys '06)*, pp. 28–41, June 2006.

[4] J. Eriksson, L. Girod, B. Hull, R. Newton, S. Madden, and H. Balakrishnan, "The Pothole Patrol: using a mobile sensor network for road surface monitoring," in *Proceedings of the 6th International Conference on Mobile Systems, Applications, and Services*, pp. 29–39, June 2008.

[5] B. Anand, A. L. Ananda, M. C. Chan et al., "Game action based power management for multiplayer online game," in *Proceedings of the ACM SIGCOMM Workshop on Networking, Systems, and Applications for Mobile Handhelds (MobiHeld '09)*, 2009.

[6] Y. Agarwal, R. Chandra, A. Wolman, P. Bahl, K. Chin, and R. Gupta, "Wireless wakeups revisited: energy management for voip over wi-fi smartphones," in *Proceedings of the 5th International Conference on Mobile Systems, Applications and Services (MobiSys '07)*, pp. 179–191, June 2007.

[7] R. K. Balan, N. Ramasubbu, K. Prakobphol, N. Christin, and J. Hong, "mFerio: the design and evaluation of a peer-to-peer mobile payment system," in *Proceedings of the 7th ACM International Conference on Mobile Systems, Applications, and Services (MobiSys '09)*, pp. 291–304, June 2009.

[8] A. G. Miklas, K. K. Gollu, K. K. W. Chan, S. Saroiu, P. K. Gummadi, and E. Lara, "Exploiting social interactions in mobile systems," in *Proceedings of the International Conference on Ubiquitous Computing (UbiComp '07)*, 2007.

[9] S. Foster, "The energy efficiency of common household battery charging systems: results and implications," *Natural Resources Defense Council*, http://www.efficientproducts.org/reports/bchargers/NRDC-Ecos_Battery_Charger_Efficiency.pdf.

[10] W. Rhines, "Mentor Graphics," http://chipdesignmag.com/lpd/blog/2009/04/15/chip-designers-scramble-for-low-power-solutions/.

[11] R. Signorelli, J. Schindall, and J. Kassakian, "Carbon nanotube enhanced double layer capacitor," in *Proceedings of the 15th International Seminar on Double Layer Capacitors and Hybrid Energy Storage Devices*, 2005.

[12] P. Banerjee, I. Perez, L. Henn-Lecordier, S. B. Lee, and G. W. Rubloff, "Nanotubular metal-insulator-metal capacitor arrays for energy storage," *Nature Nanotechnology*, vol. 4, no. 5, pp. 292–296, 2009.

[13] Weir et al., "Electrical-energy-storage unit (EESU) utilizing ceramic and integrated-circuit technologies for replacement of electrochemical batteries," US Patent 7033406, 2006.

[14] "MAXWELL Technologies," http://www.maxwell.com/products/ultracapacitors/products/k2-series.

[15] "NESSCAP Ultracapacitor Products," http://www.nesscap.com/product/edlc_large1.jsp.

[16] "Universal charger for mobile phone handsets," http://www.itu.int/newsroom/press_releases/2009/49.html.

[17] Sea-Bird Electronics, Inc, "Nickel-Metal Hydride (NiMH) Battery Charger and Battery Pack. User's Manual," http://www.seabird.com/pdf_documents/manuals/NiMH_004.pdf.

[18] J. Yun, S. Patel, M. Reynolds, and G. Abowd, "A quantitative investigation of inertial power harvesting for human-powered devices," in *Proceedings of the 10th International Conference on Ubiquitous Computing (UbiComp '08)*, pp. 74–83, September 2008.

[19] U. Maurer, A. Rowe, A. Smailagic, and D. P. Siewiorek, "eWatch: a wearable sensor and notification platform," in *Proceedings of the International Workshop on Wearable and Implantable Body Sensor Networks (BSN '06)*, pp. 142–145, April 2006.

[20] A. LaMarca, W. Brunette, D. Koizumi et al., "Making sensor networks practical with robots," in *Proceedings of the International Conference on Pervasive Computing (Pervasive '02)*, 2002.

[21] M. Matusiak, J. Paanajrvi, P. Appelqvist, M. Elomaa, M. Vainio, and A. Halme, "A novel marsupial robot society: towards long-term autonomy," in *Proceedings of the International Symposium on Distributed Autonomous Robotic Systems (DARS '08)*, 2008.

[22] A. LaMarca, D. Koizumi, M. Lease et al., "PlantCare: an investigation in practical ubiquitous systems," in *Proceedings of the 14th International Conference on Ubiquitous Computing (UbiComp '02)*, 2002.

[23] K. Lin, J. Yu, J. Hsu et al., "Heliomote: enabling long-lived sensor networks through solar energy harvesting," in *Proceedings of the International Conference on Embedded Networked Sensor Systems (SenSys '05)*, 2005.

[24] S. Guo and S. Keshav, "Fair and efficient scheduling in data ferrying networks," in *Proceedings of the 3rd International Conference on Emerging Networking EXperiments and Technologies (CoNEXT '07)*, December 2007.

[25] P. Zhang, C. M. Sadler, S. A. Lyon, and M. Martonosi, "Hardware design experiences in ZebraNet," in *Proceedings of the Second International Conference on Embedded Networked Sensor Systems (SenSys '04)*, pp. 227–238, November 2004.

[26] X. Jiang, J. Polastre, and D. Culler, "Perpetual environmentally powered sensor networks," in *Proceedings of the 4th International Symposium on Information Processing in Sensor Networks (IPSN '05)*, pp. 463–468, April 2005.

[27] C. Park and P. H. Chou, "AmbiMax: autonomous energy harvesting platform for multi-supply wireless sensor nodes," in *Proceedings of the 3rd Annual IEEE Communications Society*

on Sensor and Ad hoc Communications and Networks (Secon '06), pp. 168–177, September 2006.

[28] S. Meninger, J. O. Mur-Miranda, R. Amirtharajah, A. Chandrakasan, and J. Lang, "Vibration-to-electric energy conversion," in *Proceedings of the International Conference on Low Power Electronics and Design (ISLPED '99)*, pp. 48–53, August 1999.

[29] P. Dutta, J. Hui, J. Jeong et al., "Trio: enabling sustainable and scalable outdoor wireless sensor network deployments," in *Proceedings of the 5th International Conference on Information Processing in Sensor Networks (IPSN '06)*, pp. 407–415, April 2006.

[30] A. Kansal, J. Hsu, S. Zahedi, and M. B. Srivastava, "Power management in energy harvesting sensor networks," *ACM Transactions on Embedded Computing Systems*, vol. 6, no. 4, 2007.

[31] C. Park and P. H. Chou, "Power utility maximization for multiple-supply systems by a load-matching switch," in *Proceedings of the International Symposium on Lower Power Electronics and Design (ISLPED '04)*, pp. 168–173, August 2004.

[32] D. Lymberopoulos and A. Savvides, "XYZ: a motion-enabled, power aware sensor node platform for distributed sensor network applications," in *Proceedings of the 4th International Symposium on Information Processing in Sensor Networks (IPSN '05)*, pp. 449–454, April 2005.

[33] G. Zhou, T. He, S. Krishnamurthy, and J. A. Stankovic, "Impact of radio irregularity on wireless sensor networks," *Proceedings of the 2nd International Conference on Mobile Systems, Applications and Services (MobiSys '04)*, pp. 125–138, 2004.

[34] S. Lin, J. Zhang, G. Zhou, L. Gu, J. A. Stankovic, and T. He, "ATPC: adaptive transmission power control for wireless sensor networks," in *Proceedings of the 4th International Conference on Embedded Networked Sensor Systems (SenSys '06)*, pp. 223–236, November 2006.

[35] D. Ganesan, R. Cristescu, and B. Beferull-Lozano, "Power-efficient sensor placement and transmission structure for data gathering under distortion constraints," in *Proceedings of the 3rd International Symposium on Information Processing in Sensor Networks (IPSN '04)*, pp. 142–150, April 2004.

[36] K. Seada, M. Zuniga, A. Helmy, and B. Krishnamachari, "Energy-efficient forwarding strategies for geographic routing in lossy wireless sensor networks," in *Proceedings of the 2nd International Conference on Embedded Networked Sensor Systems (SenSys '04)*, pp. 108–121, November 2004.

[37] S. Kang, J. Lee, H. Jang et al., "SeeMon: scalable and energy-efficient context monitoring framework for sensor-rich mobile environments," in *Proceedings of the 6th International Conference on Mobile Systems, Applications, and Services (MobiSys '08)*, pp. 267–280, June 2008.

[38] S. Madden, M. Franklin, J. Hellerstein, and W. Hong, "TAG: a tiny aggregation service for Ad-Hoc sensor networks," in *Proceedings of the 9th USENIX Symposium on Operating Systems Design and Implementation (OSDI '02)*, 2002.

[39] S. Bhattacharya, H. Kim, S. Prabh, and T. Abdelzaher, "Energyconserving data placement and asynchronous multicast in wireless sensor networks," in *Proceedings of the International Conference on Mobile Systems, Applications, and Services (MobiSys '03)*, 2003.

[40] C. M. Sadler and M. Martonosi, "DALi: a communication-centric data abstraction layer for energy-constrained devices in mobile sensor networks," in *Proceedings of the 5th International Conference on Mobile Systems, Applications and Services (MobiSys '07)*, pp. 99–112, June 2007.

[41] C. C. Han, R. Kumar, R. Shea, E. Kohler, and M. Srivastava, "A dynamic operating system for sensor nodes," in *Proceedings of the 3rd International Conference on Mobile Systems, Applications, and Services (MobiSys '05)*, pp. 163–176, June 2005.

[42] T. He, S. Krishnamurthy, J. A. Stankovic et al., "Energy-efficient surveillance system using wireless sensor networks," *Proceedings of the 2nd International Conference on Mobile Systems, Applications and Services (MobiSys '04)*, pp. 270–283, 2004.

[43] X. Jiang, P. Dutta, D. Culler, and I. Stoica, "Micro power meter for energy monitoring of wireless sensor networks at scale," in *Proceedings of the 6th International Symposium on Information Processing in Sensor Networks (IPSN '07)*, pp. 186–195, April 2007.

[44] J. Flinn and M. Satyanarayanan, "Managing battery lifetime with energy-aware adaptation," *ACM Transactions on Computer Systems*, vol. 22, no. 2, pp. 137–179, 2004.

[45] H. Zeng, C. S. Ellis, and A. R. Lebeck, "Experiences in managing energy with ECOSystem," *IEEE Pervasive Computing*, vol. 4, no. 1, pp. 62–68, 2005.

[46] A. Lachenmann, P. J. Marrón, D. Minder, and K. Rothermel, "Meeting lifetime goals with energy levels," in *Proceedings of the ACM Conference on Embedded Networked Sensor Systems (SenSys '07)*, 2007.

[47] X. Liu, P. Shenoy, and M. D. Corner, "Chameleon: application-level power management," *IEEE Transactions on Mobile Computing*, vol. 7, no. 8, pp. 995–1010, 2008.

[48] J. Sorber, A. Kostadinov, M. Garber, M. Brennan, M. Corner, and E. Berger, "Eon: a language and runtime system for perpetual systems," in *Proceedings of the International Conference on Embedded Networked Sensor Systems (SenSys '07)*, 2007.

[49] N. Banerjee, A. Rahmati, M. D. Corner, S. Rollins, and L. Zhong, "Users and batteries: interactions and adaptive energy management in mobile systems," in *Proceedings of the International Conference on Ubiquitous Computing (Ubicomp '07)*, 2007.

[50] T. Zhu, Z. Zhong, Y. Gu, T. He, and Z. L. Zhang, "Leakage-aware energy synchronization for wireless sensor networks," in *Proceedings of the 7th ACM International Conference on Mobile Systems, Applications, and Services (MobiSys '09)*, pp. 319–332, June 2009.

16

Novel Complete Probabilistic Models of Random Variation in High Frequency Performance of Nanoscale MOSFET

Rawid Banchuin

Department of Computer Engineering, Siam University, 38 Petchkasem Road, Bangkok 10160, Thailand

Correspondence should be addressed to Rawid Banchuin; rawid_b@yahoo.com

Academic Editor: Jan Van der Spiegel

The novel probabilistic models of the random variations in nanoscale MOSFET's high frequency performance defined in terms of gate capacitance and transition frequency have been proposed. As the transition frequency variation has also been considered, the proposed models are considered as complete unlike the previous one which take only the gate capacitance variation into account. The proposed models have been found to be both analytic and physical level oriented as they are the precise mathematical expressions in terms of physical parameters. Since the up-to-date model of variation in MOSFET's characteristic induced by physical level fluctuation has been used, part of the proposed models for gate capacitance is more accurate and physical level oriented than its predecessor. The proposed models have been verified based on the 65 nm CMOS technology by using the Monte-Carlo SPICE simulations of benchmark circuits and Kolmogorov-Smirnov tests as highly accurate since they fit the Monte-Carlo-based analysis results with 99% confidence. Hence, these novel models have been found to be versatile for the statistical/variability aware analysis/design of nanoscale MOSFET-based analog/mixed signal circuits and systems.

1. Introduction

Nanoscale MOSFET has been adopted in many recently proposed analog/mixed signal circuits and systems such as high speed amplifier [1–3], millimeter wave components [4–6], and analog-to-digital converter [7–9]. Of course, the high frequency performances of these circuits and systems can be determined by those of their intrinsic MOSFET which can be defined by two major MOSFET's parameters, entitled gate capacitance, C_g, and transition frequency, f_T.

Obviously, imperfection in MOSFET's properties at the physical level, for example, random dopant fluctuation, line edge roughness, and so forth, causes the fluctuations in many of its characteristics such as threshold voltage, V_t, channel width, W, and channel length, L, which in turn yields the random variations in its circuit level parameters, for example, drain current, I_d, transconductance, g_m, and so forth. These variations are crucial in the statistical/variability aware design of MOSFET-based analog/mixed signal circuits and systems particularly at the nanometer regimes since their magnitudes become relatively large percentages. Hence, there are many previous researches on analytical modeling of such variations, for example, [10–13] and so forth, on which the nanoscale CMOS technology has been focused. However, they did not mention anything about variations in C_g and f_T even though they also exist and greatly affect the high frequency performances of the MOSFET-based circuits and systems. For these variations, results of studies as graphical plots and numerical measurements have been reported in previous articles, for example, [14–16] and so forth. In [17], the relationship between the variance of C_g and that of f_T has been derived. Unfortunately, formulation of each of these variations has not been mentioned. Later, the physical level oriented analytical model of random variation in C_g has been originally proposed in [18]. However, this model has been derived based on the classical model of physical level fluctuation-induced MOSFET's characteristic variation [19] which is being replaced by its more up-to-date successors such as that in [20]. Furthermore, the model in [18] is incomplete since the modeling on f_T which is as important as C_g has not been performed. To the knowledge of the author, such modeling on f_T has never been performed.

Hence, the novel probabilistic models of the random variations in nanoscale MOSFET's high frequency performances have been proposed in this paper. Beside the variation in C_g, that in f_T has also been focused on. So, these models are complete. Both random dopant fluctuation and process variation effects such as line edge roughness, which are the major causes of the random variations in the MOSFET's high frequency characteristics [16], have been taken into account. The nanoscale MOSFET equation [21] has been used as the mathematical basis similarly to [11, 12, 18]. As the precise mathematical expressions in terms of physical parameters, these models have been found to be both analytic and physical level oriented. Unlike [18], the state of the art physical level fluctuation-induced characteristic variation model of MOSFET [20] has been adopted instead of the classical one [19]. As a result, part of the models for variation in C_g is more accurate and physical level oriented than its predecessor [18]. These novel models have been verified based on the 65 nm CMOS process technology by using the Monte-Carlo SPICE simulations of the benchmark circuits and the Kolmogorov-Smirnov goodness of fit tests as very accurate since they fit the results obtained from Monte-Carlo SPICE simulations with 99% confidence. So, they have been found to be the versatile for the statistical/variability aware analysis/design of nanoscale MOSFET-based analog/mixed signal circuits and systems.

2. The Proposed Models

In this section, the proposed models will be presented. Before proceeding further, it is worthy to give some foundation on the nanoscale MOSFET's equation. It can be seen from [21] that I_d of the nanoscale MOSFET can be given with the gate oxide capacitance expressed in term of gate oxide permittivity ε_{ox} and thickness t_{ox} as

$$I_d = W\frac{\varepsilon_{ox}}{t_{ox}}\left(V_{gs} - V_t\right)v_{sat}, \tag{1}$$

where v_{sat} and V_{gs} denote the saturation velocity [20] and gate to source voltage. So, g_m of the nanoscale MOSFET is given by using (1) as follows:

$$g_m = W\frac{\varepsilon_{ox}}{t_{ox}}v_{sat}. \tag{2}$$

At this point, the derivation of the proposed models will be given by starting from finding the analytical expression of C_g and f_T. Similar to [18], C_g can be defined as [22]

$$C_g \triangleq \frac{dQ_g}{dV_{gs}}, \tag{3}$$

where Q_g denotes the gate charge [22] which can be determined as a function of I_d by using the approach adopted from [23] as

$$Q_g = -\frac{W^2 L\varepsilon_{ox}}{\alpha_d I_d V_t t_{ox}}\int_0^{V_{gs}-V_t}\left(V_{gs} - V_c - V_t\right)^2 dV_c - Q_{B,\max}, \tag{4}$$

where α_d and $Q_{B,\max}$ denote the coulomb scattering coefficient and the maximum bulk charge, respectively.

So, Q_g of the nanoscale MOSFET can be found by using (1) and (4) as

$$Q_g = -\frac{WL\left(V_{gs} - V_t\right)^2}{3\alpha_d v_{sat} V_t} - Q_{B,\max}. \tag{5}$$

As a result, C_g can be given by using (3) and (5) as follows:

$$C_g = \frac{2}{3}\frac{WL}{\alpha_d v_{sat}}\frac{V_t - V_{gs}}{V_t}. \tag{6}$$

At this point, the analytical expression of f_T can be immediately given by using (2) and (6) as follows:

$$f_T = \frac{3}{4}\frac{\alpha_d \varepsilon_{ox} v_{sat}^2}{t_{ox} L}\frac{V_t}{V_t - V_{gs}}. \tag{7}$$

By taking random dopant fluctuation and process variation effects into account, fluctuations in V_t, W, L, and so forth, denoted by ΔV_t, ΔW, ΔL, and so on, existed which yield random variations in C_g and f_T denoted by ΔC_g and Δf_T, respectively. This can be stated alternatively that ΔC_g and Δf_T are functions of ΔV_t, ΔW, ΔL, and so forth. With this in mind, ΔC_g and Δf_T can be mathematically defined as

$$\begin{aligned}\Delta C_g &\triangleq C_g\left(\Delta V_t, \Delta W, \Delta L, \ldots\right) - C_g,\\ \Delta f_T &\triangleq f_T\left(\Delta V_t, \Delta W, \Delta L, \ldots\right) - f_T,\end{aligned} \tag{8}$$

where $C_g\left(\Delta V_t, \Delta W, \Delta L, \ldots\right)$ and $f_T\left(\Delta V_t, \Delta W, \Delta L, \ldots\right)$ denoted the fluctuated C_g and f_T as functions of ΔV_t, ΔW, ΔL, and so on which are distributed in the Gaussian fashion. By using (6), (7), and (8), ΔC_g and Δf_T can be given in terms of physical level variables as follows:

$$\begin{aligned}\Delta C_g &= \frac{2}{3}\frac{(W + \Delta W)(L + \Delta L)}{\alpha_d v_{sat}}\frac{(V_t + \Delta V_t) - V_{gs}}{V_t + \Delta V_t} - \frac{2}{3}\frac{WL}{\alpha_d v_{sat}}\frac{V_t - V_{gs}}{V_t}\\ &\approx \frac{2WLV_{gs}\left(V_{FB} + 2\varphi_B - V_{tf}\right)}{3\alpha_d v_{sat}\left(V_{FB} + 2\varphi_B\right)^2},\\ \Delta f_T &= \frac{3}{4}\frac{\alpha_d \varepsilon_{ox} v_{sat}^2}{(t_{ox} + \Delta t_{ox})(L + \Delta L)}\frac{V_t + \Delta V_t}{(V_t + \Delta V_t) - V_{gs}} - \frac{3}{4}\frac{\alpha_d \varepsilon_{ox} v_{sat}^2}{t_{ox} L}\frac{V_t}{V_t - V_{gs}}\\ &\approx \frac{3\alpha_d \varepsilon_{ox} v_{sat}^2 V_{gs}\left(V_{FB} + 2\varphi_B - V_{tf}\right)}{4\pi t_{ox} L\left(V_{gs} - V_{FB} - 2\varphi_B\right)^2},\end{aligned} \tag{9}$$

where V_{FB}, V_{tf}, and φ_B denote flat band voltage, fluctuated threshold voltage, and half of surface band bending at inversion, respectively. Like other variations, Δt_{ox} denotes random variation in t_{ox}. The summation of V_{FB} and φ_B approximates the threshold voltage which its fluctuation is the dominant one. It can be seen that the deterministic behaviors of ΔC_g and Δf_T can be analytically modeled by using (9).

For the probabilistic modeling of our interested the probability density functions of ΔC_g and Δf_T, denoted by $\text{pdf}_{\Delta C_g}(\delta C_g)$ and $\text{pdf}_{\Delta f_T}(\delta f_T)$, where δC_g and δf_T stand for any sampled value of ΔC_g and that of Δf_T, respectively, must be derived. In order to do so, the up-to-date model of physical level fluctuations-induced variation in MOSFET's characteristic [20] has been adopted. At this point, $\text{pdf}_{\Delta C_g}(\delta C_g)$ and $\text{pdf}_{\Delta f_T}(\delta f_T)$ can be, respectively, derived by using (9) and the basis model as in (10) and (11), where N_{eff} and W_d denote effective doping concentration and depletion region width, respectively [20].

Consider,

$$\text{pdf}_{\Delta C_g}\left(\delta C_g\right) = \frac{4.33\ \alpha_d\ \varepsilon_{ox}\ v_{\text{sat}}\left(V_{\text{FB}} + 2\varphi_B\right)^2}{\sqrt{8\pi}\ \sqrt{q t_{\text{ox}}\ N_{\text{eff}}\ W_d\ WL\ V_{\text{gs}}}} \times \exp\left[-\frac{2.344\alpha_d^2\ \varepsilon_{\text{ox}}^2\ v_{\text{sat}}^2\left(V_{\text{FB}} + 2\varphi_B\right)^4 \delta C_g^2}{q t_{\text{ox}}\ N_{\text{eff}}\ W_d\ WL\ V_{\text{gs}}^2}\right], \quad (10)$$

$$\text{pdf}_{\Delta f_T}\left(\delta f_T\right) = \frac{\sqrt{\pi}\left(V_{\text{gs}} - V_{\text{FB}} - 2\varphi_B\right)^2}{q\alpha_d\ v_{\text{sat}}^2 V_{\text{gs}}} \sqrt{\frac{t_{\text{ox}} WL^3}{0.54 N_{\text{eff}} W_d}} \times \exp\left[-\frac{1.852\pi^2\left(V_{\text{gs}} - V_{\text{FB}} - 2\varphi_B\right)^4 t_{\text{ox}}\ WL^3 \delta f_T^2}{q^2\ \alpha_d^2\ v_{\text{sat}}^4 V_{\text{gs}}^2\ N_{\text{eff}}\ W_d}\right]. \quad (11)$$

Finally, the novel complete probabilistic models of the random variations in nanoscale MOSFET's high frequency performances have been now derived. By using these models and conventional statistical mathematics, the probabilistic/statistical behaviors of ΔC_g, Δf_T and even variations in other nanoscale MOSFET's high frequency parameters can be analytically explained in terms of related physical level variables. These models can be the mathematical tool for the statistical/variability aware optimization of MOSFET's high frequency performance and computer aid sensitivity analysis-based variability aware simulations of any circuit/system's high frequency parameters. The required computational effort has been found to be lower than that of the traditional Monte-Carlo analysis based on variation of MOSFET's characteristic which is time expensive [24]. Even though these models are oriented to variations in a single device, they can also be the modeling basis of mismatches in high frequency performances between transistors. These points will be discussed later.

Compared with [17] which proposed the relationship between the variance of C_g and that of f_T as [17]

$$\left(f_{T,\text{mean}}\right)^{-2}\sigma_{f_T}^2 = \left(g_{m,\text{mean}}\right)^{-2}\sigma_{g_m}^2 + \left(C_{g,\text{mean}}\right)^{-2}\sigma_{C_g}^2. \quad (12)$$

The models proposed in this paper, that is, $\text{pdf}_{\Delta C_g}(\delta C_g)$ and $\text{pdf}_{\Delta f_T}(\delta f_T)$ shown in (10) and (11), respectively, are more precise as they are the probability density functions in terms of physical level parameters such as W, L, and V_{FB}.

Furthermore, compared with the previous model in [18] composed of only one expression which models ΔC_g as [18]

$$\text{pdf}_{\Delta C_g}\left(\delta C_g\right) = \frac{3\alpha_d v_{\text{sat}} V_t}{2\sqrt{2\pi} A_{VT}\sqrt{WL}} \exp\left[\frac{9\alpha_d^2 v_{\text{sat}}^2 V_t^2 \delta C_g^2}{8A_{VT}^2 WL}\right]. \quad (13)$$

The models proposed in this research are more up-to-date and complete because they are derived based on newer basis model [20], and the previously overlooked Δf_T has now been modeled as $\text{pdf}_{\Delta f_T}(\delta f_T)$ in (11). Moreover, $\text{pdf}_{\Delta C_g}(\delta C_g)$ in (10) is more physical level oriented than its predecessor [18] as shown above since it contains additional physical level variables such as N_{eff}, W_d, and φ_B. The proposed $\text{pdf}_{\Delta C_g}(\delta C_g)$ is more accurate than its predecessor as can be seen from its better goodness of fit test results shown in the upcoming section in which verification of both $\text{pdf}_{\Delta C_g}(\delta C_g)$ and $\text{pdf}_{\Delta f_T}(\delta f_T)$ will be given. Hence, these novel models have been found to be the potential mathematical tool for the variability aware analysis/design of various MOSFET-based analog/mixed signal circuits and systems at the nanometer regime.

3. The Verification

Verification of the proposed models has been performed in both qualitative and quantitative aspects. In the qualitative sense, $\text{pdf}_{\Delta C_g}(\delta C_g)$ and $\text{pdf}_{\Delta f_T}(\delta f_T)$ have been graphically compared with probability distributions of ΔC_g and Δf_T obtained from the Monte-Carlo SPICE simulations of the benchmark circuits affected by normally distributed fluctuations in characteristics of MOSFET such as V_t, W, and L. For verification in the quantitative manner, goodness of fit tests have been performed to $\text{pdf}_{\Delta C_g}(\delta C_g)$ and $\text{pdf}_{\Delta f_T}(\delta f_T)$. Since it has been suspected that ΔC_g and Δf_T are normally distributed via observation of the proposed models, the appropriate goodness of fit test has been found to KS test Here, overview of the KS test and its application to this research will be given. Strategy of the KS test is to performed the comparison of the K-S test statistic (KS) and the critical value (c) where it can be stated that any model fits its target data set if and only if its KS does not exceed its c [25, 26]. For this research, KS can be defined as

$$\text{KS} = \max_{\delta x}\left\{\left|\text{CDF}'_{\Delta x}(\delta x)\right| - \left|\text{CDF}_{\Delta x}(\delta x)\right|\right\}, \quad (14)$$

where Δx can be either ΔC_g or Δf_T so δx can be either δC_g or δf_T. As a result, $\text{CDF}_{\Delta x}(\delta x)$ which represents the cumulative distribution function of any Δx can be either $\text{CDF}_{\Delta C_g}(\delta C_g)$ for $\Delta x = \Delta C_g$ and $\delta x = \delta C_g$ or $\text{CDF}_{\Delta f_T}(\delta f_T)$ for $\Delta x = \Delta f_T$ and $\delta x = \delta f_T$. Obviously, $\text{CDF}_{\Delta C_g}(\delta C_g)$ can be obtained by using $\text{pdf}_{\Delta C_g}(\delta C_g)$ as shown in (15) whereas $\text{CDF}_{\Delta f_T}(\delta f_T)$ can be derived by using $\text{pdf}_{\Delta f_T}(\delta f_T)$ as in (16).

Note that erf( ) denotes the error function which can be mathematically defined for arbitrary variable y as $\mathrm{erf}(y) = (2/\sqrt{\pi})\int_0^y \exp(-u^2)du$.

On the other hand, $\mathrm{CDF}'_{\Delta x}(\delta x)$ denotes the benchmark cumulative distribution function of any Δx which can be either ΔC_g or Δf_T, calculated by integration of its corresponding probability distribution obtained from simulations of benchmark circuits. Since the confidence level of the test is 99%, c can be given by (17), where n denotes the number of runs of Monte-Carlo SPICE analysis, which is chosen to be 3000 for this research. As a result, $c = 0.0297596$.

Consider,

$$\mathrm{CDF}_{\Delta C_g}(\delta C_g) = \int_{-\infty}^{\delta C_g} \mathrm{pdf}_{\Delta C_g}(u)\,du = \frac{1}{2}\left\{1 + \mathrm{erf}\left[\frac{\sqrt{2.344}\alpha_d \varepsilon_{ox} v_{sat}(V_{FB} + 2\varphi_B)^2 \delta C_g}{\sqrt{q t_{ox} N_{eff} W_d} W L V_{gs}}\right]\right\}, \tag{15}$$

$$\mathrm{CDF}_{\Delta f_T}(\delta f_T) = \int_{-\infty}^{\delta C_g} \mathrm{pdf}_{\Delta f_T}(v)\,dv = \frac{1}{2}\left\{1+\mathrm{erf}\left[\sqrt{\frac{1.852 t_{ox} W L^3}{N_{eff} W_d}}\frac{\pi(V_{gs}-V_{FB}-2\varphi_B)^2 \delta f_T}{q \alpha_d V_{gs} v_{sat}^2}\right]\right\}, \tag{16}$$

$$c = \frac{1.63}{\sqrt{n}}. \tag{17}$$

At this point, other issues such as verification basis, benchmark circuits, and simulation methodologies will be mentioned. As the nanoscale MOSFET is focused, the 65 nm CMOS process technology has been adopted as the verification basis. Similar to [18], parameterization of the proposed models has been performed by using the 65 nm CMOS process parameters, and the BSIM4-based benchmark circuits at 65 nm level with all necessary parameters extracted by Predictive Technology Model (PTM) [27] have been used. Of course, both NMOS and PMOS technologies have been considered.

In order to verify the accuracy of $\mathrm{pdf}_{\Delta C_g}(\delta C_g)$, the benchmark circuit and methodology used in [18] can be adopted. The core of this circuit for NMOS technological basis can be depicted in Figure 1, and C_g can be found as [18]

$$C_g = \frac{Y_{in}}{j\omega} = \frac{1}{j\omega}\frac{I_{in}}{V_{in}}, \tag{18}$$

where Y_{in}, I_{in}, and V_{in} denote the input admittance, input current, and input voltage of the benchmark circuit, respectively. Similar to [18], the Monte-Carlo SPICE simulation of this circuit gives the benchmark probability distribution of ΔC_g

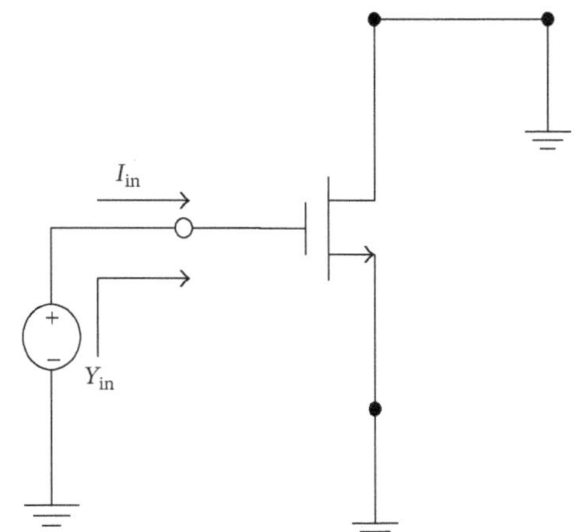

FIGURE 1: NMOS-based circuit for verifying $\mathrm{pdf}_{\Delta C_g}(\delta C_g)$ [18].

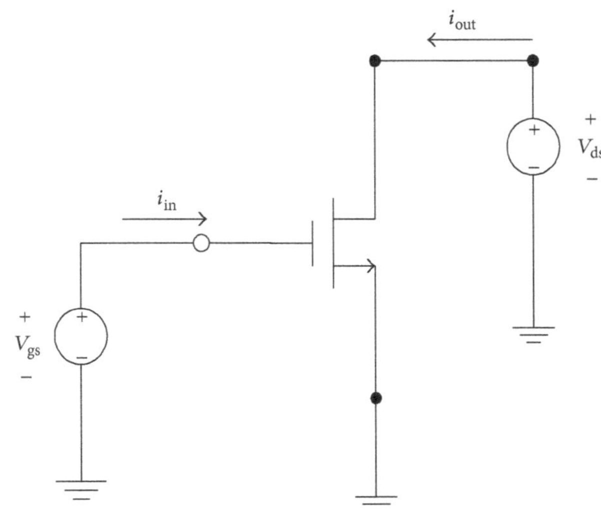

FIGURE 2: NMOS-based circuit for verifying $\mathrm{pdf}_{\Delta f_T}(\delta f_T)$.

for verification of $\mathrm{pdf}_{\Delta C_g}(\delta C_g)$ based on NMOS technology. For PMOS technology, the similar circuit with the NMOS device replaced with PMOS one and the similar methodology can be used as done in [18].

For verifying $\mathrm{pdf}_{\Delta f_T}(\delta f_T)$, core of the benchmark circuit can be depicted in Figure 2 for NMOS-based verification. From this circuit, f_T can be defined as the frequency in which the current gain is unity; that is, the magnitude of the small signal input current, $|i_{in}|$ is equal to that of the output one, $|i_{out}|$. Of course, the Monte-Carlo SPICE simulation of this circuit gives the benchmark probability distribution of Δf_T for verification of $\mathrm{pdf}_{\Delta f_T}(\delta f_T)$ based on NMOS technology. Of course, the similar circuit with the NMOS device replaced with PMOS one and the similar methodology can be used for PMOS technology-based verification. It should be mentioned here that all circuits operate under 1 V. supply voltage with $L = 60$ nm which is closed to the minimum allowable one, and $W = 1.2\,\mu$m. In the upcoming subsections, NMOS technology-based verification and PMOS-based one will be, respectively given; then it will be shown that $\mathrm{pdf}_{\Delta C_g}(\delta C_g)$ is more accurate than its predecessor.

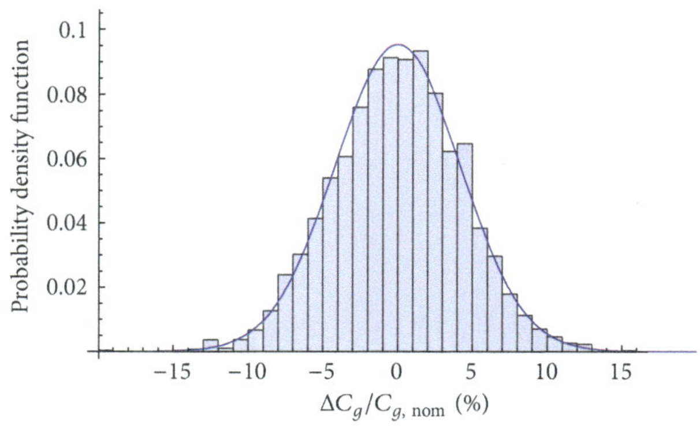

FIGURE 3: NMOS-based graphical comparison for the probability distribution of ΔC_g: $\text{pdf}_{\Delta C_g}(\delta C_g)$ (line) versus benchmark distribution from the circuit (histogram).

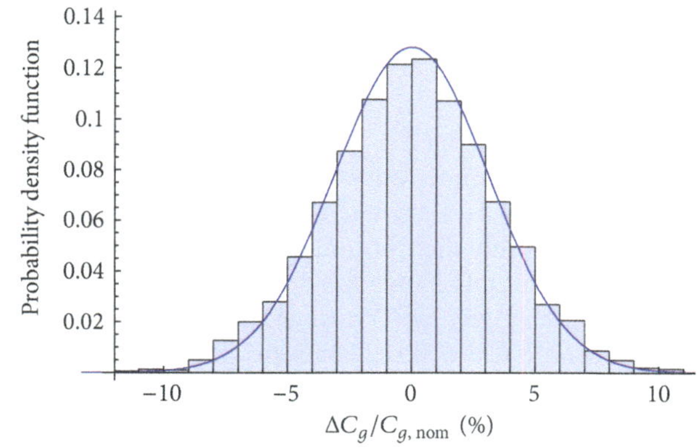

FIGURE 5: PMOS-based graphical comparison for the probability distribution of ΔC_g: $\text{pdf}_{\Delta C_g}(\delta C_g)$ (line) versus benchmark distribution from the circuit (histogram).

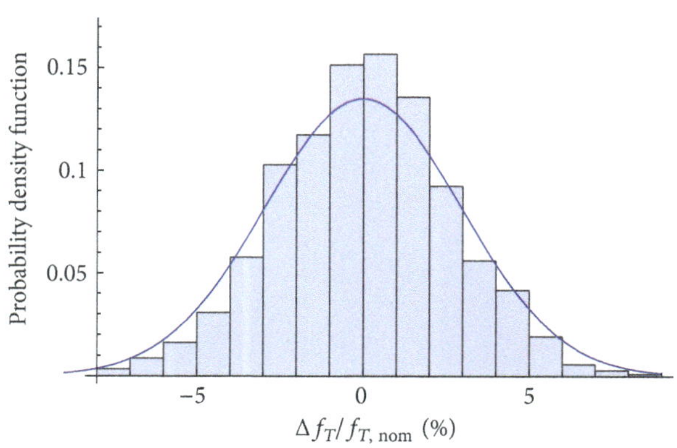

FIGURE 4: NMOS-based graphical comparison for the probability distribution of Δf_T: $\text{pdf}_{\Delta f_T}(\delta f_T)$ (line) versus benchmark distribution from the circuit (histogram).

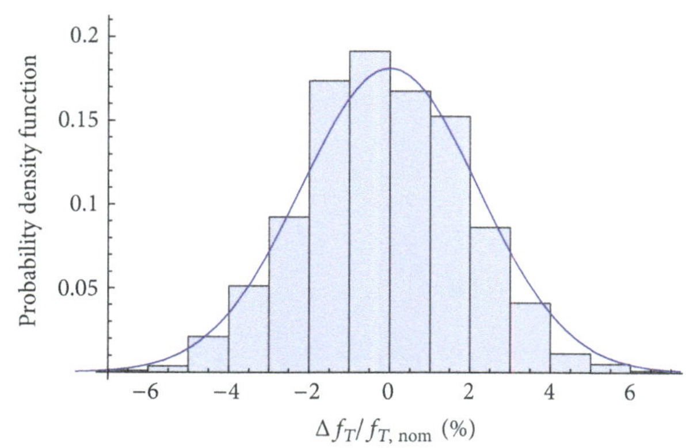

FIGURE 6: PMOS-based graphical comparison for the probability distribution of Δf_T: $\text{pdf}_{\Delta f_T}(\delta f_T)$ (line) versus benchmark distribution from the circuit (histogram).

3.1. NMOS-Based Verification. The graphical comparisons of the probability distributions are depicted in Figures 3 and 4 for verification in the qualitative sense of $\text{pdf}_{\Delta C_g}(\delta C_g)$ and that of $\text{pdf}_{\Delta f_T}(\delta f_T)$, respectively. The horizontal axes are labeled by $\Delta C_g/C_{g,\text{nom}}$ and $\Delta f_T/f_{T,\text{nom}}$ which display the amounts of ΔC_g and Δf_T as percentages of nominal C_g, $C_{g,\text{nom}}$ and nominal f_T, $f_{T,\text{nom}}$. A strong agreement between $\text{pdf}_{\Delta C_g}(\delta C_g)$ and its benchmark obtained from the Monte-Carlo simulation along with that between $\text{pdf}_{\Delta f_T}(\delta f_T)$ and its benchmark has been found. Furthermore, normal distributions of variations high frequency performances as predicted by the proposed models have been observed.

In the quantitative point of view, it can be seen that the resulting KS from the test of $\text{pdf}_{\Delta C_g}(\delta C_g)$ can be found as KS = 0.021463, and that from the test of $\text{pdf}_{\Delta f_T}(\delta f_T)$ is KS = 0.029223. These statistics do not exceed $c = 0.0297596$. This means that the proposed models can fit ΔC_g and Δf_T obtained from the Monte-Carlo SPICE simulation of NMOS-based circuits with 99% confidence. At this point, both $\text{pdf}_{\Delta C_g}(\delta C_g)$ and $\text{pdf}_{\Delta f_T}(\delta f_T)$ have been verified based on NMOS technology as highly accurate.

3.2. PMOS-Based Verification. The graphical comparisons of the probability distributions are depicted in Figures 5 and 6 for verification in the qualitative sense of $\text{pdf}_{\Delta C_g}(\delta C_g)$ and that of $\text{pdf}_{\Delta f_T}(\delta f_T)$. Strong agreements along with normal distributions of variations high frequency performances similar to those of NMOS-based verification have been found.

For verification in quantitative manner, on the other hand, KS from the test of $\text{pdf}_{\Delta C_g}(\delta C_g)$ can be found as KS = 0.016144 and that from the test of $\text{pdf}_{\Delta f_T}(\delta f_T)$ is KS = 0.029192. Similar to the previous NMOS-based case, these KS values do not exceed $c = 0.0297596$. This means that the proposed model can fit ΔC_g and Δf_T obtained from the simulation of PMOS-based circuits with 99% confidence. At this point, $\text{pdf}_{\Delta C_g}(\delta C_g)$ and $\text{pdf}_{\Delta f_T}(\delta f_T)$ are verified based on PMOS technology as highly accurate.

3.3. Higher Accuracy of $\text{pdf}_{\Delta C_g}(\delta C_g)$. In order to show that $\text{pdf}_{\Delta C_g}(\delta C_g)$ proposed in this research is more accurate than its predecessor, its KS statistics have been compared with those of such predecessor as given in [18]. Since the verification of the proposed $\text{pdf}_{\Delta C_g}(\delta C_g)$ and that of its predecessor are identical, the comparison of the obtained statistics is said to be unbiased.

Obviously, the KS values obtained from verifying the proposed $\mathrm{pdf}_{\Delta C_g}(\delta C_g)$ which are 0.025466 for NMOS-based verification and 0.016144 for PMOS based are one, respectively-lower than those of the previous one in [18] given by 0.028462 for NMOS-based verification and 0.018036 for PMOS-based one [18]. According to [25, 26], this means $\mathrm{pdf}_{\Delta C_g}(\delta C_g)$ in this research is better fit to the simulated data than its predecessor [18] according to its lower KS statistics. This can be alternatively stated that the proposed $\mathrm{pdf}_{\Delta C_g}(\delta C_g)$ is more accurate.

4. Discussion

In this section, some discussions regarding to the applications of the proposed models introduced in the previous section and some verification results will be given.

4.1. Analytical Explanation of the Probabilistic/Statistical Behaviors of ΔC_g and Δf_T and Other Variations. The probabilistic/statistical behaviors of ΔC_g, Δf_T and variations in other high frequency parameters can be analytically explained by using their probabilistic/statistical parameters which can be derived by using the proposed $\mathrm{pdf}_{\Delta C_g}(\delta C_g)$ and $\mathrm{pdf}_{\Delta f_T}(\delta f_T)$ along with traditional statistical mathematic as analytical expressions involving physical level parameters. Simple examples of these parameters are means and variances of ΔC_g and Δf_T. They can be derived by applying statistical mathematic to $\mathrm{pdf}_{\Delta C_g}(\delta C_g)$ and $\mathrm{pdf}_{\Delta f_T}(\delta f_T)$ as follows:

$$\overline{\Delta C_g} = \int_{-\infty}^{\infty} \delta C_g \mathrm{pdf}_{\Delta C_g}(\delta C_g)\, d\delta C_g = 0, \quad (19a)$$

$$\sigma^2_{\Delta C_g} = \int_{-\infty}^{\infty} \left(\delta C_g - \overline{\Delta C_g}\right)^2 \mathrm{pdf}_{\Delta C_g}(\delta C_g)\, d\delta C_g = \frac{q t_{ox} N_{\mathrm{eff}} W_d WL V_{gs}^2}{4.688\alpha_d^2 \varepsilon_{ox}^2 v_{\mathrm{sat}}^2 (V_{FB} + 2\varphi_B)^4}, \quad (19b)$$

$$\overline{\Delta f_T} = \int_{-\infty}^{\infty} \delta f_T \mathrm{pdf}_{\Delta f_T}(\delta f_T)\, d\delta f_T = 0, \quad (20a)$$

$$\sigma^2_{\Delta f_T} = \int_{-\infty}^{\infty} \left(\delta f_T - \overline{\Delta f_T}\right)^2 \mathrm{pdf}_{\Delta f_T}(\delta f_T)\, d\delta f_T = \frac{q^2 \alpha_d^2 v_{\mathrm{sat}}^4 V_{gs}^2 N_{\mathrm{eff}} W_d}{3.704\pi^2 \left(V_{gs} - V_{FB} - 2\varphi_B\right)^4 t_{ox} WL^3}, \quad (20b)$$

where $\overline{\Delta C_g}$, $\sigma^2_{\Delta C_g}$, $\overline{\Delta f_T}$, and $\sigma^2_{\Delta f_T}$ denote mean and variance of ΔC_g along with those of Δf_T, respectively. Analytically obtaining these parameters yields various benefits to the statistical/variability aware analysis/design of nanoscale MOSFET-based analog/mixed signal circuits and systems, for example, obtaining the design tradeoffs which are important issues and so forth. A simple example of such design tradeoffs is shrinking the transistor dimension reduces ΔC_g with increasing of Δf_T as a penalty. This design tradeoff can be obtained from observations of (19b) and (20b) that $\sigma^2_{\Delta C_g} \propto WL$ and $\sigma^2_{\Delta f_T} \propto 1/WL^3$, respectively. It should be mentioned here that more sophisticated parameters such as moments of various orders, skewness, and kurtosis can also be derived by applying the statistical mathematic to $\mathrm{pdf}_{\Delta C_g}(\delta C_g)$ and $\mathrm{pdf}_{\Delta f_T}(\delta f_T)$.

An interesting joint parameter of ΔC_g and Δf_T is their correlation coefficient denoted by $\rho_{\Delta C_g,\Delta f_T}$, which defines their statistical relationship. $\rho_{\Delta C_g,\Delta f_T}$ can be given in terms of $\overline{\Delta C_g}$, $\sigma^2_{\Delta C_g}$, $\overline{\Delta f_T}$, and $\sigma^2_{\Delta f_T}$ as follows:

$$\rho_{\Delta C_g,\Delta f_T} = \frac{E\left[\left(\Delta C_g - \overline{\Delta C_g}\right)\left(\Delta f_T - \overline{\Delta f_T}\right)\right]}{\sqrt{\sigma^2_{\Delta C_g}}\sqrt{\sigma^2_{\Delta f_T}}}, \quad (21)$$

where $E[\ ]$ denotes the expectation operator. By using (9) and (19a)–(20b) which can be obtained from $\mathrm{pdf}_{\Delta C_g}(\delta C_g)$ and $\mathrm{pdf}_{\Delta f_T}(\delta f_T)$, along with (21), magnitude of $\rho_{\Delta C_g,\Delta f_T}$ has been found as unity. This means that there exists a very strong statistical relationship between ΔC_g and Δf_T.

As mentioned above, the analytical explanation of the probabilistic/statistical behaviors of variations in other high frequency parameters can be performed by using the $\mathrm{pdf}_{\Delta C_g}(\delta C_g)$ and $\mathrm{pdf}_{\Delta f_T}(\delta f_T)$ as these high frequency parameters are functions of C_g and/or f_T. Let any of such high frequency parameter and its variation be denoted by X and ΔX, respectively: ΔX can be given by

$$\Delta X = \left(\frac{\partial X}{\partial C_g}\right)\Delta C_g + \left(\frac{\partial X}{\partial f_T}\right)\Delta f_T. \quad (22)$$

Obviously, ΔX is a random variable where its probabilistic/statistical behavior can be analytically explained by using its probabilistic/statistical functions/parameters which can be derived by using either the proposed $\mathrm{pdf}_{\Delta C_g}(\delta C_g)$ and $\mathrm{pdf}_{\Delta f_T}(\delta f_T)$ or related parameters which are their functions such as $\overline{\Delta C_g}$, $\sigma^2_{\Delta C_g}$, $\overline{\Delta f_T}$, $\sigma^2_{\Delta f_T}$, and so forth. Simple examples of such functions/parameters of ΔX are its mean $\overline{\Delta X}$ and variance $\sigma^2_{\Delta X}$. They can be formulated in terms of $\overline{\Delta C_g}$, $\sigma^2_{\Delta C_g}$, $\overline{\Delta f_T}$, and $\sigma^2_{\Delta f_T}$ as follows:

$$\overline{\Delta X} = \left(\frac{\partial X}{\partial C_g}\right)\overline{\Delta C_g} + \left(\frac{\partial X}{\partial f_T}\right)\overline{\Delta f_T} = 0, \quad (23)$$

$$\sigma^2_{\Delta X} = \left(\frac{\partial X}{\partial C_g}\right)^2 \sigma^2_{\Delta Cg} + \left(\frac{\partial X}{\partial f_T}\right)^2 \sigma^2_{\Delta fT} + 2\left(\frac{\partial X}{\partial C_g}\right)\left(\frac{\partial X}{\partial f_T}\right)\sqrt{\sigma^2_{\Delta Cg}\sigma^2_{\Delta fT}}. \quad (24)$$

Obviously, the fact that magnitude of $\rho_{\Delta C_g,\Delta f_T}$ is unity obtained by using $\mathrm{pdf}_{\Delta C_g}(\delta C_g)$ and $\mathrm{pdf}_{\Delta f_T}(\delta f_T)$ as stated above has been used in the formulation of (24). By using (19a)–(20b) which can be obtained from the $\text{pdf}_{\Delta C_g}(\delta C_g)$ and $\text{pdf}_{\Delta f_T}(\delta f_T)$, (23) and (24), $\overline{\Delta X}$ and $\sigma^2_{\Delta X}$ can be derived as analytical expressions in terms of physical level parameters.

More sophisticatedly, the probability density function of ΔX denoted by $\text{pdf}_{\Delta X}(\delta X)$, where δX stands for the sample variable of ΔX, can be derived by using $\text{pdf}_{\Delta C_g}(\delta C_g)$ and $\text{pdf}_{\Delta f_T}(\delta f_T)$ along with a well-known statistical mathematic methodology entitled the transformation of random variables which can be found in the usual texts such as [28, 29], as ΔX is a function of ΔC_g and/or Δf_T, since X is a function of C_g and/or f_T as aforementioned. According to [28, 29], if ΔX is function of Δu which can be either ΔC_g or Δf_T as X is function of u which can be either C_g or f_T, $\text{pdf}_{\Delta X}(\delta X)$ can be given in a straightforward manner as

$$\text{pdf}_{\Delta X}(\delta X) = \text{pdf}_{\Delta u}(\delta u(\delta X)) \left[\left[\left| \frac{\partial \delta u}{\partial \delta X} \right| \right]_{\delta u \to \delta u(\delta X)} \right]^{-1}, \tag{25}$$

where $\text{pdf}_{\Delta u}(\delta u(\delta X))$ stands for the probability density function of Δu denoted by $\text{pdf}_{\Delta u}(\delta u)$ with δu expressed in terms of δX. Since Δu can be either ΔC_g or Δf_T, $\text{pdf}_{\Delta u}(\delta u)$ can be either $\text{pdf}_{\Delta C_g}(\delta C_g)$ or $\text{pdf}_{\Delta f_T}(\delta f_T)$.

According to [29], if ΔX depends on both ΔC_g and Δf_T as X is a function of both C_g and f_T, $\text{pdf}_{\Delta X}(\delta X)$ can be given by

$$\text{pdf}_{\Delta X}(\delta X) = \int_{-\infty}^{\infty} l(\delta X, \delta Y)\, d\delta Y, \tag{26a}$$

where

$$l(\delta X, \delta Y) = \text{pdf}_{\Delta C_g, \Delta f_T}\left(\delta C_g(\delta X, \delta Y), \delta f_T(\delta X, \delta Y)\right) \times J(\delta X, \delta Y), \tag{26b}$$

$$J(\delta X, \delta Y) = \begin{vmatrix} \dfrac{\partial\left[\delta C_g(\delta X, \delta Y)\right]}{\partial \delta X} & \dfrac{\partial\left[\delta C_g(\delta X, \delta Y)\right]}{\partial \delta Y} \\ \dfrac{\partial\left[\delta f_T(\delta X, \delta Y)\right]}{\partial \delta X} & \dfrac{\partial\left[\delta f_T(\delta X, \delta Y)\right]}{\partial \delta Y} \end{vmatrix}, \tag{26c}$$

$$\delta C_g(\delta X, \delta Y) = \begin{vmatrix} \delta X & \dfrac{\partial X}{\partial f_T} \\ \delta Y & \dfrac{\partial Y}{\partial f_T} \end{vmatrix} \left[\left| \begin{matrix} \dfrac{\partial X}{\partial C_g} & \dfrac{\partial X}{\partial f_T} \\ \dfrac{\partial Y}{\partial C_g} & \dfrac{\partial Y}{\partial f_T} \end{matrix} \right| \right]^{-1}, \tag{26d}$$

$$\delta f_T(\delta X, \delta Y) = \begin{vmatrix} \dfrac{\partial X}{\partial C_g} & \delta X \\ \dfrac{\partial Y}{\partial C_g} & \delta Y \end{vmatrix} \left[\left| \begin{matrix} \dfrac{\partial X}{\partial C_g} & \dfrac{\partial X}{\partial f_T} \\ \dfrac{\partial Y}{\partial C_g} & \dfrac{\partial Y}{\partial f_T} \end{matrix} \right| \right]^{-1}, \tag{26e}$$

where δY denotes the sample variable of ΔY which stands for the variation in a conveniently chosen auxiliary function denoted by Y. Similar to X, Y is also a function of both C_g and f_T. It should be mentioned here that $\text{pdf}_{\Delta C_g, \Delta f_T}(\delta C_g(\delta X, \delta Y), \delta f_T(\delta X, \delta Y))$ can be found by applying (26d) and (26e) to $\text{pdf}_{\Delta C_g, \Delta f_T}(\delta C_g, \delta f_T)$ which stands for the joint probability density function of ΔC_g and Δf_T. This function can be found from the $\text{pdf}_{\Delta C_g}(\delta C_g)$ and $\text{pdf}_{\Delta f_T}(\delta f_T)$ by simultaneously solving the following equations:

$$\text{pdf}_{\Delta C_g}(\delta C_g) = \int_{-\infty}^{\infty} \text{pdf}_{\Delta C_g, \Delta f_T}(\delta C_g, \delta f_T)\, d\delta f_T, \tag{27a}$$

$$\text{pdf}_{\Delta f_T}(\delta f_T) = \int_{-\infty}^{\infty} \text{pdf}_{\Delta C_g, \Delta f_T}(\delta C_g, \delta f_T)\, d\delta C_g. \tag{27b}$$

The next subsection will discuss the application of $\text{pdf}_{\Delta C_g}(\delta C_g)$ and $\text{pdf}_{\Delta f_T}(\delta f_T)$ as the mathematical tool of the statistical/variability aware optimization scheme for nanoscale MOSFET's high frequency performance.

4.2. Mathematical Tool of Statistical/Variability Aware Optimization Scheme. Obviously, target of the optimization has been found to be minimizing ΔC_g and Δf_T. Simple objective functions can be obtained by using $\sigma^2_{\Delta C_g}$ and $\sigma^2_{\Delta f_T}$ which can be derived by using $\text{pdf}_{\Delta C_g}(\delta C_g)$ and $\text{pdf}_{\Delta f_T}(\delta f_T)$ as follows:

$$\begin{aligned} &\text{minimize}\left[\sigma^2_{\Delta C_g}\right], \\ &\text{minimize}\left[\sigma^2_{\Delta f_T}\right]. \end{aligned} \tag{28}$$

More sophisticatedly, the objective functions can be given in terms of $\Pr\{\Delta C_g = 0\}$ and $\Pr\{\Delta f_T = 0\}$ which stand for the probability of obtaining $\Delta C_g = 0$ and that of obtaining $\Delta f_T = 0$, respectively. These probabilities can be determined from $\text{pdf}_{\Delta C_g}(\delta C_g)$ and $\text{pdf}_{\Delta f_T}(\delta f_T)$ by using the mathematical foundation in [28] as follows:

$$\begin{aligned} \Pr\{\Delta C_g = 0\} &= \lim_{\chi \to 0}\left[\int_{-\chi}^{0} \text{pdf}_{\Delta C_g}(\delta C_g)\, d\delta C_g\right], \\ \Pr\{\Delta f_T = 0\} &= \lim_{\varphi \to 0}\left[\int_{-\varphi}^{0} \text{pdf}_{\Delta f_T}(\delta f_T)\, d\delta f_T\right], \end{aligned} \tag{29}$$

where $\chi \neq 0$ and $\varphi \neq 0$; moreover, they may or may not be equal to each other. In this case, the optimal MOSFET's parameters can be found by searching for those which maximize both $\Pr\{\Delta C_g = 0\}$ and $\Pr\{\Delta f_T = 0\}$. As a result, the objective functions can be now given by

$$\begin{aligned} &\text{maximize}\left[\Pr\{\Delta C_g = 0\}\right], \\ &\text{maximize}\left[\Pr\{\Delta f_T = 0\}\right]. \end{aligned} \tag{30}$$

If a single objective function optimization scheme is required, $\Pr\{\Delta C_g = 0, \Delta f_T = 0\}$ which is the probability to obtain $\Delta C_g = 0$ and $\Delta f_T = 0$ simultaneously, can be used. This probability can be defined as in (31), where $\text{pdf}_{\Delta C_g, \Delta f_T}(\delta C_g, \delta f_T)$ can be determined from $\text{pdf}_{\Delta C_g}(\delta C_g)$ and $\text{pdf}_{\Delta f_T}(\delta f_T)$ as shown in (27a) and (27b). In this case, the optimal MOSFET's parameters can be found by searching for those which maximize $\Pr\{\Delta C_g = 0, \Delta f_T = 0\}$. As a result, the objective function can be obtained as in (32).

Consider,

$$\Pr\{\Delta C_g = 0, \Delta f_T = 0\} = \lim_{\varphi \to 0}\left[\lim_{\chi \to 0}\left[\int_{-\varphi}^{0}\int_{-\chi}^{0} \mathrm{pdf}_{\Delta C_g,\Delta f_T}\left(\delta C_g, \delta f_T\right) d\delta C_g \, d\delta f_T\right]\right], \tag{31}$$

$$\text{maximize } \left[\Pr\{\Delta C_g = 0, \Delta f_T = 0\}\right]. \tag{32}$$

The next subsection will give a discussion on the application of $\mathrm{pdf}_{\Delta C_g}(\delta C_g)$ and $\mathrm{pdf}_{\Delta f_T}(\delta f_T)$ as the mathematical tool of the variability aware simulation of nanoscale MOSFET's high frequency performance with reduced computational effort.

4.3. Mathematical Tool for Reduced Computational Effort Variability Aware Simulation of High Frequency Parameters. Obviously, outcome of any variability aware simulation is often expressed as the standard deviation of the simulated parameter. Let any interested high frequency parameter of variability aware simulation be Z, the desired outcome in terms of σ_Z can be conventionally determined by using the Monte-Carlo simulation which requires numerous runs [24]. So, many of the computational efforts are consumed.

On the other hand, σ_Z can be computed in a more computationally efficient manner by using the sensitivity of Z with respect to C_g and that with respect to f_T. These sensitivities can be determined by using the sensitivity analysis for which the targeted circuit/system is needed to be solved only once [24, 30]; then σ_Z can be simply computed by using $\sigma^2_{\Delta C_g}$ and $\sigma^2_{\Delta f_T}$ of MOSFETs within the targeted circuit/system. It can be seen that much of the computational effort for determining σ_Z is significantly reduced by using such sensitivity analysis-based method.

At this point, the sensitivity analysis-based computation of σ_Z will be explained. Let any circuits/systems compose M MOSFETs; σ_Z can be given in terms of $\sigma^2_{\Delta C_g}$ of any ith MOSFET, $\sigma^2_{\Delta C_g,i}$, that of any jth MOSFET, $\sigma^2_{\Delta C_g,j}$, $\sigma^2_{\Delta f_T}$ of any ith MOSFET $\sigma^2_{\Delta f_T,i}$, and that of any jth MOSFET, $\sigma^2_{\Delta f_T,j}$, as

$$\begin{aligned}
\sigma_Z = \Bigg[& \sum_{i=1}^{M}\left[\left(S^Z_{C_g}\big|_i\right)^2 \sigma^2_{\Delta C_g,i} + \left(S^Z_{f_T}\big|_i\right)^2 \sigma^2_{\Delta f_T,i}\right] \\
& + \sum_{\substack{i=1 \\ i\neq j}}^{M}\sum_{j=1}^{M}\left[\left(S^Z_{C_g}\big|_i\right)\left(S^Z_{C_g}\big|_j\right)\rho_{\Delta C_g,i,\Delta C_g,j}\sqrt{\sigma^2_{\Delta C_g,i}}\sqrt{\sigma^2_{\Delta C_g,j}}\right. \\
& \left. + \left(S^Z_{f_T}\big|_i\right)\left(S^Z_{f_T}\big|_j\right)\rho_{\Delta f_T,i,\Delta f_T,j}\sqrt{\sigma^2_{\Delta f_T,i}}\sqrt{\sigma^2_{\Delta f_T,j}}\right] \\
& + 2\sum_{i=1}^{M}\sum_{j=1}^{M}\left[\left(S^Z_{C_g}\big|_i\right)\left(S^Z_{f_T}\big|_j\right)\right. \\
& \left.\times \rho_{\Delta C_g,i,\Delta f_T,j}\sqrt{\sigma^2_{\Delta C_g,i}}\sqrt{\sigma^2_{\Delta f_T,j}}\right]\Bigg]^{1/2},
\end{aligned} \tag{33}$$

where $\rho_{\Delta C_g,i;\Delta C_g,j}$, $\rho_{\Delta f_T,i;\Delta f_T,j}$ and $\rho_{\Delta C_g,i;\Delta f_T,j}$ denote correlation coefficient between ΔC_g of ith and jth MOSFET, that between Δf_T of ith and jth MOSFET, and that between ΔC_g of ith MOSFET and Δf_T of jth MOSFET, respectively. Their magnitudes can be approximated by unity when $i = j$. Furthermore, $S^Z_{C_g}|_i$, $S^Z_{C_g}|_j$, $S^Z_{f_T}|_i$, and $S^Z_{f_T}|_j$ denote sensitivity of Z to C_g of ith MOSFET, that to C_g of jth MOSFET, that to f_T of ith MOSFET, and that to f_T of jth MOSFET, respectively. These sensitivities can be determined by using the aforementioned sensitivity analysis. Finally, according to their definitions, $\sigma^2_{\Delta C_g,i}$ ($\sigma^2_{\Delta C_g,j}$) and $\sigma^2_{\Delta f_T,i}$ ($\sigma^2_{\Delta f_T,j}$) can be simply determined by substitution of the ith (jth) MOSFET's parameters into (19b) and (20b) which have been derived by using the proposed $\mathrm{pdf}_{\Delta C_g}(\delta C_g)$ and $\mathrm{pdf}_{\Delta f_T}(\delta f_T)$.

In the next subsection, the application of $\mathrm{pdf}_{\Delta C_g}(\delta C_g)$ and $\mathrm{pdf}_{\Delta f_T}(\delta f_T)$ as the mathematical basis for analytical modeling of the high frequency performances mismatches between theoretically identical nanoscale MOSFETs will be discussed.

4.4. Basis for Analytical Modeling of High Frequency Performances Mismatches. For illustration, the mismatch in C_g and that in f_T will be considered by letting such mismatches between theoretically identical devices, denoted by $\delta C_{gij} = C_{gi} - C_{gj}$ and $\delta f_{Tij} = f_{Ti} - f_{Tj}$ for mismatch in C_g and that in f_T, respectively. It should be mentioned here that C_{gi}, C_{gj}, f_{Ti}, and f_{Tj} denote C_g of ith MOSFET, C_g of jth MOSFET, f_T of ith MOSFET, and f_T of jth MOSFET.

Since the behaviors of mismatches can be conveniently modeled by their variances [31, 32], the modeling of δC_{gij} and δf_{Tij} is to derive their variances, that is, $\sigma^2_{\delta C_g ij}$ and $\sigma^2_{\delta f_T ij}$. Obviously the variances of C_{gi}, C_{gj}, f_{Ti}, and f_{Tj} which are fluctuated by the effects of random dopant fluctuation and process variation can be given by $\sigma^2_{\Delta C_g,i}$, $\sigma^2_{\Delta C_g,j}$, $\sigma^2_{\Delta f_T,i}$, and $\sigma^2_{\Delta f_T,j}$ due to (8). Of course, $\sigma^2_{\Delta C_g,i}$, $\sigma^2_{\Delta C_g,j}$, $\sigma^2_{\Delta f_T,i}$ and $\sigma^2_{\Delta f_T,j}$ can be determined by using $\mathrm{pdf}_{\Delta C_g}(\delta C_g)$ and $\mathrm{pdf}_{\Delta f_T}(\delta f_T)$ which are single devices oriented. In terms of these variances, $\sigma^2_{\delta C_g ij}$ and $\sigma^2_{\delta f_T ij}$ can be given as follows:

$$\begin{aligned}
\sigma^2_{\delta C_g ij} &= \sigma^2_{\Delta C_g,i} + \sigma^2_{\Delta C_g,j} - 2\rho_{\Delta C_g i,\Delta C_g j}\sigma_{\Delta C_g,i}\sigma_{\Delta C_g,j}, \\
\sigma^2_{\delta f_T ij} &= \sigma^2_{\Delta f_T,i} + \sigma^2_{\Delta f_T,j} - 2\rho_{\Delta f_T i,\Delta f_T j}\sigma_{\Delta f_T,i}\sigma_{\Delta f_T,j}.
\end{aligned} \tag{34}$$

Obviously, transistors are highly correlated if they are closely spaced [31, 32]. So, $\sigma^2_{\delta C_g ij}$ and $\sigma^2_{\delta f_T ij}$ are minimized if the correlations are in the positive manner and maximized if devices are negatively correlated. On the other hand,

TABLE 1: Max($\Delta C_g/C_{g,\mathrm{nom}}$) and Max($\Delta f_T/f_{T,\mathrm{nom}}$).

Max($\Delta C_g/C_{g,\mathrm{nom}}$) (%)		Max($\Delta f_T/f_{T,\mathrm{nom}}$) (%)	
NMOS	PMOS	NMOS	PMOS
14.037	10.6	7.676	6.023

TABLE 2: K-S test statistics.

From verification of $\mathrm{pdf}_{\Delta C_g}(\delta C_g)$		From Δf_T verification of $\mathrm{pdf}_{\Delta f_T}(\delta f_T)$	
NMOS	PMOS	NMOS	PMOS
0.021463	0.016144	0.029223	0.029192

correlation terms can be neglected from (34) if devices are distanced [31, 32]. In the final subsection, discussion regarding observations from the verification results will be given.

4.5. Observations from Verification Results. It can be observed from the formerly shown Monte-Carlo SPICE simulation results that the maximum percentages of ΔC_g and Δf_T denoted by Max[$\Delta C_g/C_{g,\mathrm{nom}}$] and Max[$\Delta f_T/f_{T,\mathrm{nom}}$], respectively, obtained from the NMOS-based circuits are higher than their counterparts from the PMOS ones. These maximum percentages are listed in Table 1. This means that high frequency performances of the nanoscale PMOS transistor is more robust to the random dopant fluctuation and process variation effects than that of NMOS one due to its lower Max[$\Delta C_g/C_{g,\mathrm{nom}}$] and Max[$\Delta f_T/f_{T,\mathrm{nom}}$].

Furthermore, it has been observed that the KS values obtained from the PMOS-based verification are smaller than those obtained from the NMOS-based one. These statistics are repeated in Table 2 for convenience. This means that the proposed $\mathrm{pdf}_{\Delta C_g}(\delta C_g)$ and $\mathrm{pdf}_{\Delta f_T}(\delta f_T)$ model ΔC_g and Δf_T of nanoscale PMOS transistor more accurately than they do to ΔC_g and Δf_T of NMOS device according to such smaller statistics.

5. Conclusion

The novel complete probabilistic models of the random variations in nanoscale MOSFET's high frequency performances have been proposed in this research as $\mathrm{pdf}_{\Delta C_g}(\delta C_g)$ and $\mathrm{pdf}_{\Delta f_T}(\delta f_T)$, shown in (10) and (11) as respectively. They have been found to be both analytic and physical level oriented as many physical level variables have been involved. Obviously, the formerly overlooked Δf_T has been now focused on. Both random dopant fluctuation and process variation effects have been taken into account. The nanoscale MOSFET equation [21] has been used as the mathematical basis, and the state of the art physical level fluctuation-induced characteristic variation model of MOSFET [20] has been adopted. So, $\mathrm{pdf}_{\Delta C_g}(\delta C_g)$ is more accurate and physical level oriented than its predecessor [18]. Both $\mathrm{pdf}_{\Delta C_g}(\delta C_g)$ and $\mathrm{pdf}_{\Delta f_T}(\delta f_T)$ have been verified based on the 65 nm CMOS process technology by using the Monte-Carlo SPICE simulations of the benchmark circuits and the Kolmogorov-Smirnov goodness of fit tests as very accurate since they fit the results of Monte-Carlo SPICE simulations with 99% confidence. Furthermore, they have various utilities such as analytical explanation of probabilistic/statistical behaviors of such random variations and others, as mentioned above. So, they have been found to be versatile for the statistical/variability aware analysis/design of nanoscale MOSFET-based analog/mixed signal circuits and systems.

Acknowledgment

The author would like to acknowledge Mahidol University, Thailand, for online database service.

References

[1] T. Yao, M. Gordon, K. Yau, M. T. Yang, and S. P. Voinigescu, "60-GHz PA and LNA in 90-nm RF-CMOS," in *Proceedings of the IEEE Radio Frequency Integrated Circuits Symposium*, pp. 147–150, San Francisco, Calif, USA, June 2006.

[2] M. A. Masud, H. Zirath, M. Ferndahl, and H. O. Vickes, "90 nm CMOS MMIC amplifier," in *Proceedings of the IEEE Radio Frequency Integrated Circuits (RFIC) Symposium*, pp. 201–204, Forth Worth, Tex, USA, June 2004.

[3] Y. S. Jiang, Z. M. Tsai, J. H. Tsai, H. T. Chen, and H. Wang, "A 86 to 108 GHz amplifier in 90 nm CMOS," *IEEE Microwave and Wireless Components Letters*, vol. 18, no. 2, pp. 124–126, 2008.

[4] B. Heydari, M. Bohsali, E. Adabi, and A. M. Niknejad, "Low-power mm-wave components up to 104 GHz in 90 nm CMOS," in *Proceedings of the 54th IEEE International Solid-State Circuits Conference (ISSCC'07)*, pp. 200–597, San Francisco, Calif, USA, February 2007.

[5] M. Varonen, M. Kärkkäinen, M. Kantanen, and K. A. I. Halonen, "MM-wave integrated circuits in 65-nm CMOS," *IEEE Journal of Solid-State Circuits*, vol. 43, no. 9, pp. 1991–2002, 2008.

[6] A. M. Niknejad, S. Emami, B. Heydari, M. Bohsali, and E. Adabi, "Nanoscale CMOS for mm-wave applications," in *Proceedings of the IEEE Compound Semiconductor Integrated Circuit Symposium (CSIC'07)*, pp. 1–4, Portland, Ore, USA, October 2007.

[7] D. Draxelmayr, "A 6 b 600 MHz 10 mW ADC array in digital 90 nm CMOS," in *Proceedings of the IEEE International Solid-State Circuits Conference Digest of Technical Papers (ISSCC'04)*, pp. 264–265, February 2004.

[8] Y. Shimizu, S. Murayama, K. Kudoh, H. Yatsuda, and A. Ogawa, "A 30 mW 12b 40MS/s subranging ADC with a high-gain offset-canceling positive-feedback amplifier in 90nm digital CMOS," in *IEEE International Solid-State Circuits Conference Digest of Technical Papers (ISSCC'06)*, pp. 802–811, San Francisco, Calif, USA, February 2006.

[9] A. Matsuzawa, "Design challenges of analog-to-digital converters in nanoscale CMOS," *IEICE Transactions on Electronics*, vol. E90-C, no. 4, pp. 779–785, 2007.

[10] H. Masuda, T. Kida, and S. I. Ohkawa, "Comprehensive matching characterization of analog CMOS circuits," *IEICE Transactions on Fundamentals of Electronics, Communications and Computer Sciences*, vol. E92-A, no. 4, pp. 966–975, 2009.

[11] R. Banchuin, "Process induced random variation models of nanoscale MOS performance: efficient tool for the nanoscale regime analog/mixed signal CMOS statistical/variability aware

design," in *Proceedings of the International Conference on Information and Electronic Engineering*, pp. 6–12, Bangkok, Thailand, 2011.

[12] R. Banchuin, "Complete circuit level random variation models of nanoscale MOS performance," *International Journal of Information and Electronics Engineering*, vol. 1, no. 1, pp. 9–15, 2011.

[13] K. Hasegawa, M. Aoki, T. Yamawaki, and S. Tanaka, "Modeling transistor variation using α-power formula and its application to sensitivity analysis on harmonic distortion in differential amplifier," *Analog Integrated Circuits and Signal Processing*, vol. 72, no. 3, pp. 605–613, 2012.

[14] A. R. Brown and A. Asenov, "Capacitance fluctuations in bulk MOSFETs due to random discrete dopants," *Journal of Computational Electronics*, vol. 7, no. 3, pp. 115–118, 2008.

[15] Y. Li and C. H. Hwang, "High-frequency characteristic fluctuations of nano-MOSFET circuit induced by random dopants," *IEEE Transactions on Microwave Theory and Techniques*, vol. 56, no. 12, pp. 2726–2733, 2008.

[16] M. H. Han, Y. Li, and C. H. Hwang, "The impact of high-frequency characteristics induced by intrinsic parameter fluctuations in nano-MOSFET device and circuit," *Microelectronics Reliability*, vol. 50, no. 5, pp. 657–661, 2010.

[17] H. S. Kim, C. Chung, J. Lim, K. Park, H. Oh, and H. K. Kang, "Characterization and modeling of RF-performance (fT) fluctuation in MOSFETs," *IEEE Electron Device Letters*, vol. 30, no. 8, pp. 855–857, 2009.

[18] R. Banchuin, "The probabilistic model of random variation in nanoscale MOSFET's high frequency performance," in *Proceedings of the 6th WSEAS International Conference on Computer Engineering and Applications, and Proceedings of the 2012 American Conference on Applied Mathematics*, pp. 240–245, January 2012.

[19] M. J. M. Pelgrom, A. C. J. Duinmaijer, and A. P. G. Welbers, "Matching properties of MOS transistors," *IEEE Journal of Solid-State Circuits*, vol. 24, no. 5, pp. 1433–1440, 1989.

[20] K. Kuhn, "Variability in nanoscale CMOS technology," *Science China Information Sciences*, vol. 54, no. 5, pp. 936–945, 2011.

[21] L. L. Lewyn, T. Ytterdal, C. Wulff, and K. Martin, "Analog circuit design in nanoscale CMOS technologies," *Proceedings of the IEEE*, vol. 97, no. 10, pp. 1687–1714, 2009.

[22] H. Abebe, H. Morris, E. Cumberbatch, and V. Tyree, "Compact gate capacitance model with polysilicon depletion effect for MOS device," *Journal of Semiconductor Technology and Science*, vol. 7, no. 3, pp. 131–135, 2007.

[23] R. T. Howe and C. G. Sodini, *Microelectronics: An Integrated Approach*, Prentice Hall, Upper Saddle River, NJ, USA, 1996.

[24] G. Cijan, T. Tuma, and A. Burmen, "Modeling and simulation of MOS transistor mismatch," in *Proceedings of the 6th Eurosim*, pp. 1–8, Ljubljana, Slovenia, 2007.

[25] T. Altiok and B. Melamed, *Simulation Modeling and Analysis with ARENA*, Academic Press, New York, NY, USA, 2007.

[26] S. A. Klugman, H. H. Panjer, and G. E. Willmot, *Loss Models: From Data to Decisions*, John Wiley & Sons, London, UK, 2008.

[27] Predictive Technology Model, http://ptm.asu.edu.

[28] A. H. Haddad, *Probabilistic Systems and Random Signals*, Prentice Hall, Upper Saddle River, NJ, USA, 2006.

[29] W. W. Hines, D. C. Montgomery, D. M. Goldsman, and C. M. Borror, *Probability and Statistics in Engineering*, John Wiley & Sons, London, UK, 2003.

[30] M. Conti, P. Grippa, S. Orcioni, and C. Turchetti, "Parametric yield formulation of MOS IC's affected by mismatch effect," *IEEE Transactions on Computer-Aided Design of Integrated Circuits and Systems*, vol. 18, no. 5, pp. 582–596, 1999.

[31] A. Cathignol, S. Mennillo, S. Bordez, L. Vendrame, and G. Ghibaudo, "Spacing impact on MOSFET mismatch," in *Proceedings of the IEEE Conference on Microelectronic Test Structures (ICMTS'08)*, pp. 90–95, Edinburgh, UK, March 2008.

[32] C. M. Mezzomo, A. Bajolet, A. Cathignol, R. Di Frenza, and G. Ghibaudo, "Characterization and modeling of transistor variability in advanced CMOS technologies," *IEEE Transactions on Electron Devices*, vol. 58, no. 8, pp. 2235–2248, 2011.

Permissions

The contributors of this book come from diverse backgrounds, making this book a truly international effort. This book will bring forth new frontiers with its revolutionizing research information and detailed analysis of the nascent developments around the world.

We would like to thank all the contributing authors for lending their expertise to make the book truly unique. They have played a crucial role in the development of this book. Without their invaluable contributions this book wouldn't have been possible. They have made vital efforts to compile up to date information on the varied aspects of this subject to make this book a valuable addition to the collection of many professionals and students.

This book was conceptualized with the vision of imparting up-to-date information and advanced data in this field. To ensure the same, a matchless editorial board was set up. Every individual on the board went through rigorous rounds of assessment to prove their worth. After which they invested a large part of their time researching and compiling the most relevant data for our readers. Conferences and sessions were held from time to time between the editorial board and the contributing authors to present the data in the most comprehensible form. The editorial team has worked tirelessly to provide valuable and valid information to help people across the globe.

Every chapter published in this book has been scrutinized by our experts. Their significance has been extensively debated. The topics covered herein carry significant findings which will fuel the growth of the discipline. They may even be implemented as practical applications or may be referred to as a beginning point for another development. Chapters in this book were first published by Hindawi Publishing Corporation; hereby published with permission under the Creative Commons Attribution License or equivalent.

The editorial board has been involved in producing this book since its inception. They have spent rigorous hours researching and exploring the diverse topics which have resulted in the successful publishing of this book. They have passed on their knowledge of decades through this book. To expedite this challenging task, the publisher supported the team at every step. A small team of assistant editors was also appointed to further simplify the editing procedure and attain best results for the readers.

Our editorial team has been hand-picked from every corner of the world. Their multi-ethnicity adds dynamic inputs to the discussions which result in innovative outcomes. These outcomes are then further discussed with the researchers and contributors who give their valuable feedback and opinion regarding the same. The feedback is then collaborated with the researches and they are edited in a comprehensive manner to aid the understanding of the subject.

Apart from the editorial board, the designing team has also invested a significant amount of their time in understanding the subject and creating the most relevant covers. They scrutinized every image to scout for the most suitable representation of the subject and create an appropriate cover for the book.

The publishing team has been involved in this book since its early stages. They were actively engaged in every process, be it collecting the data, connecting with the contributors or procuring relevant information. The team has been an ardent support to the editorial, designing and production team. Their endless efforts to recruit the best for this project, has resulted in the accomplishment of this book. They are a veteran in the field of academics and their pool of knowledge is as vast as their experience in printing. Their expertise and guidance has proved useful at every step. Their uncompromising quality standards have made this book an exceptional effort. Their encouragement from time to time has been an inspiration for everyone.

The publisher and the editorial board hope that this book will prove to be a valuable piece of knowledge for researchers, students, practitioners and scholars across the globe.

List of Contributors

Yuval Cohen and Yitzhak August
The Unit of Electro-Optics Engineering and the Earth and Planetary Image Facility, Ben-Gurion University of the Negev, P.O. Box 653, 84105 Beer-Sheva, Israel

Dan G. Blumberg
The Department of Geography and Environmental Development and the Earth and Planetary Image Facility, Ben-Gurion University of the Negev, P.O. Box 653, 84105 Beer-Sheva, Israel

Stanley R. Rotman
Department of Electrical and Computer Engineering and the Earth and Planetary Image Facility, Ben-Gurion University of the Negev, P.O. Box 653, 84105 Beer-Sheva, Israel

Yibo Chen and Yuan Xie
Department of Computer Science and Engineering, The Pennsylvania State University, University Park, PA 16802, USA

Yu Wang
Department of Electronics Engineering, Tsinghua University, Beijing 100084, China

Andres Takach
Design Creation and Synthesis, Mentor Graphics Corporation, Wilsonville, OR 97070, USA

Andreas G. Savva and Theocharis Theocharides
Department of Electrical and Computer Engineering, University of Cyprus, 1678 Nicosia, Cyprus

Vassos Soteriou
Department of Electrical Engineering and Information Technology, Cyprus University of Technology, 3036 Limassol, Cyprus

Mariangela Genovese, Ettore Napoli, Davide De Caro, Nicola Petra and Antonio G. M. Strollo
Department of Biomedical, Electronic and Telecommunications Engineering, University of Napoli Federico II, 80125 Napoli, Italy

Antonio Buonomo and Alessandro Lo Schiavo
Seconda Universit`a degli Studi di Napoli, Via Roma 29, 81031 Aversa, Italy

Gaspare Galati and Gabriele Pavan
Department of Electronic Engineering, Tor Vergata University, Via del Politecnico, 1-00133 Rome, Italy

Dirk Borghys
Department CISS, Royal Military Academy, 2007 Brussels, Belgium

Ingebjørg Kåsen
Land and Air Systems Division, Norwegian Defence Research Establishment (FFI), 2007 Kjeller, Norway

Véronique Achard
Theoretical and Applied Optics Department, French Aerospace Laboratory (ONERA), FR-31055 Toulouse Cedex 4, France

Christiaan Perneel
Department of Mathematics, Royal Military Academy, Brussels, Belgium

Shahaf I.Wayer and Arie Reichman
School of Engineering, Ruppin Academic Center, Emek Hefer, Israel

Sebastian Stenzel and Jürgen Freudenberger
Institute for System Dynamics, HTWG Konstanz, Brauneggerstrasse 55, 78462 Konstanz, Germany

Khaled Loukhaoukha, Jean-Yves Chouinard and Abdellah Berdai
Department of Electrical and Computer Engineering, Laval University, QC, Canada G1K 7P4

Rafael Vargas- Bernal, Gabriel Herrera-Pérez and Ma. Elena Calixto-Olalde
Departamento de Ingeniería en Materiales, Instituto Tecnológico Superior de Irapuato, Carretera Irapuato Silao Km. 12.5,
36821 Irapuato, Guanajuato, Mexico

Margarita Tecpoyotl-Torres
Centro de Investigación en Ingeniería y Ciencias Aplicadas (CIICAp), Universidad Aut´onoma del Estado de Morelos, Avenida UniversidadNo. 1001, Col. Chamilpa, 62209 Cuernavaca, Morelos, Mexico

Somaya Al-Maadeed
Department of Computer Science and Engineering, Qatar University, P.O. Box 2713, Doha, Qatar

Afnan Al-Ali and Turki Abdalla
Department of Computer Engineering, College of Engineering, University of Basrah, B.P. 49, Basrah, Iraq

Jesmin Khan and Sharif Bhuiyan
Department of Electrical Engineering, Tuskegee University, Tuskegee, AL 36088, USA

Reza Adhami
Department of Electrical and Computer Engineering, University of Alabama in Huntsville, Huntsville, AL 35899, USA

Ting Zhu
Department of Computer Science, State University of New York, Binghamton, NY 13905, USA

Chang Zhou
College of Information Engineering, China Jiliang University, Hangzhou, Zhejiang 310018, China

Rawid Banchuin
Department of Computer Engineering, Siam University, 38 Petchkasem Road, Bangkok 10160, Thailand